S0-CAH-342

Reactivity of Metal–Metal Bonds

Malcolm H. Chisholm, EDITOR
Indiana University

Based on a symposium sponsored by the Division of Inorganic Chemistry at the Second Chemical Congress of the North American Continent (180th ACS National Meeting), Las Vegas, Nevada, August 25–26, 1980.

ACS SYMPOSIUM SERIES 155

AMERICAN CHEMICAL SOCIETY
WASHINGTON, D. C. 1981

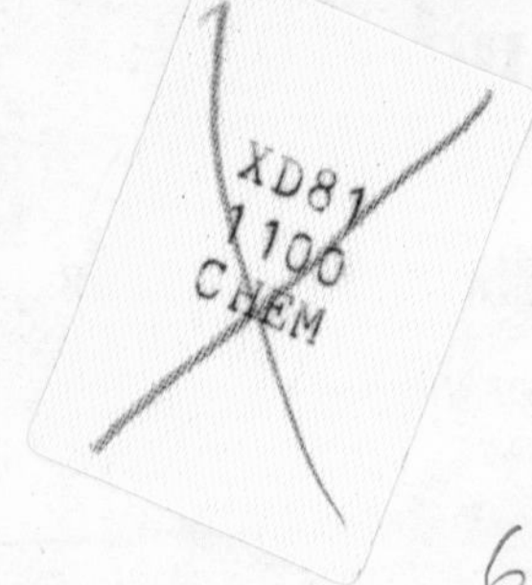

Library of Congress CIP Data

Reactivity of metal–metal bonds.
(ACS symposium series; 155 ISSN 0097-6156)

Includes bibliographies and index.

1. Metal–metal bonds—Congresses. 2. Reactivity (Chemistry)—Congresses.
I. Chisholm, Malcolm. II. American Chemical Society. Division of Inorganic Chemistry. III. Series: American Chemical Society. ACS symposium series; 155.

QD461.R38 546′.3 81-361
ISBN 0-8412-0624-4 AACR1 ASCMC 8 155
1–327 1981

PRINTED IN THE UNITED STATES OF AMERICA

ACS Symposium Series

FOREWORD

The ACS SYMPOSIUM SERIES was founded in 1974 to provide a medium for publishing symposia quickly in book form. The format of the Series parallels that of the continuing ADVANCES IN CHEMISTRY SERIES except that in order to save time the papers are not typeset but are reproduced as they are submitted by the authors in camera-ready form. Papers are reviewed under the supervision of the Editors with the assistance of the Series Advisory Board and are selected to maintain the integrity of the symposia; however, verbatim reproductions of previously published papers are not accepted. Both reviews and reports of research are acceptable since symposia may embrace both types of presentation.

CONTENTS

PREFACE

The chemistry of compounds containing metal–metal bonds is one of the most rapidly developing areas of modern coordination chemistry. At this time, virtually all the transition elements are known to form homo- or heterodinuclear compounds with metal–metal bonds that may be of integral (1, 2, 3, 4) or fractional order ($\frac{1}{2}$, $1\frac{1}{2}$, $2\frac{1}{2}$, $3\frac{1}{2}$). There are also large classes of cluster compounds ranging from polynuclear metal carbonyls and other organometallics to polynuclear metal halides, oxides, and chalconides that contain delocalized metal–metal bonds. Much of the initial interest in these compounds was devoted to the elucidation of their structures, bonding, and electronic properties. However, there is now a growing recognition that the reactivity patterns associated with these compounds will provide a rich and fruitful field of research. This volume is based on the first ACS-sponsored symposium devoted to this topic. The authors, through their research interests and findings, present a survey of the types of reactivity that presently have been established. These include metal–metal bond rupture and formation, photolysis, substitution, oxidative-addition, reductive-elimination, oligomerization, and template reactions. No doubt this group of reactions will be further elaborated upon in the future and new modes of reactivity will be discovered. Truly, compounds containing metal–metal bonds offer new dimensions and opportunities for reactivity.

I would like to thank the donors of the Petroleum Research Fund, administered by the American Chemical Society, the ACS Division of Inorganic Chemistry, and the Union Carbide Corporation for financial support used to organize this symposium.

MALCOLM H. CHISHOLM
Department of Chemistry
Indiana University
Bloomington, Indiana 47405

November 21, 1980.

1

Metal–Metal Multiple Bonds and Metal Clusters

New Dimensions and New Opportunities in Transition Metal Chemistry

F. A. COTTON

Department of Chemistry, Texas A&M University, College Station, TX 77843

The chemistry of the transition metals became a mature field, based on a sound foundation of broad, basic principles around the turn of the century as a result of the efforts of Alfred Werner, who was the first chemist to make sense out of an enormous body of experimental facts that had been painstakingly accumulated by himself, by the Danish chemist S. M. Jørgensen, and by a number of others. Working without the aid of any direct structural data, Werner was able to combine intuition, geometric reasoning and experiment to develop the central principle of his coordination theory, namely, that an ionized transition metal atom would always be surrounded by a set of neutral molecules (e.g., H_2O, NH_3) and/or anions (e.g., CN^-, Cl^-, OH^-) arranged in a geometrically well-defined pattern (square, tetrahedral, octahedral). This general conception dominated transition metal chemistry for more than six decades and even today is applicable to a vast amount of chemistry, albeit with the benefit of some further knowledge and insight, such as: (1) recognition that coordination numbers other than those emphasized by Werner (e.g., five-, seven- and eight-coordination) are important; (2) recognition that coordination geometry, especially for 5-, 7- and 8-coordination, is not always rigid; (3) much more detailed knowledge of equilibria, kinetics and spectra; (4) a quantum-mechanically sound understanding of metal-ligand bonding.

There are, however, several aspects of contemporary transition metal chemistry whose existence could not have been extrapolated from the Wernerian principles. Among these one could mention considerable areas of metal carbonyl type chemistry, much of the current field of organometallic chemistry and, most unambiguously, the chemistry of compounds containing metal-metal bonds. Although Werner dealt extensively with polynuclear complexes, these were conceived simply as two or more mononuclear complexes united only by the ligands they shared.

0097-6156/81/0155-0001$05.00/0

The notion of direct metal-metal bonds is, to the best of my knowledge, totally absent from the work of Werner, his coworkers or his direct followers. It is a modern, "non-Wernerian" concept, and it is the subject of this symposium.

Even while Werner was still alive, the first non-Wernerian compounds were discovered. In 1907 the compound "$TaCl_2 \cdot 2H_2O$" was reported and in 1913 reformulated, correctly, as $Ta_6Cl_{14} \cdot 7H_2O$ (1). During the third decade of this century polynuclear halide compounds of molybdenum were discovered (2). The true structures of these compounds could not be inferred by the available experimental methods and despite their failure to follow the general (Wernerian) patterns of behavior they attracted little attention. Even when, in 1935, C. Brosset (3) showed that the tungsten atoms in $[W_2Cl_9]^{3-}$ were very close together (*ca.* 2.5Å) and again a decade later (4) showed that the aforementioned lower halide compounds of molybdenum contained octahedral Mo_6 groups with Mo-Mo distances about equal to the interatomic distance in metallic molybdenum, no new research appeared in the literature. The presence of Ta_6 octahedra in the lower tantalum chloride was discovered in 1950 and it was explicitly concluded that metal-metal bonds were present in this and related compounds (5), but still, no one apparently was interested in looking further at such "curiosities."

It was in 1963, with yet another accidental discovery, that of the $[Re_3Cl_{12}]^{3-}$ ion (6, 7) that the field of "metal atom cluster" chemistry really had its birth, since this discovery provoked the first general discussion of the existence and probable importance of the entire class of "metal atom cluster" compounds (6).

It was in the Re_3 clusters also that the first multiple M-M bonds were explicitly and unequivocally recognized (6) and this was soon followed by the discovery of the first quadruple bond (8) and then the first triple bond (9). Since these seminal discoveries, the field of metal atom clusters and M-M multiple bonds has arisen, grown and flourished.

Survey of M-M Bonds and Clusters

The compounds under consideration here are restricted to those in which there are direct M-M bonds between transition metal atoms. We thus exclude, at least in this article, species with bonds between transition metal atoms and main group atoms, e.g., Fe-Sn, V-Au, Tl-Co, etc., bonds.

Another prefactory remark concerns the nature of proof for the existence of M-M bonds (regardless of their bond order). I shall define as first order (or *ipso facto*) evidence that a metal-metal bond exists, an internuclear distance between the metal atoms that is equal to or less than the sum of their estimated single bond radii (as defined, for example, by Pauling (10)) *when no bridging atoms or groups of atoms are*

Proof for bonding → no bridges; easy
→ bridges; harder

present. For example, in $[Re_2Cl_8]^{2-}$ or $Ir_4(CO)_{12}$ we have first order cases for the existence of M-M bonds. Whenever bridging atoms or groups are present the case for M-M bonding must be a second order one, based on calculations, measurements (or assumptions) concerning where the electrons actually are.

It is, perhaps, also worth mentioning once again that (11) "metal atom cluster compounds ... (are) ... those containing a finite group of metal atoms that are held together entirely, mainly, or at least to a significant extent, by bonds directly between the metal atoms even though some non-metal atoms may be associated intimately with the cluster." A mere polynuclear complex is not a metal atom cluster compound.

For practical purposes the field of metal-metal bonds and metal atom clusters can be divided into two broad areas. (1) Those compounds with the metal atoms in formal oxidation states of zero or close to it, including negative ones. For the most part these are polynuclear metal carbonyls, or very similar compounds. In these compounds the M-M bonds are usually long, weak and of order one. (2) Compounds with the metal atoms in low to medium positive oxidation states, and ligands of the same kinds normally found in classical Werner complexes, e.g., halide, sulfate, phosphate, carboxylate or thiocyanate ions, water, amines and phosphines. Compounds of this type include metal-metal bonds of orders ranging from about 1/2 to 4.0.

It might also be possible to define a third category of solid state systems which are closely related to those to type (2) but with some differences such as having weaker bonds or more extended arrays. However, these can also be included within type (2), and we shall not treat them separately here.

As already noted the first direct proof of the existence of a metal atom cluster was provided in 1946 by C. Brosset for the $[Mo_6Cl_8]^{4-}$ unit in two different compounds. However, this did not immediately or directly stimulate efforts to look for other such compounds. It should also be noted that since there are bridging Cl atoms, the case here for the existence of Mo-Mo bonds, while good, is only second order. This problem exists in the majority of metal atom clusters, but there are, fortunately quite a few cases in which unbridged M-M bonds exist and these provide the necessary ipso facto proof of M-M bonding. Examples are $M_2(CO)_{10}$ (M = Mn, Tc, Re), $M_3(CO)_{12}$ (M = Ru, Os) $Ir_4(CO)_{12}$, etc.

For detailed listings of the many known cluster compounds there are now numerous reviews (12-14). Suffice it to say here, that among the clusters of type (1) formed by rhodium and platinum we now have discrete arrays of thirty, or more, metal atoms and are entering into a fascinating borderline area between metal atom clusters and metals as such.

Survey of M-M Multiple Bonds

This brief survey will be confined to those multiple bonds that fit within the pattern shown in Figure 1. There are many others that do not, but space does not allow their inclusion here. As Figure 1 shows, the molecular orbital pattern of the quadruple bond, $\sigma^2\pi^4\delta^2$, where the pertinent number of electrons is eight, is a natural point of departure to give two types of bond of order 3.5, viz., $\sigma^2\pi^4\delta$ and $\sigma^2\pi^4\delta^2\delta^*$, and two types of triple bond, viz., $\sigma^2\pi^4$ and $\sigma^2\pi^4\delta^2\delta^{*2}$. All of these possibilities are to be found in real systems, as shown by the representative examples in Table 1, which also shows the extension of the scheme to still more highly electron-rich systems having bond orders of 2.5 and 1.0. For extensive reviews of the higher order bonds see refs. 14-17.

Evidence for Multiple Bonds

How does one know that there really are triple and quadruple bonds? That question is not asked very often any more, but it used to be a frequent one, especially with respect to the quadruple bond. Like any completely new idea, in the beginning the concept of the quadruple bond aroused, not unreasonably, some skepticism.

The evidence for these multiple bonds is of many kinds. It comprises an enormous collection of diverse data and computational results that could not be consistently fitted together on any other basis. The following list gives the chief points:

(1) Bond Lengths
(2) Conformations
(3) Theory
(4) Electronic and Vibronic Absorption Spectra
(5) UV Photoelectron Spectra
(6) X-ray Spectra
(7) Magnetic Anisotropy
(8) Internal Consistency

(1) Bond Lengths. The extreme shortness of triple and quadruple bonds is indicative of higher bond orders although, of course, bond order cannot be quantified in this way alone. Simply to illustrate the relevant magnitudes of the bond lengths, some of their relationships to one another and to estimated single bond distances, Figure 2 shows the observed ranges for most of the M-M bonds of various orders for clusters of the higher oxidation state (i.e., non-carbonyl) types. The vertical lines with arrowheads give the sums of Pauling single bond radii (10) for the metals concerned. There is a wealth of material represented in Figure 2 which we cannot discuss here in detail. Suffice it to say that the higher order bonds generally are so

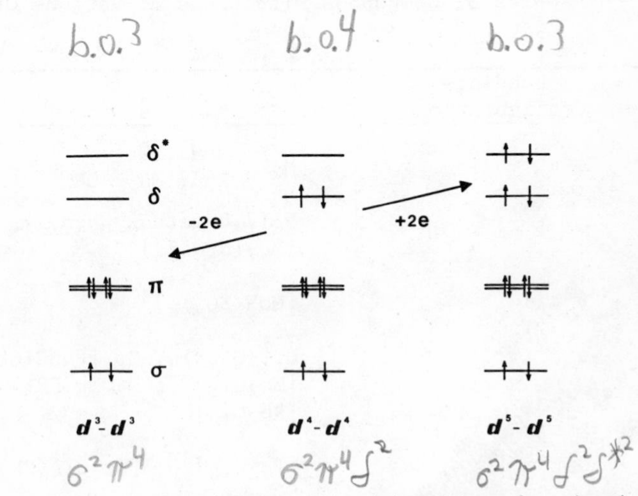

Figure 1. Diagram showing the relationship of electronic configurations for M—M bonds of orders 3 and 4

Table I

Examples of Compounds with Bonds of Various Orders

Bond Order	Bonding Configuration	Examples
3.0	$\sigma^2\pi^4$	M_2X_6 M=Mo, W; X=R, NR_2, OR V_2(2,6-methoxyphenyl)$_4$ $[Mo_2(HPO_4)_4]^-$
3.5	$\sigma^2\pi^4\delta$	$[Mo_2(SO_4)_4]^{3-}$
4.0	$\sigma^2\pi^4\delta^2$	$Cr_2(O_2CR)_4$, $Cr_2[PhNC(CH_3)O]_4$ $[Mo_2Cl_8]^{4-}$, $W_2Cl_4(PR_3)_4$ $[Re_2Cl_8]^{2-}$
3.5	$\sigma^2\pi^4\delta^2\delta^*$	$[Tc_2Cl_8]^{3-}$, $[Re_2Cl_4(PR_3)_4]^+$
3.0	$\sigma^2\pi^4\delta^2\delta^{*2}$	$Re_2Cl_4(PR_3)_4$ Os_2(2-oxopyridine)$_4Cl_2$
2.5	$\sigma^2\pi^4\delta^2\delta^{*2}\pi^*$	$Ru_2(O_2CR)_4Cl$
1.0	$\sigma^2\pi^4\delta^2\delta^{*2}\pi^{*4}$	$Rh_2(O_2CR)_4$ Rh_2(2-oxo,6-Me-pyridine)$_4$

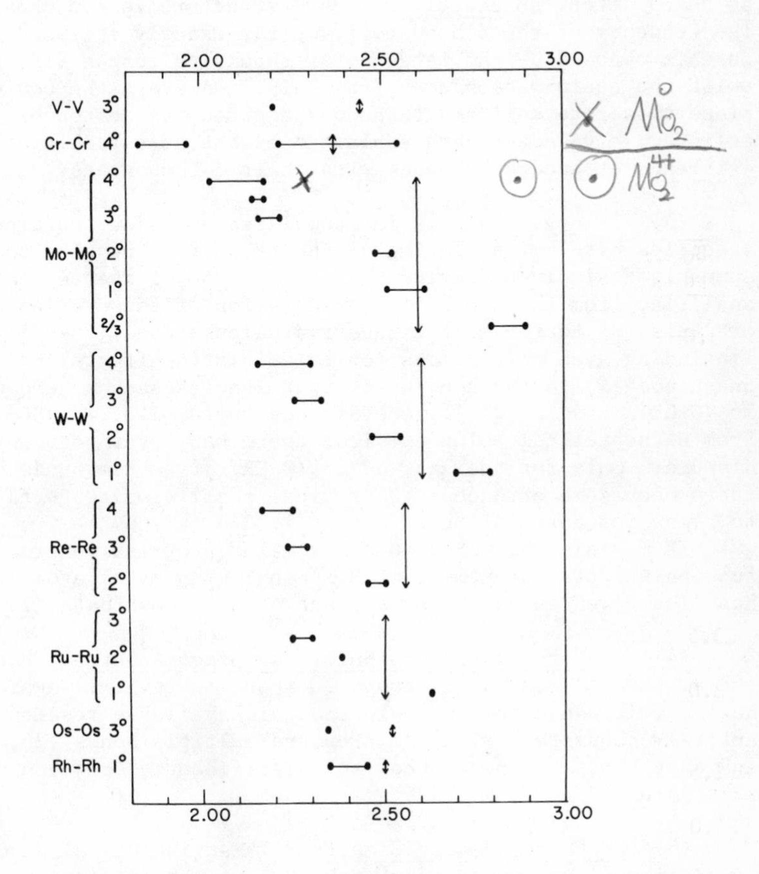

Figure 2. Bond distances for M—M bonds of various orders. The vertical double arrows for each element give twice the Pauling single-bond radii (10) for each metal.

very much shorter than the expected single bond distances that high bond orders are clearly indicated.

(2) Conformations. The eclipsed conformation in the $[Re_2Cl_8]^{2-}$ ion was one of the earliest qualitative evidences for the existence of the fourth (δ) component of the quadruple bond. Since then, this feature has consistently been observed although, as noted often, an exactly eclipsed structure is not demanded. The tendency of the δ bond to favor the exactly eclipsed structure is opposed by the tendency of repulsive forces (18, 19) to twist the conformation away from this. However, the conformations closer to eclipsed than to staggered, and often exactly eclipsed, have never been explained except as a result of the existence of quadruple bonds with their δ components.

(3) Theory. This is far too large a subject to discuss in specifics here. The bearing of theory on the question can be summarized simply by saying that every type of theoretical analysis, from the simplest consideration of the overlap of d orbitals (8) to the most exhaustive calculations (20-23) (including even corrections for relativistic effects) has led unequivocally to the conclusion that species such as $[Mo_2Cl_8]^{2-}$, $Mo_2(O_2CR)_4$ and $[Re_2Cl_8]^{2-}$ contain quadruple, i.e., $\sigma^2\pi^4\delta^2$, bonds.

From a theoretical point of view, there has never been any dispute. Only for the case of $Cr_2(O_2CR)_4$ type compounds have there been some erroneous theoretical results owing to failure to allow for correlation effects (21, 24).

It may also be noted that a similar progression from simple to sophisticated treatment of M-M bonding in metal atom clusters has also given an excellent account of all known data (25).

(4) Electronic and Vibronic Absorption Spectra. Here also, the amount of material is too voluminous to review. Again, however, it can simply be said that all available results are entirely consistent with the proposed multiple bonds (26, 27) and that there has never been any significant dissension on that point.

(5) UV Photoelectron Spectra. These in general have provided impressive support for both the qualitative ideas and the explicit quantitative results concerning the multiple M-M bonds (23, 28) and closely related ones such as the formal single bond (see Table 1) in dirhodium species (29). Such spectra have also strongly supported the accepted views on the electronic structures of many metal atom cluster compounds (30, 31).

(6) X-Ray Spectra. Although applied (32) so far only to $[Mo_2Cl_8]^{4-}$, this technique is of interest since it provides independent support for the $\sigma^2\pi^4\delta^2$ configuration.

(7) Magnetic Anisotropy. An ancillary, but interesting, form of evidence is provided by observations of diamagnetic anisotropy of multiple bonds. San Filippo was the first to discuss this possibility (33) but compounds available then did not illustrate it particularly well. In 1976 when the triply-bonded compounds $M_2(NMe_2)_6$, M = W, Mo, were described in detail, (34, 35) it was pointed out that there are large chemical shift differences (ca. 2 ppm) between the protons of the proximal and distal methyl groups and that these are attributable to the diamagnetic anisotropy of the M≡M bonds. It was also shown (34), using predictable steric effects in $Mo_2(NMeEt)_6$ that the chemical shift difference was of the correct sign for this explanation. Quite recently a quantitative interpretation (36) of this observation has shown that the Mo≡Mo and W≡W bonds have $\chi \times 10^{36}$ values of -142 and -156 m^3/molecule, which may be compared with -340 m^3/-molecules for $\chi_{C\equiv C}$; it was also estimated that for the Mo-Mo quadruple bond $\chi \times 10^{36}$ may be as high as -600 to -1000 m^3/-molecule.

(8) Internal Consistency. Only the general view of the metal-metal multiple bonding implied by Figure 1 and illustrated in Table 1 has ever been able to provide an internally consistent synthesis of all of the experimental and computational results. Quadruple and triple bonds and all their close kith and kin are as well established as any other bonds in the whole of chemistry.

Strengths of M-M Bonds

How strong are the M-M multiple bonds? This is also a very common question, but one to which a definitive answer is rather elusive. Only in the last few years have any reliable estimates become available. The first of these (37) were based on the application of the Birge-Sponer extrapolation to the many observed overtones of the M-M stretching vibration in the Raman spectra of $[Mo_2X_8]^{4-}$ and $[Re_2X_8]^{2-}$ species. The estimated bond energies were in the range of 127-190 kcal/mol for the quadruple bonds in $[Mo_2Cl_8]^{4-}$, 152±20 kcal/mol for $[Re_2Cl_8]^{2-}$ and 139±25 kcal/mol for $[Re_2Br_8]^{2-}$.

For the Mo≡Mo and W≡W triple bonds in the $M_2(NMe_2)_6$ molecules there are thermochemical data available, but because of uncertainty about how to apportion the total (known) energy of atomization among the M-M and M-N bonds, the M-M bond energies can be specified only within rather broad ranges. Thus for the Mo≡Mo and W≡W bonds the estimates are 145±45 and 185±50 kcal/mol, respectively (38). It is encouraging that the ranges estimated thermochemically for these bonds are consistent with the results of the earlier Birge-Sponer extrapolations.

An attempt to estimate the Re-Re quadruple bond energy in the $[Re_2Br_8]^{2-}$ ion using the heat of formation of the cesium salt and estimates of the lattice energy and other ancillary quantities has lead to a result of 100±12 kcal/mol (39).

One other experimental number that is relevant is a value of 96.5±0.5 kcal/mol for the dissociation energy of the Mo_2(g) molecule to ground state atoms (40). The good thing about this number is that it is precise; the problem with it is that its' relationship to the desired bond energies in compounds containing Mo-Mo triple and quadruple bonds is not selfevident. The bond in Mo_2 has contributions from six pairs of electrons, but two of these contributions are apparently very weak and another is at best small (41). Thus, this bond might be expected to have the strength of only a triple bond, or perhaps a little more. In addition, because of the greater diffuseness and consequent poorer overlap of the d orbitals on two neutral Mo atoms, as compared with the situation for Mo ions in the formal oxidation states +2 and +3, even the relatively strong $d\sigma$ and $d\pi$ components of the bond in Mo_2 may be less strong than their counterparts in species such as $Mo_2(NR_2)_6$, $Mo_2(O_2CCH_3)_4$ or $[Mo_2Cl_8]^{4-}$. The bond dissociation energy of 96.5 kcal/mol in the Mo_2 molecule may thus be regarded as a definite lower limit for the Mo-Mo bond energies in these other species. This view is quite consistent with the previously cited experimental results.

Finally, there have been two credible attempts to estimate M-M multiple bond energies from theory. For the Mo_2H_6 model compound , a SCF-HF-CI calculation, suitably scaled to results for other multiple bonds of known energy (H_2 and P_2), the Mo≡Mo bond dissociation energy was estimated to be 125±15 kcal/mol (42). In a less secure procedure involving scaling to the Mo_2 molecule, which may be invalid, bond dissociation energies of 73 and 97 kcal/mol have been proposed for $[Mo_2Cl_8]^{4-}$ and $Mo_2(O_2CCH)_4$ (20).

Comparison of Clusters and Multiple Bonds

There is a formal electronic relationship between cluster formation and M-M multiple bond formation. For a metal atom having *n* electrons and *n* d orbitals available after formation of metal-ligand bonds, there is a range of possibilities for M-M interactions. If *n* = 4, as with the square (or pyramidal) $ReCl_4^-$ or $MoCl_4^{2-}$ unit, we can envision all possibilities from the union of just two of these to form one bond of order 4, through intermediate stages such as a set of three with double bonds between pairs of metal atoms to a large cluster in which each metal atom forms four single M-M bonds. As shown in Figure 3, *all* these possibilities are actually known in Nature.

There is more than just the foregoing purely formal relation between dinuclear multiple bonded species and metal atom clusters. It has been found in several cases that interconversion can be carried out experimentally. The earliest example was the preparation (Equation 1) of $[Re_2Cl_8]^{2-}$ from

Re_3Cl_9 in molten $[H_2NEt_2]Cl$ (43). More recently, Walton (44) has shown that the dinuclear, quadruply-bonded carboxylato species can be almost quantitatively converted to the Re_3X_9 species (Equation 2). Wilkinson (45) has lately shown that the

$$Re_3Cl_9 \xrightarrow[\text{molten}]{(NH_2Et_2)Cl} [Re_8Cl_8]^{2-} \qquad (1)$$

$$3Re_2(O_2CCH_3)_4X_2 \xrightarrow[X=Cl,\ Br]{HX(gas)} 2Re_3X_9 \qquad (2)$$

trinuclear $Re_3Cl_3(CH_2SiMe_3)_6$ reacts under quite mild conditions with PMe_3 to give a mixture of isomers of $Re_2Cl_2(CH_2SiMe_3)_2$-$(PMe_3)_4$ (Equation 3), and that $Re_3(CH_3)_9$ reacts, again under

$$Re_3Cl_3R_6 \xrightarrow{PMe_3} Re_2Cl_2(CH_2SiMe_3)_2(PMe_3)_4 \qquad (3)$$

very mild conditions, with PMe_3 to give a dinuclear species (Equation 4). It is to be noted that reaction (3) involves reduction as well as transformation of Re_3 to Re_2.

$$Re_3(CH_3)_9 \xrightarrow{PMe_3} Re_2(CH_3)_6(PMe_3)_2 \qquad (4)$$

With molybdenum, Sattelberger (46) has shown how to convert $[Mo_2Cl_8]^{4-}$ to $[Mo_6Cl_8]^{4+}$ (Equation 5).

$$3[Mo_2Cl_8]^{4-} \xrightarrow[\text{molten}]{NaAlCl_4} [Mo_6Cl_8]^{4+} \qquad (5)$$

Relevance and Uses

Under this heading come two kinds of relevance: first, relevance to the rest of chemistry as a pure science, and, second, relevance in a utilitarian or technological sense.

The field of metal atom clusters and M-M multiple bonds is no island separated from the rest of chemistry, but is related as a peninsula to the mainland. It, therefore, has considerable relevance and, indeed, importance to the rest of chemistry. This may be illustrated by the following observations.

Compounds containing clusters or M-M multiple bonds often provide synthetic routes to conventional complexes that are superior to others and, in some cases afford products not yet accessible in other ways. This is particularly the case when species with M-M multiple bonds are treated with strong π-acceptor ligands. Since this subject will be discussed in detail later in this symposium by R. A. Walton, no more will be said here.

If species with M-M multiple bonds are to be desirable starting materials for syntheses, they must themselves be readily accessible. Although in some cases the first preparative methods to be discovered were exotic or inconvenient, for the majority of the common species convenient, routine preparations are now known. As an illustration, let us consider preparation of compounds of the $[Mo_2Cl_8]^{4-}$ ion. This was first reached by the following process, requiring the use of the relatively expensive

$$Mo(CO)_6 + \text{excess } HOAc \longrightarrow Mo_2(OAc)_4$$

$$Mo_2(OAc)_4 + HCl(aq) \longrightarrow [Mo_2Cl_8]^{4-}$$

carbonyl as a starting material. It has now been shown (47) how to proceed from readily available MoO_3 using aqueous chemistry to $[Mo_2Cl_8]^{4-}$ as follows:

$$MoO_3 \text{ in } HCl \xrightarrow{+e^-} MoCl_6{}^{3-} \longrightarrow Mo_2Cl_9{}^{3-} \xrightarrow{Zn} [Mo_2Cl_8]^{4-}$$

A recent development of particular interest is the recognition that the aquo ion of Mo^{IV} is presumably a trinuclear metal atom cluster compound, and that it reacts with a variety of ligands (e.g., oxalate (48), EDTA (48), SCN^- (49), F^- (59)) to to give crystalline complexes containing an equilateral Mo_3 triangle with Mo-Mo single bonds (*ca.* 2.50Å) (51). These observations are unusually interesting because they show how metal clusters may be involved in the most basic, simple, benchtop chemistry of an element, and also because they extend the range of oxidation states for metal atom clusters into a higher range, +4, than ever before.

It is also noteworthy that both molybdenum and tungsten carbonyls react with carboxylic acids to give excellent yields of another type of trinuclear cluster with equilateral triangles of metal atoms, in which the triangles are capped above and below by single oxygen atoms (52).

Another interesting example of the application of clusters to a problem in a different area of chemistry is the use of the $Mo_2{}^{4+}$ unit to complex peptides in such a way that they retain their natural conformations, which can then easily be studied by X-ray crystallography (53).

In the area of technological application there are two major topics: the (real or postulated) involvement of metal atom clusters in catalysis and the remarkable superconducting properties of the Chevrel phases. The former topic has been abundantly covered in current review literature (54), although there are cautions to be recognized (55), and will not be further discussed here. Even the $Rh_2{}^{4+}$ complexes have been reported to be catalytically active. Thus, $Rh_2(O_2CCF_3)_4$ exhibits catalytic activity, toward cyclopropanation of alkenes with alkyl

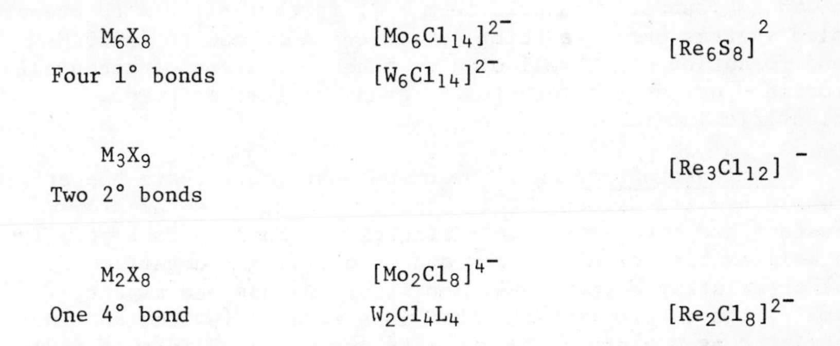

M_6X_8
Four 1° bonds
$[Mo_6Cl_{14}]^{2-}$
$[W_6Cl_{14}]^{2-}$
$[Re_6S_8]^{2}$
M_3X_9
Two 2° bonds
$[Re_3Cl_{12}]^{-}$
M_2X_8
One 4° bond
$[Mo_2Cl_8]^{4-}$
$W_2Cl_4L_4$
$[Re_2Cl_8]^{2-}$

Figure 3. Schematic of how larger clusters are formally related to diatomic multiply bonded clusters

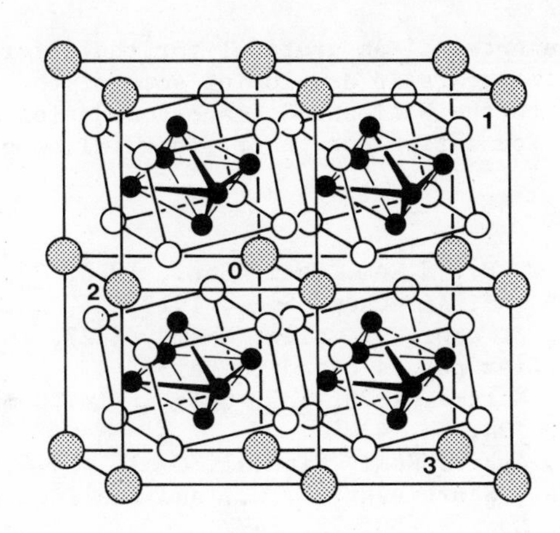

Figure 4. Structure of one of the Chevrel compounds $PbMo_6S_8$. This drawing was kindly supplied by M. J. Sienko of Cornell University. ((●) Mo; (○) S; (⊛) Pb)

diazoacetates (56) and toward the oxidation of cyclohexene when combined with a vanadium cocatalyst (57).

The extraordinary superconductance (58) of the Chevrel type compounds of which $PbMo_6S_8$, Figure 4, is a representative constitute a remarkable example of how new and valuable physical as well as chemical, properties may be found in the field of M-M bonded compounds. It would seem that further efforts to prepare solid systems under reducing conditions designed to favor M-M bond formation might well open up other new areas of technical promise - or, at the very least, areas of further purely scientific interest.

Concluding Statement. Over the past years there has arisen a whole new transition metal chemistry, that of metal atom clusters and metal-metal multiple bonds. This is conceptually as well as historically a new and revolutionary departure from the preexisting Werner type chemistry, and has one aspect, namely, the quadruple bond, that is a totally new concept in chemistry as a whole. The activity and progress in this field over the past 17 years has been phenomenal and shows every sign of continuing unabated for some time. Very probably important technical applications of compounds with M-M bonds will soon become more numerous.

Acknowledgements. I am grateful for the efforts of many coworkers over the years in developing some of the ideas discussed here and to the National Science Foundation and The Robert A. Welch Foundation for their financial support.

Literature Cited

1. Cf. Chapin, W. H., J. Am. Chem. Soc., 1910, 32, 327; Harned, H. S., J. Am. Chem. Soc., 1913, 35, 1078.
2. Lindner, K., Z. anorg. u. allg. Chem., 1927, 162, 203 and numerous earlier papers cited therein.
3. Brosset, C., Nature, 1935, 135, 874; Arkiv. Kemi, Mineral. Geol., 1935, 12A, No. 4.
4. Brosset, C., Arkiv. Kemi, Mineral. Geol., 1946, A22, No. 11.
5. Vaughn, P. A., Sturtivant, J. H., and Pauling, L., J. Am. Chem. Soc., 1950, 72, 5477.
6. Bertrand, J. A., Cotton, F. A., and Dollase, W. A., J. Am. Chem. Soc., 1963, 85, 1349; Inorg. Chem., 1963, 2, 1166.
7. Robinson, W. T., Fergusson, J. E., and Penfold, B. R., Proc. Chem. Soc., 1963, 116.
8. Cotton, F. A., Curtis, N. F., Harris, C. B., Johnson, B.F.G., Lippard, S. J., Mague, J. T., Robinson, W. R. and Wood, J.S., Science, 1964, 145, 1305; Cotton, F. A., Inorg. Chem., 1965, 4, 334.

9. Bennett, M. J.; Cotton, F. A; and Walton, R. A., J. Am. Chem. Soc., 1966, 88, 3866; Proc. Roy. Soc., London, Ser. A, 1968, 303, 175.
10. Pauling, L., "The Nature of the Chemical Bond"; 3rd Ed., Cornell University Press, 1960, page 403 and related text.
11. Cotton, F. A., Quart. Rev., 1966, 20, 389.
12. Chini, P.; Longoni, G.; and Albano, V. G., Adv. Organometal. Chem., 1976, 14, 285.
13. King, R. B., Prog. Inorg. Chem., 1972, 15, 287.
14. Cotton, F. A. and Wilkinson, G., "Advanced Inorganic Chemistry"; 4th Ed., John Wiley, N. Y., 1980, Chapter 26.
15. Cotton, F. A., Accts. Chem. Res., 1978, 11, 225.
16. Chisholm, M. H; and Cotton, F. A., Accts. Chem. Res., 1978, 11, 356.
17. Templeton, J. L., Prog. Inorg. Chem., 1979, 26, 211.
18. Cotton, F. A.; Gage, L. D.; Mertis, K.; Shive, L. W. and Wilkinson, G., J. Am. Chem. Soc., 1976, 98, 6922.
19. Cotton, F. A.; Fanwick, P. E.; Fitch, J. W.; Glicksman, H. D. and Walton, R. A., J. Am. Chem. Soc., 1979, 101, 1752.
20. Norman, J. G., Jr.; and Ryan, P. B., J. Computa. Chem., 1980, 1, 59 and references to earlier work by J.G.N. therein.
21. Benard, M., J. Am. Chem. Soc., 1978, 100, 2354.
22. Hay, P. J., J. Am. Chem. Soc., 1978, 100, 2897.
23. Bursten, B. E.; Cotton, F. A.; Green, J. C.; Seddon, E. A.; and Stanley, G. G., J. Am. Chem. Soc., 1980, 102, 4579.
24. Cotton, F. A. and Stanley, G. G., Inorg. Chem., 1977, 16, 2668; Benard, M., J. Chem. Phys., 1979, 71, 2546.
25. Bursten, B. E.; Cotton, F. A. and Stanley, G. G., Israel J. Chem., 1980, 19, 132.
26. Trogler, W. C. and Gray, H. B., Accts. Chem. Res., 1978, 11, 232.
27. Cotton, F. A., J. Mol. Struct., 1980, 59, 97.
28. Bursten, B. E.; Cotton, F. A.; Cowley, A. H.; Hanson, B. E.; Lottman, M. and Stanley, G. G., J. Am. Chem. Soc., 1979, 101, 6244.
29. Berry, M.; Garner, C. D.; Hillier, I. H.; Mac Dowell, A. and Clegg, W., J.C.S. Chem. Comm., 1980, 494.
30. Trogler, W. C.; Ellis, D. E. and Berkowitz, J., J. Am. Chem. Soc., 1979, 101, 5896.
31. Bursten, B. E.; Cotton, F. A.; Green, J. C.; Seddon, E. A. and Stanley, G. G., J. Am. Chem. Soc., 1980, 102, 955.
32. Haycock, D.; Urch, D. S.; Garner, C. D., and Hillier, I. H. J. Electron. Spect. Related Phenom., 1979, 17, 345.
33. San Filippo, J., Inorg. Chem., 1972, 11, 3140.
34. Chisholm, M. H.; Cotton, F. A.; Frenz, B. A.; Reichert, W. T.; Shive, L. W.; and Stults, B. R., J. Am. Chem. Soc., 1976, 98, 4469.
35. Chisholm, M. H.; Cotton, F. A.; Extine, M. and Stults, B. R. J. Am. Chem. Soc., 1976, 98, 4477.

36. McGlinchey, M. J., Inorg. Chem., 1980, 19, 1392.
37. Trogler, W. C.; Cowman, C. D.; Gray, H. B. and Cotton, F. A.; J. Am. Chem. Soc., 1977, 99, 2993.
38. Connor, J. A.; Pilcher, G.; Skinner, H. A.; Chisholm, M. H. and Cotton, F. A., J. Am. Chem. Soc., 1978, 100, 7738.
39. Morss, L. R.; Porcja, R. J.; Nicoletti, J. W.; San Filippo, J., and Jenkins, H.D.B., J. Am. Chem. Soc., 1980, 102, 1923.
40. Gupta, S. K.; Atkins, R. M. and Gingerich, K. A., Inorg. Chem., 1978, 17, 3211.
41. Bursten, B. E. and Cotton, F. A., Faraday Symposium No. 14, in press; Bursten, B. E.; Cotton, F. A.; and Hall, M. B., J. Am. Chem. Soc., 1980, 102, 6348.
42. Hall, M. B., J. Am. Chem. Soc., 1980, 102, 2104.
43. Bailey, R. A. and McIntyre, J. A., Inorg. Chem., 1966, 5, 1940.
44. Glicksman, H. D.; Hamer, A. D.; Smith, T. J. and Walton, R. A., Inorg. Chem., 1976, 15, 2205.
45. Edwards, P.; Mertis, K.; Wilkinson, G.; Hursthouse, M. B.; and Abdul Malik, K. M., J.C.S. Dalton, 1980, 334.
46. Sattelberger, A. P., personal communication.
47. Bino, A.; and Gibson, D., J. Am. Chem. Soc., 1980, 102, 4277.
48. Bino, A.; Cotton, F. A. and Dori, Z. J. Am. Chem. Soc., 1978, 100, 5252; 1979, 101, 3842.
49. Murmann, R. K., and Shelton, M. E., J. Am. Chem. Soc., 1980, 102, 3984.
50. Müller, A. Bielefeld University, Private Communication.
51. Jostes, S.; Müller, A.; and Cotton, F. A., Angew. Chem. in press.
52. Bino, A.; Cotton, F. A.; Dori, Z.; Koch, S.; Küppers, H. Millar, M., and Sekutowski, J. C. Inorg. Chem., 1978, 17, 3245.
53. Bino, A; and Cotton, F. A. J. Am. Chem. Soc., 1980, 102, 3014.
54. Smith, A. K.; and Bassett, J. M. J. Mol. Catal., 1977, 2, 229; Muetterties, E. L., Science, 1977, 196, 839; Muetterties, E. L.; Rhodin, T. N.; Band, E.; Brucker, C. F.; and Pretzer, W. R. Chem. Rev., 1979, 79, 91; Ugo, R. "Aspects of Homogeneous Catalysis, Wiley,"Vols. 1, 2, 1970, 1976.
55. Moskovits, M. Accts. Chem. Res., 1979, 12, 229.
56. Hubert, A. J.; Noels, A. F.; Anciaux, A. F.; and Teyssie, P. Synthesis, 1976, 600.
57. Noels, A. F.,; Hubert, A. J.; and Teyssie, P. J. Organomet. Chem., 1979, 166, 79.
58. See Delk, II, F. S.; and Sienko, M. J. Inorg. Chem., 1980, 19, 1352, for extensive references.

RECEIVED November 21, 1980.

2

Anything One Can Do, Two Can Do, Too—And It's More Interesting

MALCOLM H. CHISHOLM
Department of Chemistry, Indiana University, Bloomington, IN 47405

I should like to propose that all the types of reactions which have been established for mononuclear transition metal complexes will also occur for dinuclear transition metal complexes, and furthermore, that the latter will show additional modes of reactivity which are uniquely associated with the metal-metal bond. In this article, I shall support this proposal by illustrations taken from the reactions of dinuclear compounds of molybdenum and tungsten, two elements which enter into extensive dinuclear relationships (1).

Coordination Numbers and Geometries

For any transition element in a given oxidation state and d^n configuration, there is generally a fairly well defined coordination chemistry. Ligand field stabilization energies are often important, if not dominant, and readily account for the fact that mononuclear complexes of Cr(3+), Co(3+) and Pt(4+) are almost invariably octahedral, while those of Cr(2+) and high spin Co(2+) may be either 4- or 6-coordinate. The effect of the charge on the metal and attainment of an 18 valence shell of electrons are also two strong forces in determining preferred coordination numbers. High coordination numbers (7 and 8) are common for the early transition elements in their high oxidation states (4+ and 5+), while the latter transition elements in their low oxidation states often have low coordination numbers

0097-6156/81/0155-0017$05.75/0

(2, 3 and 4). Similar factors control the coordination preferences of the dinuclear compounds of molybdenum and tungsten.

$(M{\equiv}M)^{4+}$ Complexes (2). In scores of dinuclear molybdenum compounds, and to a lesser extent ditungsten compounds, a central M_2^{4+} unit has a M-M quadruple bond of configuration $\sigma^2\pi^4\delta^2$. The metal atomic orbitals involved in the M≣M bond are $d_{z^2}(\sigma)$, d_{xz}, d_{yz} (π) and d_{xy} (δ). The remaining metal atomic orbitals are available for use in metal ligand bonding giving rise to two types of compounds shown schematically in I and II below.

I II

In compounds of type I, four metal ligand bonds are formed in a plane using metal atomic s, p_x, p_y, $d_{x^2-y^2}$ orbitals. The overall geometry about the dimetal center is eclipsed as a result of the formation of the M-M δ bond. In these compounds, the metal atoms attain a 16 valence shell of electrons. In compounds of type II, two additional bonds are formed along the M-M axis. These utilize the p_z atomic orbitals of each metal atom and, in this way, the EAN rule is satisfied. At this time, the majority of compounds are of type I: axial coordination to give II is generally weak as evidenced by long Mo-axial ligand distances in the solid state. An immediate analogy is seen with the coordination chemistry of Pt(2+) which is most often square planar with coordination number 4, but sometimes 5 coordination is observed giving 16 and 18 valence shell electronic configurations, respectively. The absence of a M≣M bond between two trigonally coordinated metal atoms, e.g. as is common for compounds with M≡M bonds, can be understood: a trigonal field leaves the $d_{x^2-y^2}$ and d_{xy} orbitals degenerate and so a $\sigma^2\pi^4\delta^2$ configuration would be paramagnetic with the two unpaired electrons residing in the δ orbitals.

$(M{\equiv}M)^{6+}$ Compounds (3). A large number of compounds containing a central M_2^{6+} unit have recently been synthesized for molybdenum and tungsten in which a central M≡M bond arises from the formation of a σ bond (d_{z^2}-d_{z^2}) and a degenerate pair of π bonds (d_{xz}-d_{xz}, d_{yz}-d_{yz}) (4). In these compounds, the coordination number of the metal atoms is commonly 3 or 4, but examples of 5 and 6 are also known. For coordination number 3, the three ligand atoms bonded to each metal atom lie in a plane, giving rise to ground state structures which are either staggered, IIIa, or eclipsed, IIIb, depending upon the requirements of the ligands. Similarly, for coordination number 4, the four ligand atoms bonded to each metal atom lie in a plane, and the conformation about the M-M bond depends on the requirements of the ligands giving rise to staggered IVa, eclipsed IVb (shown below) or intermediate situations.

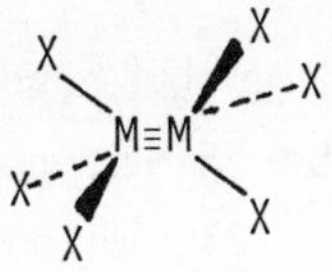

IIIa 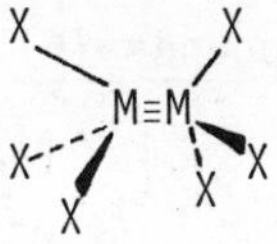 IIIb

IVa  IVb

For coordination numbers 5 and 6, which are less common, the structures of $W_2(O_2CNEt_2)_4Me_2$ and $W_2(O_2CNMe_2)_6$, shown in V and VI below, reveal that five equivalent bonds can readily be formed in a pentagonal plane (using metal s, p_x, p_y, d_{xy} and

$d_{x^2-y^2}$ atomic orbitals), but the sixth bond which must be formed along the M-M axis (using the p_z atomic orbital) is formed only weakly (5).

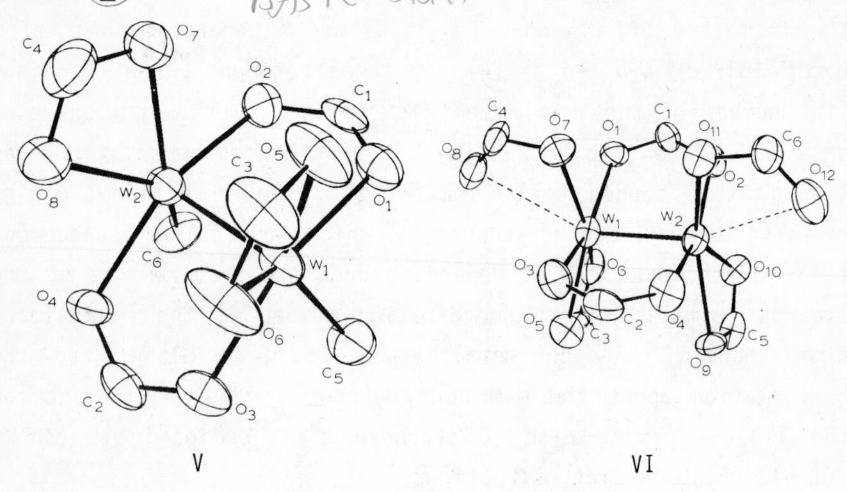

V VI

$(M{=}M)^{8+}$ Compounds (6). Compounds containing M=M bonds are, at present, not common. The coordination number 5 is, however, seen in both $Mo_2(OPr^i)_8$ (7) and $Mo_2(OBu^t)_6(CO)$ (8), which have, with respect to each metal atom, trigonal bipyramidal and square based pyramidal structures, respectively. See VII and VIII below.

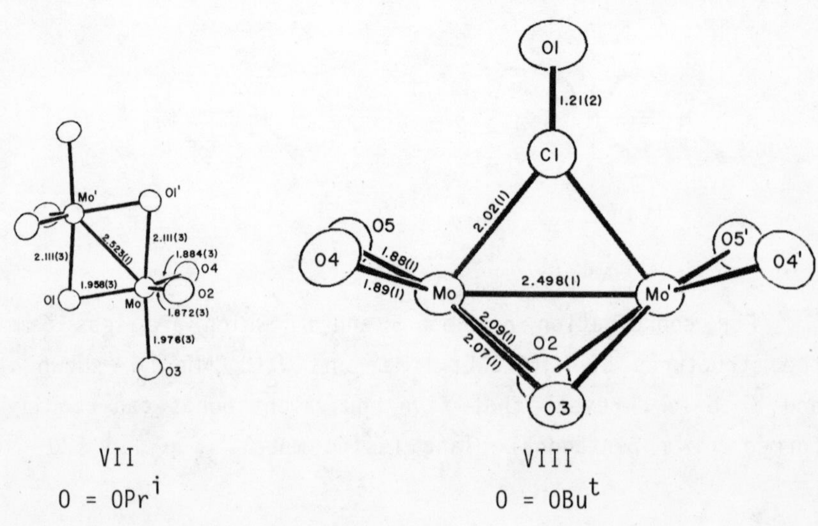

VII
$O = OPr^i$

VIII
$O = OBu^t$

$(M\text{-}M)^{10+}$ Compounds. Single M-M bonds formed by molybdenum and tungsten in the +5 oxidation state are dependent on the nature of the bridging ligands (9). For example, Mo_2Cl_{10} is paramagnetic (10) and does not show structural evidence (11) for a M-M bond. Oxygen ligands, however, appear to favor the M-M bond (12) as is seen in the recent structural characterizations of $Mo_2X_4(OPr^i)_6$ compounds which have the structure shown in IX below (13) with Mo-to-Mo distances of 2.73 Å. The coordination number 6 is seen in IX. The M-M bond arises from the interactions of metal d orbitals which have their lobes directed inbetween the bridging ligands. An octahedral geometry for each metal is thus well suited for this type of d^1-d^1 M-M single bond.

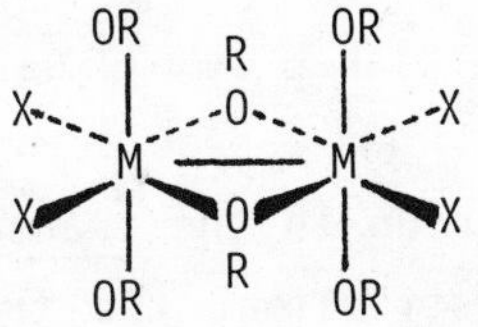

X = Br or Cl

IX

For the sake of brevity, I have restricted my attention here to M-M bonds of integral order and have considered only the cases where all the available d^n electrons are used to form M-M bonds. There are, however, dinuclear compounds having M-M bonds of fractional order (14) and dinuclear compounds in which not all the available d^n electrons contribute to M-M bonding. Well known examples of the latter are $Cp_2Mo_2(CO)_4$ (M≡M) (15) and $Cp_2Mo_2(CO)_6$ (M-M) (16) compounds which both contain molybdenum atoms in oxidation state number +1 (formally they are d^5-d^5 dimers), but by considerations of the EAN rule and the observed M-M distances (2.448(1) and 3.235(1) Å) are commonly considered to have M-M triple and single bonds, respectively.

Ligand Substitution Reactions (17)

Mononuclear compounds can be broadly classified as kinetically labile or inert toward ligand substitution reactions: e.g. Cr(3+), Co(3+) and Pt(4+) are inert, while Cr(2+) and high spin Co(2+) are labile. These dramatic differences in kinetic lability are easily rationalized by ligand field considerations. Other factors which are important in determining rates of ligand substitution are the size of the ligands, the formal positive charge and the number of valence electrons on the metal. Many complexes which satisfy the EAN rule are substitutionally "inert" and undergo substitution by an initial loss of a ligand. Group 6 transition metal carbonyl compounds provide good examples of the latter phenomenon. On the other hand, the coordinatively unsaturated square planar complexes of the group 8 transition elements react by ligand association reactions (S_N2). The labilizing effect of a group in the trans position to the group which is undergoing substitution, the trans-effect phenomenon, is well documented for square planar and octahedral complexes and can allow kinetic control in the isolation of isomers of octahedral and square planar compounds.

All of these considerations carry over into the dinuclear chemistry of molybdenum and tungsten. Ligand substitution reactions around the $(M\equiv M)^{4+}$ and $(M\equiv M)^{6+}$ moieties are well documented (6). Kinetically, they are slower than the nmr time-scale. Thus, $Mo_2R_4(PMe_3)_4$ compounds (M≡M) will not exchange coordinated PMe_3 with excess PMe_3 to give a single PMe_3 resonance, but addition of a different phosphine will lead to rapid scrambling on the synthetic time-scale.

In the reaction between anti-$W_2Cl_2(NEt_2)_4$ and alkyllithium reagents LiR (2 equiv), substitution of Cl-by-R occurs with retention of configuration (18). Kinetically anti-$W_2R_2(NEt_2)_4$ is formed which then isomerizes to a mixture of anti and gauche rotamers, with the latter being the favored rotamer. This substitution reaction can be viewed as an example of an S_F2 process in which the new bond is formed as the old bond is broken within a square plane (19). Formation of the anti rotamer arises because the cogging effect of NR_2 groups produces an energy barrier to rotation about the M≡M bond (E_{Act} = ca. 24 Kcal mol^{-1}) (20). Ligand substitution is thus kinetically faster than anti ⇌ gauche isomerization. Another example of kinetic control in ligand substitution at a dimolybdenum center is seen in the following. Hexane solutions of 1,2-$Mo_2Br_2(CH_2SiMe_3)_4$ (M≡M) react with $LiNMe_2$ and $HNMe_2$ to give 1,1- and 1,2-$Mo_2(NMe_2)_2$-$(CH_2SiMe_3)_4$, respectively, which once formed do not isomerize readily (21). Addition of Bu^tOH to 1,1- and 1,2-$Mo_2(NMe_2)_2$-$(CH_2SiMe_3)_4$ yields 1,1- and 1,2-$Mo_2(OBu^t)_2(CH_2SiMe_3)_4$, respectively, whereas with CO_2 (1 atmos, 25°C), the 1,1-isomer yields 1,1'-$Mo_2(NMe_2)(O_2CNMe_2)(CH_2SiMe_3)_4$, while the 1,2-isomer does not react. Clearly, a rich substitution chemistry surrounds these dinuclear compounds and remains to be explored and exploited.

Stereochemical Lability

Three types of stereochemical lability have been observed: (1) rotations about M-M bonds, (2) cis ⇌ trans isomerizations at each metal center and (3) bridge ⇌ terminal ligand exchange processes.

Rotations about the M≣M bond are not observed (on the nmr time-scale) and are not expected since this would rupture the δ bond. In compounds containing the central $(M{\equiv}M)^{6+}$ moiety, which have the $\sigma^2\pi^4$ configuration, rotation appears to be only restricted by the steric properties of the ligands. The compounds 1,1- and 1,2-$Mo_2(NMe_2)_2(CH_2SiMe_3)_4$ show E_{Act} for M-M rotation of ca. 15 Kcal mol^{-1}. For $Mo_2Me_2(CHSiMe_3)_4$, the barrier is less than 9 Kcal mol^{-1} and conformers have not been frozen out on the nmr time-scale (22).

The compounds $Mo_2(OR)_6L_2$, where R = alkyl or $SiMe_3$ and L = a neutral nitrogen donor ligand, contain three oxygen atoms and one nitrogen atom coordinated to each molybdenum atom (23,24,-25). Thus, there are a pair of mutually trans OR ligands and one which is trans to the N atom. At low temperatures in toluene-d_8, the ^1H nmr spectra are consistent with the expected 2:1 ratio of OR ligands described above. Upon raising the temperature, all the OR ligands become equivalent on the nmr time-scale. For $Mo_2(OPr^i)_6(NCNMe_2)$, the molybdenum atoms are different because of an asymmetric central $Mo_2(\mu\text{-}NCNMe_2)$ moiety (26). Here the low temperature limiting spectrum reveals four types of OR ligands in the expected integral ratio 1:1:2:2. Upon raising the temperature, there is a collapse of a pair of these signals to give at ca. +35°C, and 220 MHz, three OPr^i resonances in the integral ratio 3:2:1, consistent with the view that cis $\rightleftharpoons$ trans OPr^i exchange is rapid at one molybdenum, but not at the other. These exchange processes have been shown to be intramolecular and thus, parallel the common lability associated with five coordinate mononuclear complexes (27). For the dinuclear compounds, however, the fifth coordination site is the other metal atom.

The compounds $Mo_2(OPr^i)_8$ (M=M) (7) and $Mo_2(OBu^t)_6(\mu\text{-}CO)$ (M=M) (8), which contain bridging OR ligands, show rapid bridge $\rightleftharpoons$ terminal group exchange on the nmr time-scale. Similarly, the compounds $W_2Me_2(O_2CNEt_2)_4$ and $W_2(O_2CNMe_2)_6$ show exchange between bridging and terminal carbamato ligands on the nmr time-scale (5).

Organometallic Reactions

Tolman (28) has suggested that all commonly occurring organometallic reactions, including those that are important in catalysis, can be classified by five named reactions, each with its microscopic reverse: (1) Lewis base association and dissociation, (2) Lewis acid association and dissociation, (3) insertion and deinsertion (ligand migration reactions), (4) oxidative addition and reductive elimination, and (5) oxidative coupling and reductive decoupling. With the exception of the simple Lewis acid association-dissociation reactions, we have studied examples of all of the aforementioned reactions.

Lewis Base Association and Dissociation. The dinuclear alkoxides $M_2(OR)_6$ reversibly add donor ligands to give $M_2(OR)_6L_2$ compounds (23,24): the position of equilibrium is dependent on the bulkiness of R and L. In these reversible Lewis base addition reactions, the M-M distances are essentially unchanged, c.f. (23,24) Mo-to-Mo = 2.222(1) Å in $Mo_2(OCH_2CMe_3)_6$ with Mo-to-Mo = 2.242(1) Å in $Mo_2(OSiMe_3)_6(HNMe_2)_2$, since the metal atoms do not attain an 18 valence shell electronic configuration. However, for compounds containing M≡M bonds in which the metal atoms have a completed valence shell of electrons, e.g. Cp_2M_2-$(CO)_4$ compounds (M = Mo and W), the formation of the two new metal-ligand bonds will occur only with a reduction in M-M bond order, from three to one, c.f. the Mo-to-Mo distances of 2.448(1) Å (15) and 3.235(1) Å (16) found for $Cp_2Mo_2(CO)_4$ and $Cp_2Mo_2(CO)_6$, respectively.

Insertion-Deinsertion Reactions. An example of a facile reversible insertion-deinsertion reaction is seen in the reactions between $M_2(OR)_6$ compounds and CO_2 which give $M_2(OR)_4$-$(O_2COR)_2$ compounds (29). These reactions were shown to proceed by a direct insertion mechanism (i.e. not by a mechanism involving catalysis by traces of alcohols) with energies of activation of not greater than 20 Kcal mol^{-1}.

Oxidative-Addition Reactions. Chuck Kirkpatrick has been studying oxidative additions to $Mo_2(OPr^i)_6$ (M≡M) (30). Addition of Pr^iOOPr^i leads to $Mo_2(OPr^i)_8$ (M=M) (7). Addition of each of the halogens Cl_2, Br_2 and I_2 proceeds to give the compounds $Mo_2(OPr^i)_6X_4$ (M-M) where X = Cl, Br or I. In these oxidative-additions, a stepwise change in M-M bond order, from three to two to one, is achieved. A large number of related additions have been noted, but await detailed structural characterizations.

The formation of $W_4(\mu\text{-H})_2(OPr^i)_{14}$ in the reaction between $W_2(NMe_2)_6$ and Pr^iOH (excess) can also be viewed as an oxidative-addition: $W_2(OPr^i)_6 + Pr^iO\text{-}H \rightarrow \frac{1}{2}[W_2(\mu\text{-H})(OPr^i)_7]_2$ (31). An ORTEP view of the central $W_4(\mu\text{-H})_2O_{14}$ skeleton is shown below. The molecule is centrosymmetric with alternating short (2.446(1) Å) and long (3.407(1) Å) W-to-W distances consistent with the presence of W-to-W double and non-bonding distances, respectively.

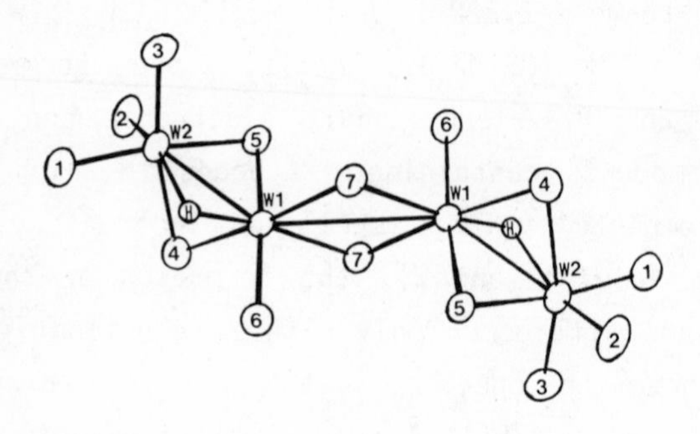

Curtis and coworkers (32) have also documented a number of oxidative-addition reactions in their studies of the reactivity of $Cp_2Mo_2(CO)_4$.

Reductive Eliminations. The synthesis of an extensive series of 1,2-$M_2R_2(NMe_2)_4$ compounds (33,34) from the reaction between 1,2-$M_2Cl_2(NMe_2)_4$ (M≡M) and alkyllithium reagents, LiR (2 equiv), where R = CH_3, Et, i-Pr, n-Bu, sec-Bu, t-Bu, CH_2CMe_3 and CH_2SiMe_3, affords the opportunity of studying the decomposition pathways of alkyl groups coordinated to dimetal centers. Some particularly interesting comparisons can be made with mononuclear σ-alkyl complexes of the transition elements, which have been the subject of much investigation (35).

The structure of the $Mo_2Et_2(NMe_2)_4$ molecule is shown in Figure 1 (34). This view emphasizes the virtual C_2 axis of symmetry which exists for gauche 1,2-$M_2X_2(NMe_2)_4$ compounds. All the hydrogens were located and refined, and in Figure 2 a stereoview of the molecule is shown which shows the orientations of the C-H bonds. Two important points can be seen: (1) the conformation about the C-C bonds of the ethyl groups is, within the limits of experimental error, perfectly staggered. Thus, as is shown in the Newman projection below, two β-hydrogen atoms are equidistant from the molybdenum atom to which the ethyl ligand is σ-bonded. The β-H-to-Mo distances are 3.25(5) Å, which, taken together with the staggered conformation about the C-C bond, indicate the absence of any significant β-H---Mo interaction (36). (2) The $MoNC_2$ units, which are planar, are aligned along the M-M axis and one hydrogen atom from each methyl group is contained within this plane. This introduces relatively short CH---Mo distances as is shown in Figure 3. The shortest distances are between the molybdenum atoms and the hydrogen atoms on the distal methyl groups, but only a little longer are the distances involving proximal methyl hydrogens and the other molybdenum atoms across the Mo≡Mo bond. This leads one to wonder, "what are the closest natural CH---Mo distances in molecules of this type?" This question is readily answered: keeping the central $Mo_2N_4C_2$ unit rigid, but allowing rotations about Mo-N, Mo-C and C-C bonds produces CH---Mo distances of 2.3 to 2.4 Å, as shown in Figure 4, between the methyl protons of the ethyl ligand and the molybdenum atom to which it is not directly bonded. The methyl groups of the ethyl ligands thus provide β and γ hydrogens to the two molybdenum atoms and the H---Mo distances across the Mo≡Mo bond are the shortest.

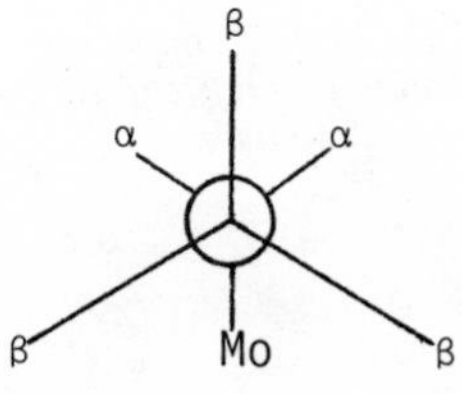

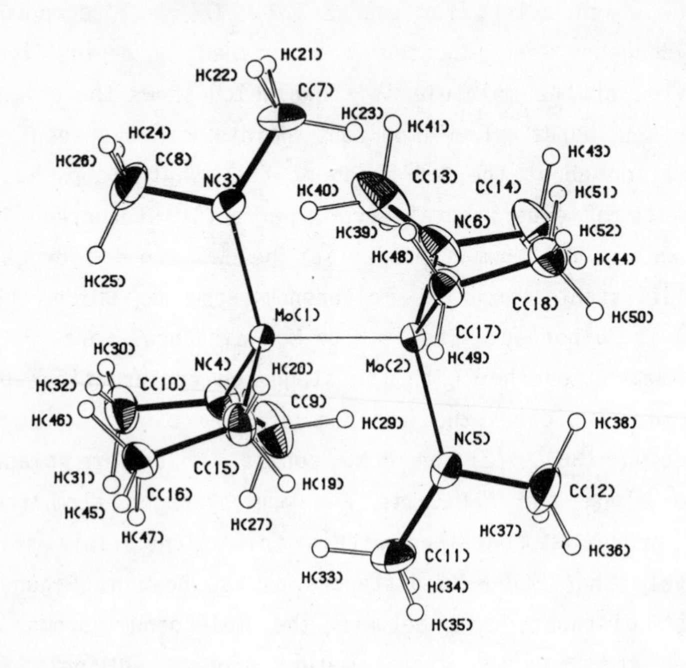

Figure 1. An ORTEP view of the $Mo_2Et_2(NMe_2)_4$ molecule emphasizing the virtual C_2 axis of symmetry. Pertinent structural parameters are Mo – – – Mo (M≡M) = 2.206(1) Å, Mo—N = 1.96 Å (av), Mo—C = 2.18 Å (av), Mo—Mo—N angle = 103° (av) and Mo—Mo—C angle = 101° (av).

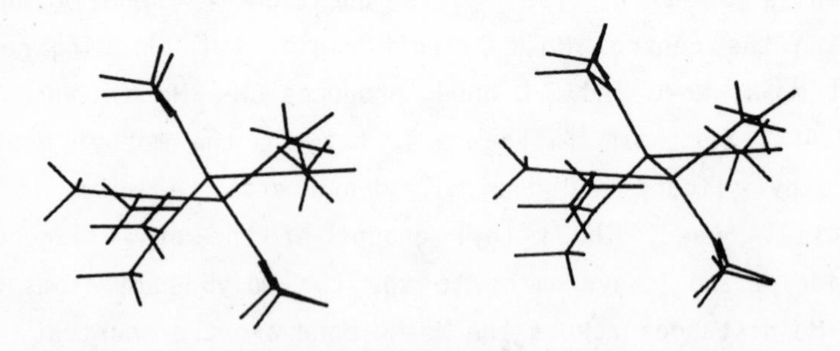

Figure 2. Stereoview of the $Mo_2Et_2(NMe_2)_4$ molecule looking down the Mo—Mo bond. This stick view of the molecule emphasizes the positions of all the hydrogen atoms.

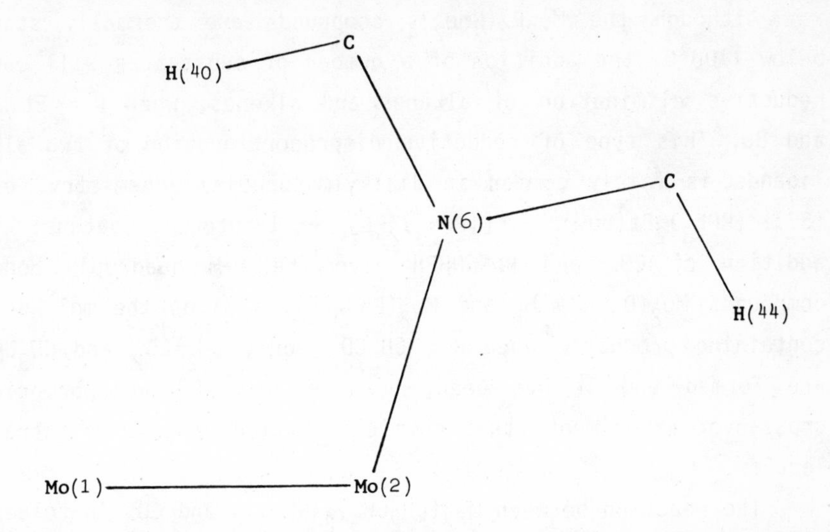

Figure 3. A line drawing of one Mo_2NC_2 fragment showing the Mo --- HC distances that arise to the two hydrogen atoms that are confined in the plane of the Mo—NC_2 unit. (Mo(2) --- H(44) 2.77 Å; Mo(2) --- H(40) 3.18 Å; Mo(1) --- H(40) 2.96 Å)

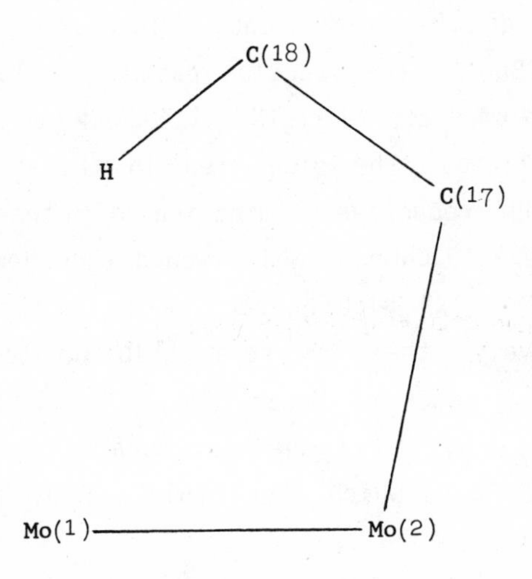

Figure 4. A line drawing of one Mo_2-ethyl fragment showing the short CH --- Mo distance that arises from rotations about Mo—C and C—C bonds (Mo(1) --- H 2.36 Å; Mo(2) --- H 2.97 Å)

Although the $Mo_2R_2(NMe_2)_4$ compounds are thermally stable below +100°C, the addition of a number of substrates will cause reductive elimination of alkanes and alkenes, when R = Et, Pr and Bu. This type of reductive disproportionation of two alkyl ligands is fairly common in dialkylmononuclear chemistry, e.g. (37) $(PPh_3)_2Pt(Bu^n)_2 \rightarrow [(PPh_3)_2Pt]_x$ + 1-butene + butane. The addition of CO_2 and PhNNNHPh give the M-M quadruply bonded compounds $Mo_2(O_2CNMe_2)_4$ and $Mo_2(PhN_3Ph)_4$ (38) as the molybdenum containing products. When R = CH_2CD_3, only $CH_2{=}CD_2$ and CD_3CH_2D are formed and it has been shown by use of the appropriate cross-over experiments that this elimination process is intramolecular.

The reaction between $Mo_2(CH_2CH_3)_2(NMe_2)_4$ and CO_2 in toluene-d_8 has been followed at -30°C by ^1H nmr spectroscopy. The build-up of an intermediate, namely $Mo_2Et_2(NMe_2)_2(O_2CNMe_2)_2$ can be seen. The spectrum corresponding to this intermediate is shown in Figure 5 and is entirely consistent with the view that the molecule has a virtual C_2 axis of symmetry passing through the mid-point of the Mo-Mo bond, i.e. as is found (29) for $Mo_2(OBu^t)_4(O_2COBu^t)_2$. The reaction pathway following the formation of the intermediate $Mo_2Et_2(NMe_2)_2(O_2CNMe_2)_2$ poses two interesting possibilities. The slow step in the reaction could be involved in the reductive-elimination of ethane and ethylene from $Mo_2Et_2(NMe_2)_2(O_2CNMe_2)_2$ which would then generate a species "$Mo_2(NMe_2)_2(O_2CNMe_2)_2$" highly reactive to further reaction with CO_2. Alternatively, the slow step could involve CO_2 insertion into the Mo-NMe_2 bonds of $Mo_2Et_2(NMe_2)_2(O_2CNMe_2)_2$ to give say, $Mo_2Et_2(O_2CNMe_2)_4$, c.f. (5) the structurally characterized compound $W_2Me_2(O_2CNEt_2)_4$ which then rapidly eliminates ethane and ethylene.

$$Mo_2(NMe_2)_6 + ArNNNHAr\ (excess) \rightarrow Mo_2(NMe_2)_4(ArN_3Ar)_2\ (M{\equiv}M) + Me_2NH \qquad (1)$$

$$Mo_2R_2(NMe_2)_4 + ArNNNHAr \text{ (excess)} \rightarrow Mo_2R_2(NMe_2)_2(ArN_3Ar)_2 \text{ (M≡M)} + Me_2NH \quad (2)$$

$$Mo_2R_2(NMe_2)_4 + ArNNNHAr \text{ (excess)} \rightarrow Mo_2(ArN_3Ar)_4 \text{ (M≣M)} + \text{alkane} + \text{1-alkene} + HNMe_2 \quad (3)$$

We have as yet been unable to distinguish between these pathways, though we are inclined toward favoring the former pathway based on the analogy with the reactions observed with triazenes shown in 1, 2 and 3 above. The difference between (2) and (3) rests solely on the nature of R. When R = CH_3 or CH_2CMe_3, reaction 2 occurs. When R = Et and Bu^n, reaction 3 occurs. The molecular structure of the compound $Mo_2Me_2(NMe_2)_2$-$(ArN_3Ar)_2$, where Ar = p-tolyl, is shown in Figure 6. The molecule has a crystallographically imposed C_2 axis of symmmetry relating the two 4-coordinate molybdenum atoms. It is not unreasonable to suppose that reaction 3 proceeds via initial formation of the $Mo_2R_2(NMe_2)_2(ArN_3Ar)_2$ compounds, which then undergo elimination of alkane and alkene generating the coordinatively unsaturated molecules $Mo_2(NMe_2)_2(ArN_3Ar)_2$ which react rapidly with the excess triazine to give $Mo_2(ArN_3Ar)_4$ (M≣M).

At this time, nothing definitive can be said concerning the intimate mechanism of elimination beyond noting the facts: it is intramolecular, quantitative and irreversible - no scrambling of labels is observed.

A plausible interpretation of these results is that elimination occurs by a rate determining step which is akin to a cyclometallation reaction across the M-M bond. This is suggested by the close CH---Mo distance shown in Figure 4. Once the Mo-H bond is formed, elimination of alkane is rapid. Rather interestingly, the elimination of alkene from the dinuclear center may then be a slower final step: analogous reactions involving $W_2R_2(NMe_2)_4$ and either CO_2 or triazines yield 1 mole of alkane

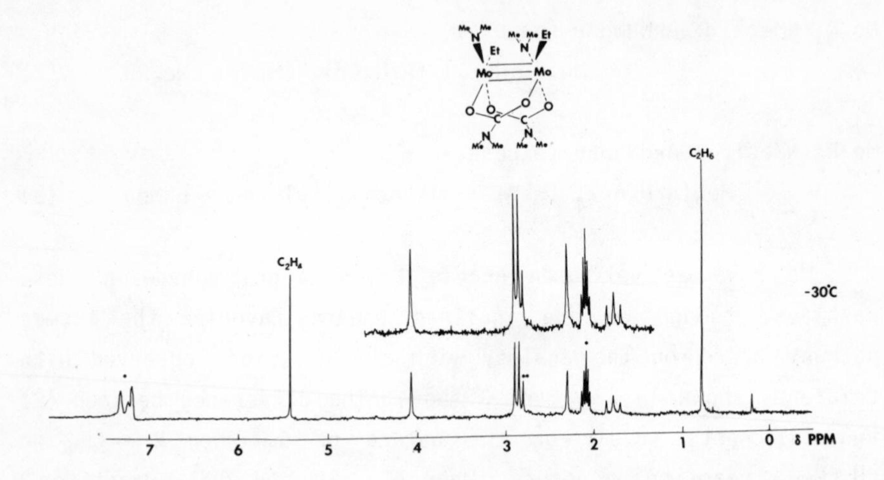

Figure 5. *¹H NMR spectrum recorded at −30°C, 100 MHz, during the reaction between $Mo_2Et_2(NMe_2)_4$ and CO_2, showing the formation of the intermediate $Mo_2Et_2(NMe_2)_2(O_2CNMe_2)_2$, along with the products C_2H_4, C_2H_6, and $Mo_2(O_2$-$CNMe_2)_4$. The solvent is toluene-d_8 and the protio impurities are indicated by (*). The signal noted by (**) corresponds to $Mo_2(O_2CNMe_2)_4$ which, because of its very low solubility, is mostly precipitated as it is formed.*

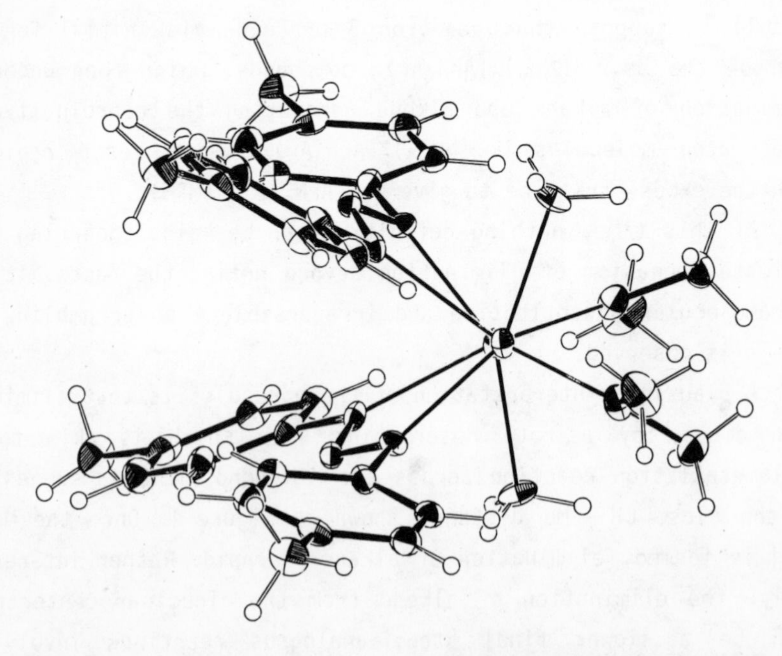

Figure 6. *An ORTEP view of the $Mo_2Me_2(NMe_2)_2(C_7H_8N_3C_7H_8)_2$ molecule (C_7H_8 = p-tolyl) viewed down the Mo≡Mo bond emphasizing the C_2 axis of molecular symmetry which relates the two 4-coordinate molybdenum atoms. The Mo – – – Mo distance is 2.175(1) Å and Mo—C = 2.193(4) Å.*

and as yet uncharacterized tungsten compounds which retain the elements of the alkene.

It is clear from this brief summary that dinuclear reductive elimination sequences have pathways which arise uniquely from the presence of the dimetal center, in addition to those accessible to mononuclear complexes (39).

Oxidative-Coupling and Reductive-Decoupling. Examples of oxidative-coupling and reductive-decoupling of ligands about a dimetal center are seen in the work of Stone, Knox and their coworkers (40) and are not discussed here.

Direct Attack on the M-M Multiple Bond

A number of small unsaturated molecules or functional groups have been found to add across the M-M triple bond in $Cp_2Mo_2(CO)_4$ to give adducts $Cp_2Mo_2(CO)_4(un)$, where un = acetylenes (41), allenes (42), cyanamides (43) and thioketones (44). In all of these adducts, the unsaturated fragments act as 4-electron donors to the dimetal center, thus reducing the M-M bond order from three to one.

The same group of unsaturated molecules have been found to react with $Mo_2(OPr^i)_6$, but as yet none of the adducts has been characterized by a full X-ray study. The spectroscopic data observed (45) for $Mo_2(OPr^i)_6(NCNMe_2)$ are entirely consistent, however, with a mode of addition directly analogous to that observed above.

Carbon monoxide has been shown (8) to reversibly add across the M≡M bond of $Mo_2(OBu^t)_6$ to give $Mo_2(OBu^t)_6(\mu\text{-CO})$ (M=M). See VIII before. We have previously speculated (46) that this carbene-like addition to a M≡M bond could be one of general class of reactions that are common to dinuclear compounds. Similar additions of CO across Pd-Pd and Pt-Pt bonds have been reported (47).

Addition of nitric oxide to $M_2(OR)_6$ compounds gives $[M(OR)_3\text{-}NO]_2$ compounds. The latter do not contain M-M bonds: the structure of $[Mo(OPr^i)_3NO]_2$ is closely related to that of $Mo_2(OPr^i)_8$, shown in VII, but has a Mo-to-Mo distance of 3.325(1) Å (48). In a formal sense, the M≡M bond in $Mo_2(OPr^i)_6$ has been replaced by two metal-ligand triple bonds (M≡N→O). The absence of any direct M-M bond in $[M(OR)_3NO]_2$ compounds is also supported (in addition to the structural evidence cited above) by the fact that the alkoxy bridged dimers are readily cleaved by donor ligands such as pyridine. This has been structurally confirmed for $W(OBu^t)_3$-(NO)(py) (49).

Walton and coworkers (50) have found that compounds containing M≡M bonds are also cleaved by the addition of certain unsaturated molecules such as isocyanides. Very recently we have found (51) that $Mo_2(OBu^t)_6$ (M≡M) reacts with each of molecular oxygen (2 equiv) and aryl azides (>4 equiv) to give $Mo(O)_2$-$(OBu^t)_2$ and $Mo(NAr)_2(OBu^t)_2$, respectively. In both reactions, the metal-metal triple bond is cleaved and the elements of $2Bu^tO$ are eliminated from the metal center. Though the addition of aryl azides to low valent mononuclear transition metal complexes is known to give arylimido compounds (52), the ease of cleavage of the M≡M bond seen in these reactions, once again, emphasizes the lability of the dinuclear center toward addition reactions.

Conclusions

The time is ripe for truly exciting developments in the reactivity of dinuclear transition metal compounds. The potential for cyclic sequences of reactions, as is required for catalytic reactions, has already been realized. (1) It has been shown, by Muetterties, et al. (53), that alkynes can be selectively hydrogenated to alkenes (cis 2H-addition) by $Cp_2Mo_2(CO)_4$: the rate determining step involves CO dissociation from the acetylene adducts $Cp_2Mo_2(CO)_4(R_2C_2)$. (2) We have found that

$W_4(\mu\text{-}H)_2(OPr^i)_{14}$ will isomerize olefins: it selectively takes 1-butene to cis-2-butene, for example (54). While one can but speculate about the ultimate impact of dinuclear transition metal chemistry, it is surely fair to say that the elucidation of the intimate mechanisms of reactions at dinuclear metal centers will prove more challenging and fascinating than did their analogues at mononuclear centers.

Acknowledgements

I thank Research Corporation, the Petroleum Research Fund administered by the American Chemical Society, the Office of Naval Research and the National Science Foundation for their financial support of various aspects of this work. I am also particularly grateful to the Alfred P. Sloan Foundation and the Camille and Henry Dreyfus Teacher-Scholar Grant Program for awards which have allowed me additional degrees of freedom in promoting my research in this area. Finally, I acknowledge my long time collaborations with F. Albert Cotton and coworkers at Texas A&M University, as well as my own group whose contributions are cited in the references.

Abstract

All of the reaction chemistry surrounding mononuclear transition metal compounds could just as easily be carried out at a bimetallic center. Indeed, there should be additional types of reactions uniquely associated with the M-M bond. This general premise is discussed in light of the recent reactivity found for dimolybdenum and ditungsten compounds. For a given M_2^{x+} center, the preferred coordination geometries, ligand substitution behaviors and dynamical properties (fluxional behavior) are noted. The ability of these compounds to undergo Lewis base association and dissociation reactions, reversible insertions or ligand migrations, oxidative addition and reductive elimination reactions, as well as direct additions across the M-M bond are discussed. Finally, the possible mechanisms for one of these reactions, reductive elimination of alkane and alkene from a dimolybdenum center, is considered in detail.

Literature Cited

1. Cotton, F.A.; Wilkinson, G. "Advanced Inorganic Chemistry", 4th Edition, 1980, Wiley-Interscience.
2. For recent reviews, see (a) Cotton, F.A., Chem. Soc. Rev., 1975, 4, 27; (b) Cotton, F.A. Acc. Chem. Res., 1978, 11, 225; (c) Templeton, J.A. Prog. Inorg. Chem., 1979, 26, 211.
3. For recent reviews, see (a) Chisholm, M.H.; Cotton, F.A. Acc. Chem. Res., 1978, 11, 356 and (b) Chisholm, M.H. Faraday Society Symposium, 1980, 14, xxx.
4. For detailed calculations, see (a) ref. 2b; (b) Cotton, F.A.; Stanley, G.G.; Kalbacher, B.; Green, J.C.; Seddon, E.; Chisholm, M. H. Proc. Natl. Acad. Sci., U.S.A., 1977, 74, 3109; (c) Hall, M.B. J. Am. Chem. Soc., 1980, 102, 2104; (d) Bursten, B.E.; Cotton, F.A.; Green, J.C.; Seddon, E.A.; Stanley, G.G. J. Am. Chem. Soc., 1980, 102, 4579.
5. Chisholm, M.H.; Cotton, F.A.; Extine, M.W.; Stults, B.R. Inorg. Chem., 1977, 16, 603.
6. Chisholm, M.H. Transition Metal Chemistry, 1978, 3, 321.
7. Chisholm, M.H.; Cotton, F.A.; Extine, M.W.; Reichert, W.W. Inorg. Chem., 1978, 17, 2944.
8. Chisholm, M.H.; Cotton, F.A.; Extine, M.W.; Kelly, R.L. J. Am. Chem. Soc., 1979, 101, 7645.
9. Cotton, F.A. Acc. Chem. Res., 1969, 2, 240.
10. Klemm, W.; Steinberg, H. Z. Anorg. Allgem. Chem., 1936, 227, 193.
11. Sands, D.E.; Zalkin, A. Acta Cryst., 1956, 12, 723.
12. See references to other studies cited in ref. 9.
13. Chisholm, M.H.; Kirkpatrick, C.C.; Huffman, J.C. J. Am. Chem. Soc., submitted.

14. Cotton, F.A.; Frenz, B.A.; Webb, T.R. J. Am. Chem. Soc., 1973, 95, 4431.
15. Klinger, R.J.; Butler, W.; Curtis, M.D. J. Am. Chem. Soc., 1975, 97, 3535; idem, ibid, 1978, 100, 5034.
16. Adams, R.D.; Collins, D.M.; Cotton, F.A. Inorg. Chem., 1974, 13, 1086.
17. For general discussion, see (a) Basolo, F.; Pearson, R.G. "Inorganic Reaction Mechanisms", 2nd Ed., 1968, John Wiley Publishers; (b) Tobe, M.L. "Inorganic Reaction Mechanisms", 1972, T. Nelson Publishers; (c) Wilkins, R.G. "The Study of Kinetics and Mechanism of Reactions of Transition Metal Complexes", Allyn and Bacon Publishers, 1974.
18. Chisholm, M.H.; Extine, M.W. J. Am. Chem. Soc., 1976, 98, 6393; Chisholm, M.H.; Cotton, F.A.; Extine, M.W.; Millar, M.; Stults, B.R. Inorg. Chem., 1977, 16, 320.
19. See, Basolo, F.; Pearson, R.G. in "The Mechanisms of Inorganic Reactions", 2nd Ed., 1968, John Wiley and Sons Publishers, page 126.
20. Chisholm, M.H.; Cotton, F.A.; Extine, M.W.; Millar, M.; Stults, B.R. Inorg. Chem., 1976, 15, 2244.
21. Chisholm, M.H.; Rothwell, I.P. J.C.S. Chem. Commun., 1980, xxx.
22. Chisholm, M.H.; Rothwell, I.P. J. Am. Chem. Soc., 1980, 102, 5950.
23. Chisholm, M.H.; Cotton, F.A.; Murillo, C.A.; Reichert, W.W. Inorg. Chem., 1977, 16, 1801.
24. Chisholm, M.H.; Cotton, F.A.; Extine, M.W.; Reichert, W.W. J. Am. Chem. Soc., 1978, 100, 153.
25. See also the solid state structure and dynamical solution behavior of $W_2(OPr^i)_6(py)_2$. Akiyama, M.; Chisholm, M.H.; Cotton, F.A.; Extine, M.W.; Haitko, D.A.; Little, D.; Fanwick, P.E. Inorg. Chem., 1979, 18, 2266.
26. Chisholm, M.H.; Kelly, R.L. Inorg. Chem., 1979, 18, 2321.
27. Muetterties, E.L. Acc. Chem. Res., 1970, 3, 266.

28. Tolman, C.A. Chem. Soc. Rev., 1972, 1, 337.
29. Chisholm, M.H.; Cotton, F.A.; Extine, M.W.; Reichert, W.W. J. Am. Chem. Soc., 1978, 100, 1727.
30. Chisholm, M.H.; Kirkpatrick, C.C.; Huffman, J.C. Inorg. Chem., in press.
31. Akiyama, M.; Little, D.; Chisholm, M.H.; Haitko, D.A.; Cotton, F.A.; Extine, M.W. J. Am. Chem. Soc., 1979, 101, 2504.
32. See Chapter 12 in this volume.
33. Chisholm, M.H.; Haitko, D.A. J. Am. Chem. Soc., 1979, 101, 6784.
34. Chisholm, M.H.; Haitko, D.A.; Huffman, J.C. J. Am. Chem. Soc., submitted.
35. Kochi, J.K. "Organometallic Mechanisms and Catalysis", Academic Press Publishers, 1978, Ch. 12 and references therein.
36. This contrasts with other significant and short CH---Mo interactions: Cotton, F.A.; Day, V.W. J.C.S. Chem. Commun., 1974, 415; Cotton, F.A.; LaCour, T.; Stanislowski, A.G. J. Am. Chem. Soc., 1974, 96, 754.
37. Whitesides, G.M.; Gaasch, J.F.; Stedronsky, E.R. J. Am. Chem. Soc., 1972, 94, 5258.
38. Cotton, F.A.; Rice, G.W.; Sekutowski, J.C. Inorg. Chem., 1979, 18, 1143.
39. Reductive elimination: $L_nM(H)R \rightarrow L_nM + R\text{-}H$ is generally much faster than by C-C bond formation. See, Norton, J.R. Acc. Chem. Res., 1979, 12, 139.
40. Knox, S.A.R.; Stansfield, R.F.D.; Stone, F.G.A.; Winter, M.J.; Woodward, P. J.C.S. Chem. Commun., 1978, 221. See also, Chapter 13 in this volume.
41. Bailey, W.I.; Chisholm, M.H.; Cotton, F.A.; Rankel, L.A. J. Am. Chem. Soc., 1978, 100, 5764.
42. Bailey, W.I.; Chisholm, M.H.; Cotton, F.A.; Murillo, C.A.; Rankel, L.A. J. Am. Chem. Soc., 1978, 100, 802.

43. Chisholm, M.H.; Cotton, F.A.; Extine, M.W.; Rankel, L.A. J. Am. Chem. Soc., 1978, 100, 807.
44. Alper, H.; Silavwe, N.D.; Birnbaum, G.I.; Ahmed, F.R. J. Am. Chem. Soc., 1979, 101, 6582.
45. Chisholm, M.H.; Kelly, R.L. Inorg. Chem., 1979, 18, 2321.
46. Chisholm, M.H. Advances in Chemistry Series, 1979, 173, 396.
47. See Chapters 8 and 9 in this volume and references therein.
48. Chisholm, M.H.; Cotton, F.A.; Extine, M.W.; Kelly, R.L. J. Am. Chem. Soc., 1978, 100, 3354.
49. Chisholm, M.H.; Cotton, F.A.; Extine, M.W.; Kelly, R.L. Inorg. Chem., 1979, 18, 116.
50. Walton, R.A. Chapter 11 this volume.
51. Chisholm, M.H.; Kirkpatrick, C.C.; Raterman, A, results to be published.
52. Hillhouse, G.L. Ph.D. Thesis, Indiana University, 1980.
53. Muetterties, E.L.; Slater, S. Inorg. Chem., in press.
54. Akiyama, M.; Chisholm, M.H.; Cotton, F.A.; Extine, M.W.; Haitko, D.A.; Leonelli, J.; Little, D. J. Am. Chem. Soc., in press.

RECEIVED November 21, 1980.

3

Structure and Reactivity of Some Unusual Molybdenum and Tungsten Cluster Systems

R. E. McCARLEY, T. R. RYAN, and C. C. TORARDI

Ames Laboratory and Department of Chemistry, Iowa State University, Ames, IA 50011

One aspect of the reactivity of metal-metal bonds concerns the approaches to rational synthesis of metal cluster species of increasing nuclearity. A commonly desired approach involves the addition of a metal atom, metal ligand fragment or metal cluster fragment to an existing metal cluster species such that units of increasing size and complexity can be built up in a controlled stepwise manner. This approach has been quite successful when applied to organometallic clusters (1), but not nearly so effective in the synthesis of cluster species to which only classical ligands are attached. In the latter category the elements providing the most numerous examples of metal-metal bonded structures are those found early in the 4d and 5d transition series (2). Molybdenum is particularly outstanding in the number, diversity and robustness of its cluster compounds (2). Thus it is natural to examine the reactivity of molybdenum clusters selected because of their availability, well-characterized structural features, and particular metal-metal bond type or electronic arrangement.

The work reported here concerns the addition of quadruply bonded dimers of the type $Mo_2Cl_4L_4$ according to eqn. 1,

$$2\ Mo_2Cl_4L_4 = Mo_4Cl_8L_4 + 4L \tag{1}$$

whereby new tetranuclear clusters of rectangular geometry are generated. Finally, a glimpse of exciting chemistry yet to come from ternary oxide compounds is provided by the new metal-metal bonded structures $NaMo_4O_6$ and $Ba_{1.13}Mo_8O_{16}$.

Experimental

Details of the experimental work and structure determinations will be given in separate papers to be published in regard to the separate aspects of work summarized here.

0097-6156/81/0155-0041$05.00/0

Rectangular Clusters

Molybdenum Clusters of the Type $Mo_4Cl_8L_4$. The chemistry described here was initiated with the notion that it should be possible to observe addition of molecular species X-Y across multiple metal-metal bonds, for example as in the process

$$L_nM{\equiv}ML_n + X\text{-}Y \longrightarrow \underset{\substack{|\\X}}{L_nM}{\equiv}\underset{\substack{|\\Y}}{ML_n}$$

Such a process was considered likely because in the quadruply bonded dimers the metal atoms have vacant coordination sites to accept the ligand atoms of X and Y, and the X-Y bond energy plus the weak δ-bond of the dimer are compensated by formation of M-X and M-Y bonds. On the further assumption that unfavorable steric crowding of ligands about the metal atoms account for the lack of addition chemistry previously observed for these dimers we sought to prepare compounds having weakly bound ligands. Dissociation in solution as in eqn. 2

$$Mo_2Cl_4L_4 = Mo_2Cl_4L_3 + L \tag{2}$$

would then provide species more reactive towards addition. To this end we attempted the preparation of $Mo_2Cl_4(P\phi_3)_4$, which because of the lower donor strength and large size of triphenylphosphine was expected to meet the desired requirements. Although many derivatives of $Mo_2Cl_4(PR_3)_4$, where R = alkyl, had been reported (3), $Mo_2Cl_4(P\phi_3)_4$ had not.

All attempts to prepare $Mo_2Cl_4(P\phi_3)_4$ thus far have failed, but in the course of this work a molecule having the desired reactivity was obtained via eqn. 3

$$(NH_4)_5Mo_2Cl_9{\cdot}H_2O + 2\phi_3P + 2Ch_3OH \xrightarrow[\text{excess } \phi_3P]{CH_3OH}$$

$$Mo_2Cl_4(P\phi_3)_2(CH_3OH)_2 + 5NH_4Cl + H_2O \tag{3}$$

The compound is crystallized from methanol as the solvate $Mo_2Cl_4(P\phi_3)_2(CH_3OH)_2{\cdot}nCH_3OH$ ($n{\sim}2.2$). A structure determination (4) revealed the molecular configuration shown in Fig. 1. A noteworthy feature of the structure is the long Mo-O bond, 2.211(5)Å, to the coordinated methanol ligands; other features of the structure are not unusual. Based on the bond radius for Mo derived from the average Mo-Cl distance, 2.404(2)Å, the Mo-O bonds are ca. 0.14Å longer than the sum of covalent radii. This feature of the structure suggests easy loss of the methanol ligands in accordance with the reaction subsequently explored.

Some reactions of $Mo_2Cl_4(P\phi_3)_2(CH_3OH)_2$ are given in Scheme I. Upon dissolving this compound in benzene at 25° a

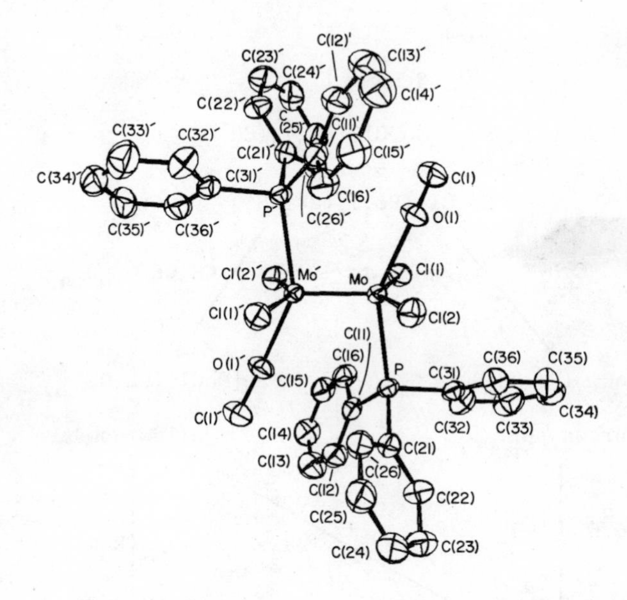

Figure 1. Molecular structure of $Mo_2Cl_4(P\phi_3)_2(CH_3OH)_2$

Scheme I

Reactions of $Mo_2Cl_4(P\phi_3)_2(CH_3OH)_2$

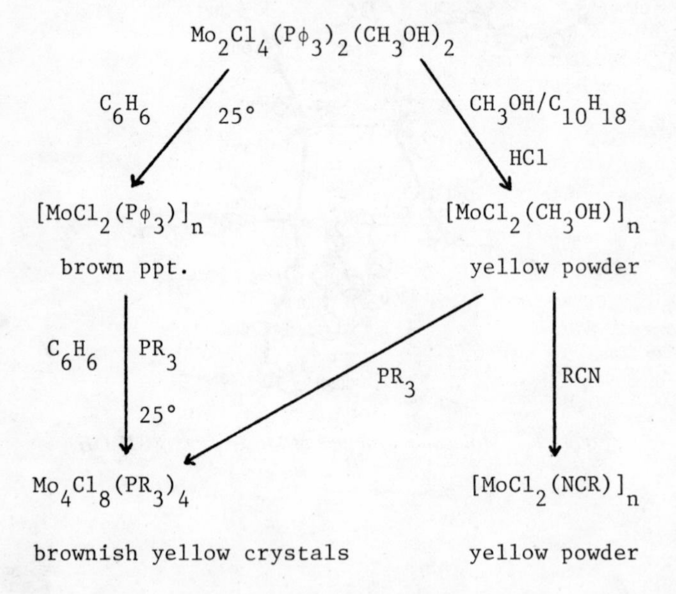

reaction leading to the precipitation of a brown solid is complete within *ca.* 30 min. This compound is difficult to obtain in pure form by this reaction, because of secondary reaction with methanol.

Analytical data and i.r. spectra indicated however that all methanol ligands had been lost from the starting material and the brown product consisted of $[MoCl_2(P\phi_3)]_n$. Insolubility of this product prevented definitive characterization, hence efforts were made to secure more tractable derivatives. Reaction of the brown $[MoCl_2(P\phi_3)]_n$ with tributylphosphine at 25° in benzene led to displacement of $P\phi_3$ and formation of a brownish yellow solution, from which brown-yellow crystals were obtained. A structure determination revealed a tetramer of the composition $Mo_4Cl_8(PBu^n_3)_4$, but satisfactory refinement of the structure could not be achieved. Refinement of the structure of the triethylphosphine derivative $Mo_4Cl_8(PEt_3)_4$ was more successful; the structure of the molecule is shown in Fig. 2.

We notice immediately that the molecule contains four Mo atoms in a rectangular array, bridged on each long edge by two Cl atoms. Within the rectangle the Mo-Mo bond distances are 2.211(3)Å on the short edges and 2.901(2)Å on the long edges. The long distance falls well within the range of *ca.* 2.5 to 3.2Å known for Mo-Mo single bonds (5). The short bond distances are in excellent agreement with those known for compounds with Mo-Mo triple bonds (6), e.g. $Mo_2(NMe_2)_6$, 2.214(3)Å; $Mo_2(OCH_2CMe_3)_6$, 2.222(2)Å; $Mo_2(OSiMe_3)_6(HNMe_2)_2$, 2.242(1)Å. Thus the structural data clearly indicate the rectangle consists of single bonds on the long edges and triple bonds on the short edges. Remembering that the rectangle is formed from two quadruply bonded dimers we may schematically represent the reaction as a 2 + 2 addition process

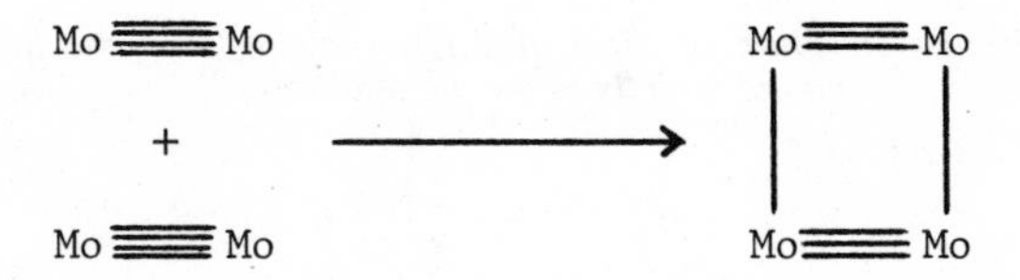

If this view is valid we might expect that the δ-bonding orbitals in the dimers are recast in the tetramer to form the long σ-bonds.

After characterization of the $Mo_4Cl_8(PR_3)_4$ derivatives it naturally was of interest to establish more general synthetic routes to the rectangular clusters of this type. Poor overall yields of *ca.* 20 percent based on conversion of $Mo_2Cl_4(P\phi_3)_2(CH_3OH)_2$ to $Mo_4Cl_8(PR_3)_4$ as discussed above presented an impediment to further chemistry of the tetramers. Also we wished to know if the dimer $Mo_2Cl_4(P\phi_3)_2(CH_3OH)_2$ was unique in leading to formation of the rectangular units, a result which seemed rather unlikely. Indeed, subsequent work proved that much more efficient syntheses could be obtained from more convenient starting

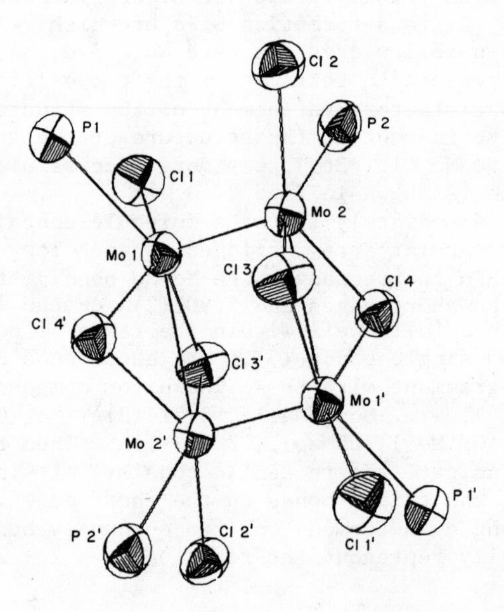

Figure 2. Molecular structure of $Mo_4Cl_8(PEt_3)_4$ showing only the ligand atoms bound directly to the metal atoms

materials. The governing factor proved to be simply stoichiometric control of the added ligand/metal ratio. Some successful synthetic reactions are enumerated in Scheme II.

Reactions 5 and 6 are particularly convenient since the most common starting material for synthesis of quadruply bonded dimers of molybdenum, viz. $Mo_2(O_2CCH_3)_4$, is converted directly into the rectangular clusters in high yield. The products of reactions 7, 8 and 9 are formulated as tetramers and assumed to consist of rectangular cluster units because their yellow color and electronic absorption maxima are quite similar to $Mo_4Cl_8(PBu_3)_4$, and each is easily converted into $Mo_4Cl_8(PEt_3)_4$ or $Mo_2Cl_4(PEt_3)_4$ upon reaction at 25° with the stoichiometric amount of PEt_3. In the presence of excess PEt_3 all of these tetramers are converted back into dimers:

$$Mo_4Cl_8L_4 + 8PEt_3 \longrightarrow 2Mo_2Cl_4(PEt_3)_4 + 4L$$

The contrast in color between the dimers $Mo_2Cl_4L_4$, generally red to blue, and the tetramers, brown to yellow, is quite striking. The intense band assigned as the $\delta \rightarrow \delta^*$ transition in the dimers (7), at ca. 500-600nm, disappears upon formation of the tetramers and is replaced by new bands at 375-440nm(strong) and 600-700nm(weak). Absorption maxima for these bands obtained from reflectance spectra of several of the derivatives are given in Table I. The assignment of these transitions must await a thorough study of the spectra. However one possibility for their

Table I. Band Maxima ($cm^{-1} \times 10^{-3}$) for $Mo_4Cl_8L_4$

L	PBu_3	$P\phi_3$	EtCN	MeOH	
ν_1	14.3	14.5	15.6	16.7	weak
ν_2	22.7	23.1	24.4	26.7	strong

assignment is based on the Mo's which can be derived from the set of four d-orbitals appropriate for δ-bonding in the short direction, and σ-bonding in the long direction of the rectangular cluster. In the idealized C_{2h} symmetry of the cluster found in the structure of $Mo_4Cl_8(PEt_3)_4$ these MO's consist of the bonding set $a_g(\sigma-\delta)$, $b_u(\sigma-\delta^*)$, and the antibonding set $a_u(\sigma^*-\delta)$, $b_g(\sigma^*-\delta^*)$, with relative energies $a_g<b_u<a_u<b_g$. Assuming these MO's comprise the HOMO's and LUMO's of the cluster, the ground state electron configuration should be $a_g^2b_u^2$. On this basis the transition $b_u^2 \longrightarrow b_u^1\ b_g^1$ is dipole allowed and may correspond to the strong band ν_2 given in Table I, while the dipole forbidden transition $b_u^2 \longrightarrow b_u^1\ a_u^1$ may correspond to the weak band ν_1. In these terms the striking change in color on going from dimer to rectangular tetramer may be attributed to stabilization of the δ-orbitals through σ-bond formation.

Scheme II

$$2K_4Mo_2Cl_8 + 4\ PEt_3 \xrightarrow[\text{reflux}]{CH_3OH} \underset{60\%}{Mo_4Cl_8(PEt_3)_4} + 8KCl \quad (4)$$

$$2Mo_2(O_2CCH_3)_4 + 8(CH_3)_3SiCl + 4PEt_3 \xrightarrow[\text{reflux}]{THF} \underset{50\%}{Mo_4Cl_8(PEt_3)_4} + 8(CH_3)_3SiOCOCH_3 \quad (5)$$

$$2Mo_2(O_2CCH_3)_4 + 4AlCl_3 + 4PEt_3 \xrightarrow[25°]{THF} \underset{50\%}{Mo_4Cl_8(PEt_3)_4} + 4AlCl(O_2CCH_3)_2 \quad (6)$$

$$2Mo_2Cl_4(P\phi_3)_2(CH_3OH)_2 \xrightarrow[\text{HCl, }60°]{C_6H_{12}} \underset{90\%}{Mo_4Cl_8(CH_3OH)_4} + 4P\phi_3 \quad (7)$$

$$Mo_4Cl_8(CH_3OH)_4 + 4C_2H_5CN \xrightarrow[25°]{EtCN} \underset{93\%}{Mo_4Cl_8(C_2H_5CN)_4} + 4CH_3OH \quad (8)$$

$$Mo_4Cl_8(C_2H_5CN)_4 + 4P\phi_3 \xrightarrow[25°]{THF} Mo_4Cl_8(P\phi_3)_4 + 4C_2H_5CN \quad (9)$$

The Tungsten Cluster $W_4Cl_8(PBu^n_3)_4$. An entry to synthesis of the related tungsten clusters, $W_4Cl_8L_4$ was provided by the recent success by Sharp and Schrock (8) in the synthesis of the quadruply bonded dimers $W_2Cl_4(PR_3)_4$ (R = methyl, ethyl, propyl, etc.). For the synthesis of $W_4Cl_8(PBu_3)_4$, this procedure was altered as shown in Scheme III. A structure determination of this compound confirmed the rectangular cluster whose basic configuration is depicted in Fig. 3. Here it is notable that the idealized symmetry, D_2, differs from that of $Mo_4Cl_8(PEt_3)_4$, C_{2h}. However essential features of the bonding in the two cases are quite similar. A comparison of some important bond distances and angles is provided in Table II. The most interesting comparison concerns the metal-metal bond distances, where it can be seen that the long bonds of the rectangle are shorter in the tungsten case, and the short bonds are longer. Based on a comparison of known distances in quadruply bonded dimers the differences observed in the rectangular clusters were not surprising. Since the W-W distances in quadruply bonded dimers range from ca. 0.1 to 0.15Å longer than the Mo-Mo distances in related compounds (9), the δ-orbitals of the tungsten dimers should be more readily available for W-W σ-bonding upon formation of tetramer. In other words, the σ-bonds on the long edges are strengthened by the reduced δ-bonding on the short edges. The slightly more acute $M-Cl_b-M$ and more open Cl_b-M-Cl_b bond angles in the tungsten case are consistent with this view.

A major difference in reactivity of the tungsten cluster as compared to the molybdenum clusters is evidenced by the extreme sensitivity of the former to decomposition in air. Further work on the synthesis and reactions of the tungsten clusters is in progress.

New Ternary Molybdenum Oxides

Our relatively innocent venture into the chemistry of molybdenum oxide systems has led to the discovery of a whole new class of ternary oxide compounds dominated in their structural features by strong metal-metal interactions. Interest in the question of whether the trinuclear clusters in the well known compound $Zn_2Mo_3O_8$ (10, 11) could accept more than 6 electrons in the metal-metal bonding orbitals prompted our initial experiments. Indeed we succeeded in synthesis of the first compound $LiZn_2Mo_3O_8$ containing 7 electrons in the monocapped M_3O_{13} cluster type (12) found in $Zn_2Mo_3O_8$ without otherwise altering the ligands bound to the cluster. A structure determination of $LiZn_2Mo_3O_8$ was completed, but because of the low x-ray scattering factor and possible disorder of the Li^+ ions the positions of the latter could not be determined. The synthesis of $NaZn_2Mo_3O_8$ was next attempted with the expectation that the Na positions could be more easily

Scheme III

Preparation of $W_4Cl_8(PBu^n_3)_4$

$$2\ WCl_4 + 2\ Na(Hg) \xrightarrow{THF} W_2Cl_6(THF)_4 + 2\ NaCl$$

$$W_2Cl_6(THF)_4 + 2\ PBu^n_3 \xrightarrow{THF} W_2Cl_6(THF)_2(PBu^n_3)_2 + 2\ THF$$

red-orange solution

$$W_2Cl_6(THF)_2(PBu^n_3)_2 + 2\ Na(Hg) \xrightarrow{THF} \text{green solution} + 2\ NaCl$$

green solution → (− THF) → green residue

green residue → (brief hexane xtn.) → undissolved brown solid

undissolved brown solid → (hexane, slow xtn.) → brown crystals

brown crystals $W_4Cl_8(PBu^n_3)_4$ 18%

Table II

A Comparison of Bond Distances(Å) and Angles(deg.) in $W_4Cl_8(PBu^n_3)_4$ and $Mo_4Cl_8(PEt_3)_4$

Type bond or angle	$W_4Cl_8P_4$	$Mo_4Cl_8P_4$	
short M-M	2.309(2)	2.211(3)	
long M-M	2.840(1)	2.901(2)	
M-Cl(bridge)	2.396(5) 2.417(5)	2.381(6) 2.417(6)	2.373(5) 2.422(6)
M-Cl(terminal)	2.400(5)	2.425(5)	2.421(6)
M-P	2.530(5)	2.558(6)	2.556(7)
$M-Cl_b-M$	72.3(1)	75.2(2)	73.7(2)
Cl_b-M-Cl_b	102.8(2)	100.5(2)	100.4(2)
M-M-M	89.93(3)	90.6(1)	89.4(1)

determined. Attack of fused silica at high temperature by alkali metal salts dictated use of alternate material for the reaction vessel. A molybdenum tube seemed most feasible for this purpose. Hence the reaction was attempted with a mixture of Na_2MoO_4, ZnO, MoO_2 and Mo sealed in a Mo-tube at 1100°C. Upon opening the tube after a few days gleaming metallic needles were observed growing out from the inner walls of the tube.

The Structure of $NaMo_4O_6$. The subsequent structure determination of one of these needles established the formula of the new compound as $NaMo_4O_6$ (13). Subsequently it was found that single phase preparations could be achieved by the reaction.

$$Na_2MoO_4 + 4MoO_2 + 3Mo \xrightarrow[\text{Mo-tube}]{1100°} 2NaMo_4O_6 \qquad (10)$$

This remarkably stable compound with Mo in the net 2.75 oxidation state is resistant to oxidation in air up to 350° and unaffected by dilute aqueous acids. At 25° the resistivity of a single crystal along the needle axis is ca. 10^{-4} ohm cm, a value indicative of metallic character.

The structure of $NaMo_4O_6$ proved to be most remarkable. As shown in Fig. 4 the tetragonal unit cell (a = 9.559(3), c = 2.860(1)Å) contains octahedral cluster units each joined to adjacent units in the a b plane through Mo-O-Mo bonding. Along the c-axis the octahedral units are fused at opposite edges to

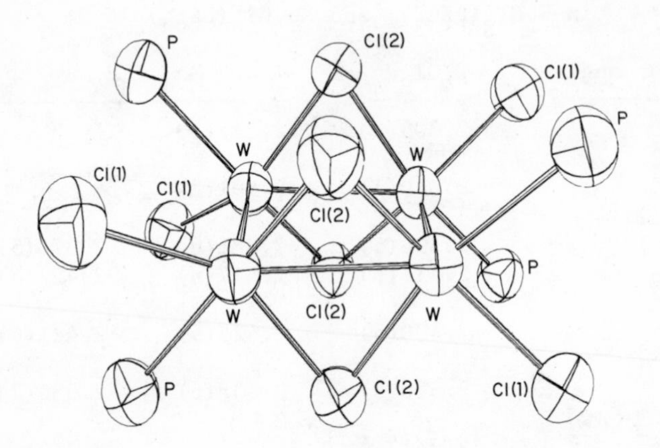

Figure 3. Structure of the metal–metal and metal–ligand framework of $W_4Cl_8(PBu_3^n)_4$. For clarity the butyl groups of the tributylphosphine ligands have been omitted.

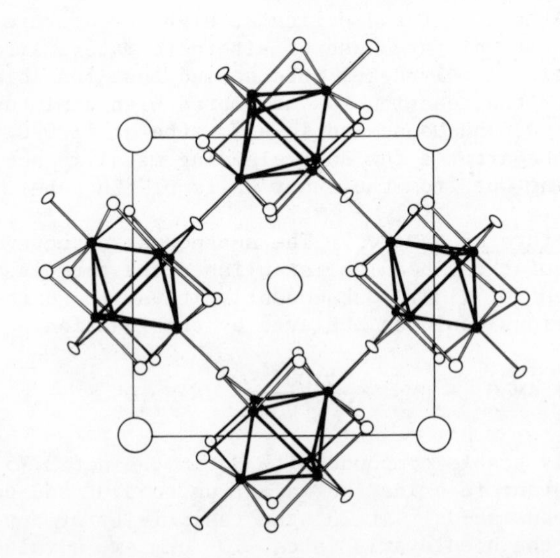

Figure 4. View of the unit cell of $MaMo_4O_6$ along the c-axis ((●) Mo; (○) 0; (○) Na)

create infinite linear chains, with the outward edges bridged by O-atoms. The compound then may be viewed as a metallic infinite chain polymer crosslinked by Mo-O-Mo bonds. A view of one chain perpendicular to the chain axis is provided in Fig. 5. Notable features include the repeat distance between units of 2.860(1)Å; this distance thus constitutes both the Mo-Mo(bonded) and O-O (nonbonded) distance along the direction of chain propagation. The strongest Mo-Mo bonds are those between pairs of waist atoms perpendicular to the chain, 2.751(3)Å, and those joining waist to apex atoms, 2.778(2)Å. Within the repeat unit there are 13 Mo-Mo bonds of average distance 2.801Å, which is only 0.076Å greater than the distance between nearest neighbors in bcc molybdenum metal. Each O-atom is strongly bonded to three Mo-atoms at distances of 2.014(8) to 2.067(8)Å. The shortest of these involve the O atoms having trigonal planar hybridization bridging between the chains.

A final three dimensional perspective of the structure is provided in Fig. 6. From this view it is readily discerned that the Na^+ ions are stacked along the c-axis in tunnels created by crosslinking of the metal oxide cluster chains. Each Na^+ ion occupies a site of tetragonal coordination symmetry surrounded by eight O atoms at a distance of 2.740(8)Å. When these structural features are taken into consideration the formula $NaMo_4O_6$ may be rewritten as $Na^+[(Mo_2Mo_{4/2}O_{8/2}O_{2/2})O^-_{2/2}]$ to reflect the connectivity relations within and between the chains.

Within the chains each repeat unit Mo_4O_6 has available 13 electrons to participate in Mo-Mo bonding, and there are 13 bonds per repeat unit. Thus if all 13 electrons participate in bonding orbitals the average Mo-Mo bond order is 0.5. Use of the Pauling relation (14) $D_n = D_1 - 0.6 \log n$, with D_1(Mo-Mo) = 2.614, permits calculation of the bond order n associated with each Mo-Mo bond of length D_n in the repeat unit. (The value D_1 = 2.614 has been corrected from the value given by Pauling (14) to account for the body centered structure of molybdenum metal.) When averaged over all Mo-Mo bonds the result is n(ave.) = 0.49, which confirms that all 13 electrons indeed reside in Mo-Mo bonding orbitals. Since each apical Mo is bonded to 6 metal neighbors, and each waist Mo is likewise bound to 7, we can see that the structure is dominated and probably determined by the metal-metal interactions. Although this structure is unique among metal oxide compounds, and probably presages many others, several metal halide compounds with related structures have recently been discovered, e.g. Gd_2Cl_3 (15, 16), Sc_5Cl_8 (17), and Y_2Br_3 (18). It is not by accident that the halide examples are found among the metals with fewer valence electrons and larger metallic radii. In these structures the M-M and X-X distances must be equal along the chain, consequently the larger spacing required by Cl-Cl or Br-Br nonbonded contacts dictate lower M-M bond orders.

In the $NaMo_4O_6$ structure (13) the Na^+ ions were found to exhibit an unusually large thermal parameter, an understandable

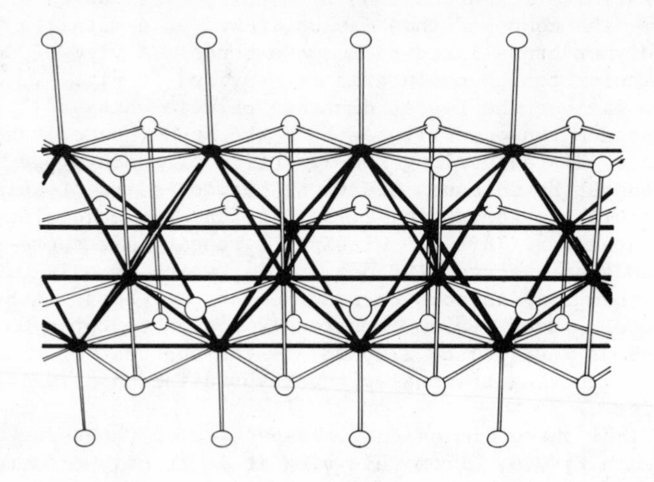

Figure 5. Mo_4O_6 cluster chain viewed perpendicular to the chain direction ((●) Mo; (○) O)

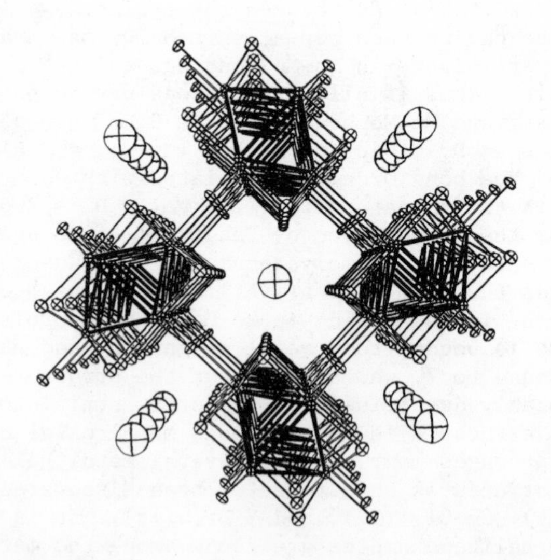

Figure 6. Three-dimensional view of the $NaMo_4O_6$ structure as seen along the c-axis.

fact since the Na-O distances are <u>ca</u>. 0.39Å longer than the sum of ionic radii. Indeed the cation sites appear of sufficient size to accommodate ions as large as K^+ or Ba^{2+}. Reactions were thus performed to prepare KMo_4O_6 and $BaMo_4O_6$. In both cases new compounds were formed and work is still in progress to identify their composition and structure. From the reaction of $BaMoO_4$, MoO_2 and Mo at 1100° several crystalline phases were observed. A single crystal of one phase exhibiting columnar growth habit was secured and the structure determined. The composition of this compound, $Ba_{1.13}Mo_8O_{16}$, was established from the structure refinement (R = 0.033, R_w = 0.042), which also showed the crystal was triclinic with lattice constants a = 7.311(1), b = 7.453(1), c = 5.726(1)Å, α = 101.49(2)°, β = 99.60(2)°, γ = 89.31(2)°, Z = 1, P1.

<u>The Structure of $Ba_{1.13}Mo_8O_{16}$</u>. The structure of $Ba_{1.13}Mo_8O_{16}$ is extremely interesting for several reasons. First, the structure again involves infinite chains of metal cluster units bound together in such a way that tunnels are constructed for the Ba^{2+} ions. This feature of the structure is remarkably like that of $NaMo_4O_6$ as shown in Fig. 7, which is a view down the c-axis, also the tunnel axis. The repeat distance along the c-axis is almost exactly twice that in $NaMo_4O_6$, a result of the fact that the Mo-Mo separations are alternately long and short, though the O-O spacings are roughly equal at 2.86Å. Secondly, the infinite chains are of two types: one which consists of nearly regular clusters $Mo_4O_8 = Mo_4O_2O_{8/2}O_{6/3}$ as shown in Fig. 8, and the other having clusters of the same basic structure but with two elongated Mo-Mo bonds as shown in Fig. 9. Finally, the positions occupied by the Ba^{2+} ions within the tunnels are of three kinds, each with only fractional occupation number, viz. BaI, 0.86; BaII, 0.13; BaIII, 0.13. The arrangement is such that sites I and II, or II and III cannot be occupied simultaneously because of their close proximity.

A view perpendicular to one of the chains is shown in Fig. 10. It is seen that the metal atoms fill all octahedral sites between two oxide layers, each three atoms wide by infinite length. The discrete cluster units are formed by displacement of the metal atoms along a line which is colinear with the chain, such that the Mo-Mo spacing is alternately long and short, e.g. 3.16 and 2.57Å. All O-atoms capping triangular faces above and between cluster units along the chain are trigonal pyramidal, and those bound on the edges are trigonal planar.

Metal-metal bonding within the clusters of regular geometry may be understood as resulting from 10 electrons in bonding σ-orbitals directed along the 5 bonded edges. This indicates the net oxidation states of 3+ for the Mo atoms on the shared edge of the two triangles, and 4+ for the two atoms on the outer apices. Each regular cluster unit then must assume anionic charge $Mo_4O_8^{2-}$. It is difficult to reason why the distortion in the

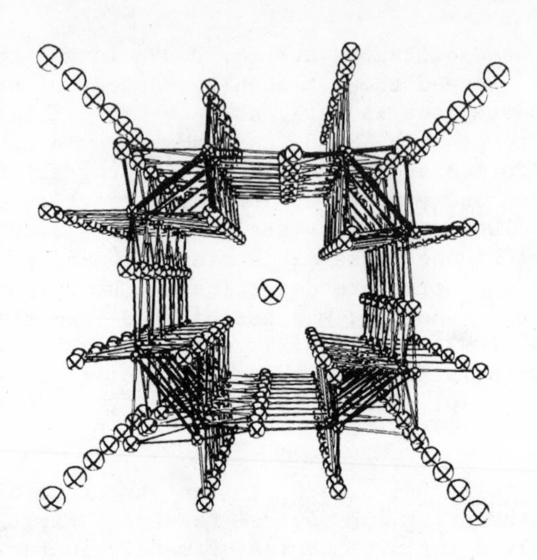

Figure 7. Three-dimensional view of the $Ba_{1.13}Mo_8O_{16}$ structure as seen along the c-axis

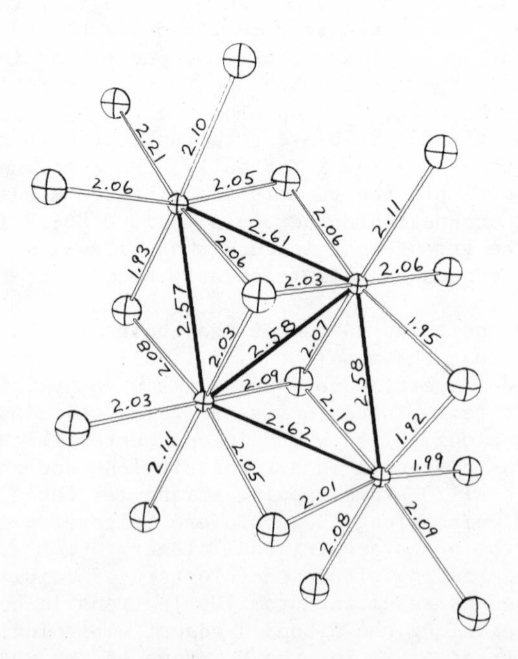

Figure 8. Structure and bond distances of the regular cluster unit in $Ba_{1.13}Mo_8O_{16}$

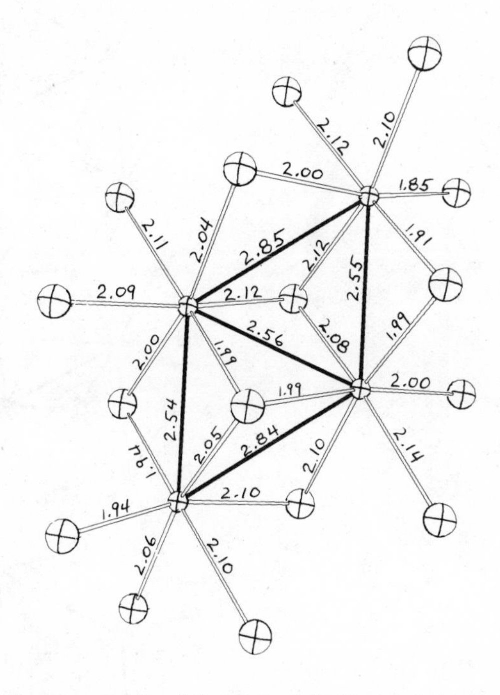

Figure 9. Structure and bond distances of the distorted cluster unit in $Ba_{1.13}Mo_8O_{16}$

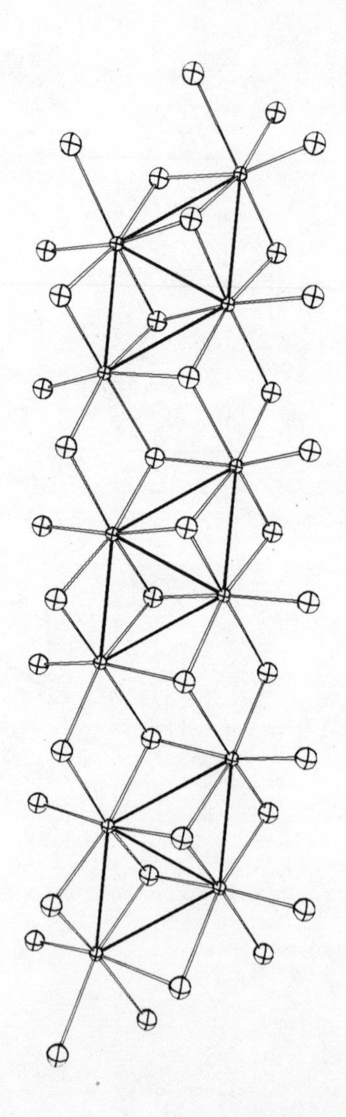

Figure 10. View showing the interlinking of cluster units within the infinite chains of $Ba_{1.13}Mo_8O_{16}$

"irregular" cluster unit occurs exactly as it does. From the Mo-Mo distances in these units it appears the three short bonds are of order 1.0 and the two elongated bonds are approximately of order 0.5. Since the atoms on the shared edge are involved in ca. 2.5 bonds each, their net oxidation state is approximately 3.5+, whereas the atoms on the outer apices are involved in only 1.5 bonds and have approximate state 4.5+. In both cases the shortest Mo-O distances are those about the outer apical atoms, which lends support to their assessment of a higher net oxidation state. Based on these considerations the compound perhaps is better formulated as $Ba_{1.13}(Mo_4O_8^{2-})(Mo_4O_8^{0.26-})$.

The astute reader will recognize that this compound is a metal-metal bonded adaptation of the well-known hollandite structure (19). Further, the metal cluster configuration is like that in $CsNb_4Cl_{11}$ (20), though the latter compound possesses a double layer structure rather than the tunnel structure observed here. Finally these new structures provide only a glimpse of what is possible, and an excitement about new compounds of unusual structure and properties yet to come from studies of highly reduced ternary and quaternary refractory metal oxide systems.

Abstract

Synthesis, structure and properties of the new tetranuclear clusters $Mo_4Cl_8L_4$ (L = PBu_3, PEt_3, EtCN and MeOH) and $W_4Cl_8(PBu_3)_4$ are described. These compounds result from condensation of quadruply bonded dimers in a 2 + 2 cycloaddition-ligand elimination process and exhibit rectangular geometry of the M_4 cluster. Metal-metal bond distances of 2.211(3) and 2.901(2) for $Mo_4Cl_8(PEt_3)_4$, and 2.309(2), 2.840(1)Å for $W_4Cl_8(PBu_3)_4$ were found for the short and long edges of the rectangular cluster units. The short distances agree well with known M-M triple bond lengths, and the long distances fall within the range known for M-M single bonds. Also described are the synthesis and structure of two new ternary molybdenum oxides with metal-metal bonded cluster units, $Ba_{1.13}Mo_8O_{16}$ and $NaMo_4O_6$. In each case the structure features infinite chains of metal clusters crosslinked by M-O-M bonds in such a way that unidirectional tunnels are formed for accommodation of the Na^+ or Ba^{2+} ions. In $Ba_{1.13}Mo_8O_{16}$ the clusters consist of Mo_4O_{16} units with the connectivity relation $Mo_4O_2O_{8/2}O_{6/3}$ and the coplanar metal atoms at the vertices of two triangles sharing a common edge. The overall structural features (both Mo-Mo and Mo-O bonding) suggest the formulation $Ba^{2+}_{1.13}(Mo_4O_8^{2-})(Mo_4O_8^{0.26-})$. The extremely interesting structure of $NaMo_4O_6$ consists of infinite chains of octahedral clusters Mo_6 fused on opposite (parallel) edges and bridged on the outwardly exposed edges by O-atoms. Within the chains the connectivity is $Mo_2Mo_{4/2}O_2O_{8/2}$; if the crosslinking between chains is included the connectivity becomes $Na^+[(Mo_2Mo_{4/2}O_{2/2}O_{8/2})O_{2/2}^-]$. The compound is tetragonal and exhibits metallic conductivity.

Literature Cited

1. See for example Vahrenkamp, H., Structure and Bonding, 1977, 32, 1-56.
2. Cotton, F. A.; Wilkinson, G., Advanced Inorganic Chemistry, 4th Ed., John Wiley and Sons, New York, 1980, p. 1094.
3. San Filippo, Jr., J., Inorg. Chem., 1972, 11, 3140.
4. McGinnis, R. N.; Ryan, T. R.; McCarley, R. E., J. Am. Chem. Soc., 1978, 100, 7900.
5. Cotton, F. A., J. Less-Common Met., 1977, 54, 3.
6. Chisholm, M. H.; Cotton, F. A., Acc. Chem. Res., 1978, 11, 356.
7. Trogler, W. C.; Gray, H. B., Acc. Chem. Res., 1978, 11, 232.
8. Sharp, P. R.; Schrock, R. R., J. Am. Chem. Soc., 1980, 102, 1430.
9. Cotton, F. A.; Felthouse, T. R.; Lay, D. G., J. Am. Chem. Soc., 1980, 102, 1431.
10. McCarroll, W. H.; Katz, L.; Ward, R., J. Am. Chem. Soc., 1957, 79, 5410.
11. Ansell, G. B.; Katz, L., Acta Cryst., 1966, 21, 482.
12. Cotton, F. A., Inorg. Chem., 1964, 3, 1217.
13. Torardi, C. C.; McCarley, R. E., J. Am. Chem. Soc., 1979, 101, 3963.
14. Pauling, L., The Nature of the Chemical Bond, 3rd Ed., Cornell University Press, 1960, p. 400.
15. Lokken, D. A.; Corbett, J. D., Inorg. Chem., 1973, 12, 556.
16. Simon, A.; Holzer, N.; Mattausch, Hj., Z. Anorg. Allg. Chem., 1979, 456, 207.
17. Poeppelmeier, K. R.; Corbett, J. D., J. Am. Chem. Soc., 1978, 100, 5039.
18. Mattausch, Hj.; Hendricks, J. B.; Eger, R.; Corbett, J. D.; Simon, A., Inorg. Chem., 1980, 19, 2128.
19. Wells, A. F., Structural Inorganic Chemistry, 4th Ed., Clarendon Press, Oxford, 1975, p. 459.
20. Broll, A.; Simon, A.; von Schnering, H. G.; Schafer, H., Z. Anorg. Allg. Chem., 1969, 367, 1.

(The Ames Laboratory is operated for the U.S. Department of Energy by Iowa State University under Contract No. W-7405-Eng-82. This research was supported by the Assistant Secretary for Energy Research, Office of Basic Energy Sciences, WPAS-KC-02-03.)

RECEIVED November 21, 1980.

4

Rhodium Carbonyl Cluster Chemistry Under High Pressures of Carbon Monoxide and Hydrogen

Polynuclear Rhodium Carbonyl Complexes and Their Relationships to Mononuclear and Binuclear Rhodium Carbonyl Complexes[1]

JOSÉ L. VIDAL, R. C. SCHOENING, and W. E. WALKER

Union Carbide Corporation, P.O. Box 8361, South Charleston, WV 25303

The complexity of the chemistry of rhodium carbonyl clusters has been responsible for the slow development of this field of chemistry.[1] We have been involved for some time in the study of these species under high pressures of carbon monoxide and hydrogen.[2] The study of these systems by high pressure infrared spectroscopy showed the relatively low fingerprint ability of this technique[3] and the possibility that some species could remain undetected as a consequence of an overlapping of spectral features. The importance of determining the presence in those systems of rhodium carbonyl complexes of nuclearity lower even than that of such simple clusters as $[Rh_5(CO)_{15}]^-$ is rooted in the relatively recent proposals that mononuclear transition metal complexes of Co[4] and Ru[5] are able to hydrogenate carbon monoxide and of mechanisms for this behavior.[6] In contrast, polynuclear complexes have been mentioned as probably enhancing this reactivity.[7] Thus, it was of interest to determine the presence of mono- and binuclear-rhodium complexes in a system showing the catalytic formation of polyalcohols from synthesis gas. The participation of mono- and binuclear anionic, cationic and neutral complexes in the aggregation and fragmentation reactions of rhodium carbonyl clusters under ambient or low pressure conditions has been described before. For instance, $[Rh(CO)_4]^-$ has been shown[1] to participate in condensation reactions that increase

[1] This is the fourth part of a series.

0097-6156/81/0155-0061$05.75/0

stepwise the size and degree of reduction of the clusters (equations 1-3). It was also reported that the formation of this anion could originate in the fragmentation of the clusters by carbon monoxide facilitated by a basic medium.

$$Rh_4(CO)_{12} + [Rh(CO)_4]^- \rightleftharpoons [Rh_5(CO)_{15}]^- \quad (1)$$

$$[Rh_5(CO)_{15}]^- + [Rh(CO)_4]^- \rightleftharpoons [Rh_6(CO)_{15}]^{2-} + 4CO \quad (2)$$

$$[Rh_6(CO)_{15}]^{2-} + [Rh(CO)_4]^- \rightleftharpoons [Rh_7(CO)_{16}]^{3-} + 3CO \quad (3)$$

Another simple rhodium carbonyl complex also known to be involved in the fragmentation and aggregation reactions of clusters is $Rh_2(CO)_8$. This species has been shown to participate in the reactions of neutral rhodium carbonyl species in either matrixes(8) or solutions(9) (equation 4), but it has not yet been implicated in the chemistry of large anionic clusters.

$$2Rh_6(CO)_6 + 24CO \rightleftharpoons 3Rh_4(CO)_{12} \rightleftharpoons 6Rh_2(CO)_8 - 12CO \quad (4)$$

A second neutral species that has not previously been shown to be involved in the reactions of clusters is $HRh(CO)_4$. This complex has recently been produced under high pressures of carbon monoxide and hydrogen by means of the reactions of $Rh_4(CO)_{12}$ with CO and H_2 (equations 5) or by bringing protonic acid in contact with $[Rh(CO)_4]^-$ under 1000 atm. of $CO:H_2$ (equation 6).

$$Rh_4(CO)_{12} + 4CO \rightleftharpoons 2Rh_2(CO)_8$$
$$Rh_2(CO)_8 + H_2 \rightleftharpoons 2HRh(CO)_4 \quad (5)$$

$$[Rh(CO)_4]^- + H^+ \rightleftharpoons HRh(CO)_4 \quad (6)$$

The potential participation of $HRh(CO)_4$ in the reactions of high nuclearity rhodium carbonyl clusters could be analogous to the formation of $[Ni_2(CO)_6H]^-$ upon fragmentation of $[Ni_{12}(CO)_{21}H_2]^{2-}$ by carbon monoxide (equation 7)(10)

$$[Ni_{12}(CO)_{21}H_2]^{2-} + 23CO \rightleftharpoons 8Ni(CO)_4 + 2\,[Ni_2(CO)_6H]^- \quad (7)$$

We have already observed(3,9) that simple anionic, e.g., $[Rh(CO)_4]^-$, and neutral, e.g, $HRh(CO)_4$ or $Rh_2(CO)_8$, rhodium complexes could be involved in the fragmentation and aggregation reactions of polynuclear species. In addition, the cationic moiety "$Rh(CO)_2^+$" has been found to formally coordinate to $[Rh_{14}(CO)_{25}]^{4-}$ to form $[Rh_{15}(CO)_{27}]^{3-}$ (equation 8), or to get detached from the latter cluster by halide ligands (equation 9)(11) under ambient conditions.

$$[Rh_{14}(CO)_{25}]^{4-} + [Rh(CO)_2(CH_3CN)_2]^+ \rightleftharpoons [Rh_{15}(CO)_{27}]^{3-} + 2CH_3CN \quad (8)$$

$$[Rh_{15}(CO)_{27}]^{3-} + 2Br^- \rightleftharpoons [Rh_{14}(CO)_{25}]^{4-} + [Rh(CO)_2Br_2]^- \quad (9)$$

The potential catalytic ability of polynuclear complexes for the conversion of $CO:H_2$ into chemicals(7) is in contrast with recent results involving mononuclear species(4,6) and the proposal of a mechanism for such conversion.(6)

Our interest in understanding the behavior of rhodium carbonyl clusters in systems which catalytically convert $CO:H_2$ into alcohols(3) prompted us to test the potential presence of mononuclear and binuclear rhodium carbonyl complexes in these systems. A positive characterization of these species under these circumstances would show a previously unknown behavior of rhodium carbonyl clusters under high pressure of carbon monoxide. It could also show the existence of a parallel behavior between the chemistry of these species under ambient and high pressures of carbon monoxide, and it may shed some light on the catalytic reactions occurring in those systems.(3)

Because of our previous success in applying Fourier-transform infrared spectroscopy to the study of the rhodium carbonyl clusters under high pressures of carbon monoxide and hydrogen(2,3), we have applied the same technique and equipment in this work.(3) The temperature has been kept in all these experiments below 200° with maximum pressures of 832.0 atm to maximize the trend towards fragmentation of clusters. The absence of bases, e.g., salts or amines, in the systems under evaluation in this work was desirable to eliminate the ambiguity that would result from the enhancement of the fragmentation of clusters by carbon monoxide in a basic medium.(1)

RESULTS AND DISCUSSION

Rhodium carbonyl clusters with encapsulated heteroatoms, e.g., $[Rh_6(CO)_{15}C]^{2-}$, $[Rh_9P(CO)_{21}]^{2-}$ and $[Rh_{17}S_2(CO)_{32}]^{3-}$, have been observed to be more resistant to fragmentation under 500-1000 atm of $CO:H_2$ than homometallic rhodium clusters. Polynuclear species account for 95-98 and 80-85 percent, respectively, of the total rhodium content of those systems, while the greatest catalyst activity has been noted with the latter complexes. $[Rh(CO)_4]^-$ is the only fragmentation product detected in both instances.(3)

The reasons above directed our attention to the role of $[Rh(CO)_4]^-$ in the fragmentation-aggregation reactions of clusters. This anion is generated during the fragmentation of $[Rh_7(CO)_{16}]^{3-}$ under 600 atm of $CO:H_2$ (Figure 1, equation 10). $[Rh(CO)_4]^-$ is also formed under similar conditions from other clusters in the presence of bases, e.g., $[Rh_{12}(CO)_{30}]^{2-}$ (equation

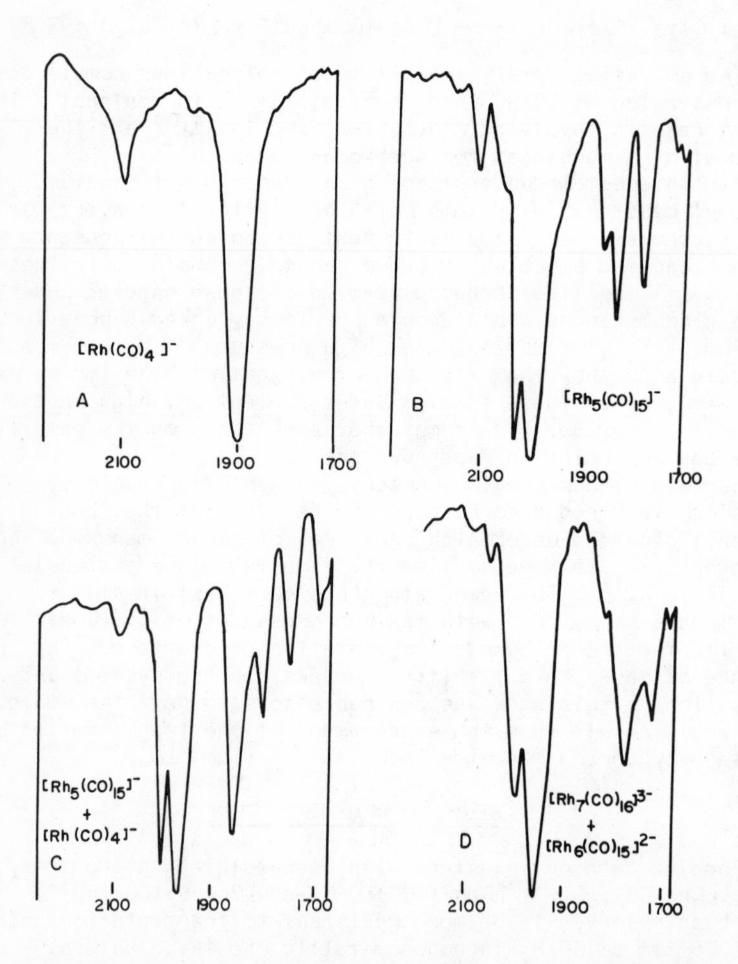

Figure 1. IR spectra under 537 atm of $CO{:}H_2$ (1:1) and 125°C of: (a) a solution of $[(Ph_3P)_2N][Rh(CO)_4]$ in 18-crown-6; (b) a solution of $[(Ph_3P)_2N]_2[Rh_{12}(CO)_{30}]$ in 18-crown-6; (c) a solution of $[(Ph_3P)_2N]_3[Rh_7(CO)_{16}]$ in 18-crown-6; (d) solution (c) at atmospheric conditions. Assignments of the spectra shown in the figures.

11). The same behavior has been observed under ambient conditions.[1]

$$[Rh_7(CO)_{16}]^{3-} + 7CO \rightleftharpoons [Rh_5(CO)_{15}]^- + 2[Rh(CO)_4]^- \quad (10)$$

$$[Rh_{12}(CO)_{30}]^{2-} + R_3N + CO \rightleftharpoons 2[Rh_5(CO)_{15}]^- + 2[Rh(CO)_4]^- \quad (11)$$

The parallel between the aggregation reactions of clusters with $[Rh(CO)_4]^-$ under ambient and high pressure conditions is more clearly shown by the comparison of the ability of these reactions to bring about the growth of clusters one atom and one negative charge at a time under both conditions. This behavior (equations 2 and 3) has been also noted under 537-840 atm (equation 11, Figure 2).[12]

$$[Rh_9P(CO)_{21}]^{2-} + [Rh(CO)_4]^- \rightleftharpoons [Rh_{10}P(CO)_{22}]^{3-} + 3CO \quad (12)$$

The fragmentation reactions of $[Rh_{13}(CO)_{24}H_3]^{2-}$ and $[Rh_{13}(CO)_{24}H_2]^{3-}$ have not been previously studied, although is has been reported that these clusters are formed together with $[Rh_{14}(CO)_{25}]^{4-}$ and $[Rh_{15}(CO)_{27}]^{3-}$ during the reaction of $[Rh_{12}(CO)_{30}]^{2-}$ with hydrogen at 80° and 1 atm (equation 13)[11,13]

$$[Rh_{12}(CO)_{30}]^{2-} + H_2 \longrightarrow [Rh_{13}(CO)_{24}H_x]^{(5-n)-} + [Rh_{14}(CO)_{25}]^{4-} + [Rh_{15}(CO)_{27}]^{3-} \quad (13)$$

We also examined the fragmentation and aggregation reactions of $[Rh_{13}(CO)_{24}H_x]^{(5-n)-}$ under high pressures of CO and H_2.[3] It was of interest in this respect to determine which rhodium carbonyl complexes result from the former reaction and whether the parallel formation of any organic products derived from the hydrogenation of carbon monoxide is also occurring. The latter possibility was considered because of the presence of hydrides in the cluster and the involvement of hydrido carbonyl complexes in the hydrogenation of carbon monoxide.[4] Unfortunately, it was not possible to detect any organic products formed from this reaction even after the cyclic repetition of the transformations below (equation 14).

$$\begin{array}{ccc} [Rh_{13}(CO)_{24}H_3]^{2-} & \xrightarrow[\text{10 ATM CO}]{\text{BASE}} & [Rh_{13}(CO)_{24}H_2]^{3-} \\ \text{CO} \downarrow\uparrow \text{H}_2 & & \text{CO} \downarrow\uparrow \text{H}_2 \\ [Rh_5(CO)_{15}]^- & & [Rh_5(CO)_{15}]^- + [Rh(CO)_4]^- \end{array} \quad (14)$$

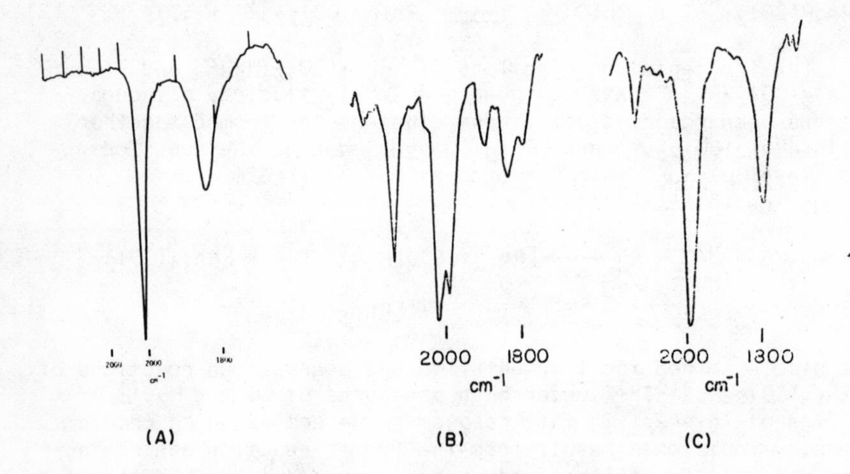

Figure 2. IR spectra under 537 atm of $CO:H_2$ (1:1) and 180°C: (a) a solution of $[Cs(18\text{-crown-}6)_n]_2[Rh_9P(CO)_{21}]$ in 18-crown-6; (b) a solution of $[Cs(18\text{-crown-}6)_n]_3[Rh_{10}P(CO)_{22}]$ in 18-crown-6; (c) same solution as in (b) before high-pressure treatment.

The products detected and the relative conditions required for their formation (equation 14) are consistent with the trend towards increased fragmentation by carbon monoxide upon increasing the negative charge: metal atom ratio, e.g., $Rh_6(CO)_{14}^{4-}$, $Rh_6(CO)_{15}^{2-}$, $Rh_6(CO)_{16}$.(1) The presence of $[Rh(CO)_4]^-$ in the basic system above could have been ascribed to partial fragmentation of $[Rh_{13}(CO)_{24}H_2]^{3-}$ but a quantitative infrared study of that solution showed a molar ratio of $[Rh_5(CO)_{15}]^-$ to $[Rh(CO)_4]^-$ of ca., 2:3. This, together with our inability to detect any hydrogenation product of carbon monoxide and with the high acidity of $HRh(CO)_4$, was indicative of the potential formation of this mononuclear hydride during the fragmentation of $[Rh_{13}(CO)_{24}H_3]^{2-}$. A comparison of the infrared spectrum of $[Rh_5(CO)_{15}]^-$ and that recently reported(9) for $HRh(CO)_4$ (Figure 1 and Table I) shows the overlapping of the main features of both spectra. Thus, substraction of the bands of $[Rh_5(CO)_{15}]^-$ from the spectrum of the solution obtained after reaction of $[Rh_{13}(CO)_{24}H_3]^{2-}$ with carbon monoxide leaves the pattern expected for $HRh(CO)_4$ (Figure 3). We have concluded that the fragmentation and aggregation reactions of $[Rh_{13}(CO)_{24}\text{-}H_x]^{(5-x)-}$ (x = x,3) could be written as below (equations 15 and 16)

$$[Rh_{13}(CO)_{24}H_3]^{2-} + 18CO \rightleftharpoons 2[Rh_5(CO)_{15}]^- + 3HRh(CO)_4 \quad (15)$$

$$[Rh_{13}(CO)_{24}H_2]^{3-} + 18CO + 2R_3N \rightleftharpoons 2[Rh_5(CO)_{15}]^- + 3[Rh(CO)_4]^- + 2[R_3NH]^+ \quad (16)$$

The characterization of $HRh(CO)_4$ was essential to our studies. This species was expected to be analogous to $HM(CO)_4$(M=Co, Ir) previously reported by Whyman,(14) but it was not possible to detect its formation upon reacting $Rh_4(CO)_{12}$ with $CO:H_2$.(15) In addition, its presence could not be confirmed during the protonation of $[Rh(CO)_4]^-$. Instead, $Rh_6(CO)_{16}$ was formed when that reaction was carried out at ambient conditions.(16)

We looked at the reaction of $Rh_4(CO)_{12}$ with $CO:H_2$ in tetraglyme and dodecane. The formation of soluble side-products, e.g., $[Rh_5(CO)_{15}]^-$, in tetraglyme, precluded a clear detection of $HRh(CO)_4$, but it was possible to remove these impurities by using dodecane. The interference caused by the presence of unconverted $Rh_4(CO)_{12}$ when the reaction was conducted at 25° and 1320 atm. of $CO:H_2$ (1:2) was avoided by Fourier-substraction of the spectrum of this cluster from the one observed for the products of the reaction. The remaining features are assigned to $HRh(CO)_4$ (equation 5) by comparison with those of $HCo(CO)_4$ and $HIr(CO)_4$, prepared also by reaction of the neutral carbonyls, $Co_2(CO)_8$ and $Ir_4(CO)_{12}$, with CO and H_2 in dodecane (Table I).

TABLE I. Comparison of the Infrared Spectra for $HM(CO)_4$

Compound	Absorptions (intensity) (cm^{-1} + 1)			Reference
$HCo(CO)_4$	1992(w)	2030(5)	2075(m)	This work (dodecane)
	1996(vw)	2029(5)	2052(m)	R. Whyman[1] (heptane)
$HRh(CO)_4$	2008(vw)	2039(5)	2070(m)	This work "
$HIr(CO)_4$	1995(w)	2030(5)	2070(m)	This work "
	1999(w)	2031(5)	2054(m)	R. Whyman[1]

(1) J. Organometal. Chem., 81, 97 (1974).

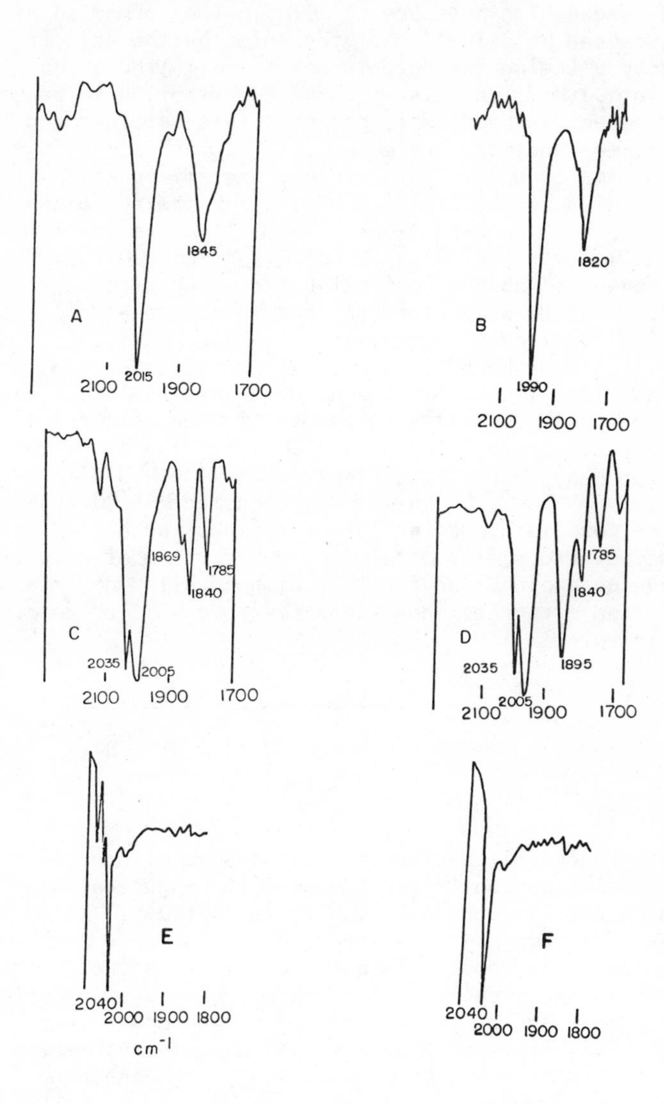

Figure 3. IR spectra of 18-crown-6 solutions of: (a) $[Cs(18\text{-crown-}6)_n]_2[Rh_{13}(CO)_{24}H_3]$ under 10 atm of CO; (b) solution in (a) after addition of N-*methylmorpholine; (c) solution in (a) under 140 atm of CO and 80°C; (d) solution in (b) under 140 atm of CO and 40°C; (e) solution in (c) after Fourier-subtraction of the pattern of $[Rh_5(CO)_{15}]^-$; (f) $HRh(CO)_4$ obtained from $Rh_4(CO)_{12}$ and $CO{:}H_2$.*

Another pathway for the preparation of $HRh(CO)_4$ was found to be the acidification of solutions of the conjugate base $[Rh(CO)_4]^-$ under high pressure of $CO:H_2$. The formation of $Rh_6(CO)_{16}$ caused by oxidation of the anion by the acid is precluded by bringing the acid in contact with the anion under 1000 atm. of $CO:H_2$. In this way, the coordination to carbon monoxide present in rhodium tetracarbonyl is retained and the tendency to aggregation is precluded. A comparison with the basicity of the other two tetracarbonyl metallate anions (Co and Ir) as indicated by the stoichiometries for their protonation to $HM(CO)_4$ under similar conditions (Table II) shows the following trend of basicity: $[Ir(CO)_4]^- > [Co(CO)_4]^- > [Rh(CO)_4]^-$. In fact, we have been able to establish the order of acidities of the conjugate acids by preforming them in dodecane solutions by means of the reaction of $CO:H_2$ with $Co_2(CO)_8$, $Rh_4(CO)_{12}$ and $Ir_4(CO)_{12}$, followed by addition of controlled amounts of amines of different basicities. It was noted (Table III) that while HIr(CO)4 does not react with an excess of N-methylmorpholine (pka (25°) = 7.4), $HCo(CO)_4$ and $HRh(CO)_4$ reacted quantitatively. On the other hand, $HCo(CO)_4$ did not react with N,N-dimethylaniline (pKa (25°) = 4.8) but it deprotonated $HRh(CO)_4$.

All the results above indicate a scale of acidities in which the second row transition metal complex is the most acidic. We do not have an explanation for this at hand although it appears that a similar order has been found for some similar complexes in the iron triad.[17]

	Order of Acidity
Cobalt Triad	$HRh(CO)_4 > HCo(CO)_4 > HIr(CO)_4$
Iron Triad	Ru > Fe > Os

The other neutral species we have studied with respect to its potential participation in the fragmentation-aggregation reactions of rhodium carbonyl cluster is $Rh_2(CO)_8$. This species has not been previously implicated in these reactions in the case of anionic rhodium carbonyl clusters, although its involvement in such reactions for neutral clusters (equation 4) has already been shown. An indication of the presence of this species in these types of reactions could be the formation of $Rh_6(CO)_{16}$ in the reaction of $[Rh_5(CO)_{15}]^-$ with carbon monoxide, as observed by Chini, et al.,[18a] and by us[18b] under ambient conditions

$$3[Rh_{12}(CO)_{30}]^{2-} + 16CO \longrightarrow 6[Rh_5(CO)_{15}]^- + Rh_6(CO)_{16} \qquad (17)$$

TABLE II. Comparative Protonation Studies of $[M(CO)_4]^-$ with Protonic Acids Under High Pressures of $CO:H_2$ (1:1) in Tetraglyme-Toluene (6:1)

Anion (mmol)	Acid (mmol)	Results	Conditions
$[Co(CO)_4]^-$ (0.95)	H_3PO_4 (50.0)	$[Co(CO)_4]^-$	690 atm.; 50°
	H_2SO_4 (29.9)	$HCo(CO)_4$	
$[Rh(CO)_4]^-$ (0.90)	H_3PO_4 (52.1)	$[Rh(CO)_4]^-$	" ; 100°
	H_2SO_4 (29.4)	$HRh(CO)_4$	
$[Ir(CO)_4]^-$ (0.38)	H_3PO_4 (4.12)	$HIr(CO)_4$	690 atm.; 50°

TABLE III. Comparative Reactivity of $HM(CO)_4$ with Amines Under High Pressures of $CO:H_2$ (1:1)

Initial Complex (mmol metal)	Amine (mmole)	pKa (25°)	Results (mmol)	Conditions
$HCo(CO)_4$ (1.0)	N,N-dimethylaniline (1.0)	4.8	$[Co(CO)_4]^-$ (0.35)	Dodecane; 690 atm; 100°
	N-methylmorpholine (1.0)	7.4	$[Co(CO)_4]^-$ (1.0)	
(2.34)	N-methylmorpholine (1.0)		$[Co(CO)_4]^-$ and $HCo(CO)_4$	Tetraglyme-Toluene " 50°
(2.34)	N,N-dimethylaniline (1.0)		$HCo(CO)_4$	
$HIr(CO)_4$ (1.5)	N-methylmorpholine (10.0)		$HIr(CO)_4$ (1.0)	Dodecane; " "
(1.0)				Tetraglyme-Toluene; " "
$Rh_4(CO)_{12}$; $Rh_2(CO)_8$; $HRh(CO)_4$ (2.73)	N,N-dimethylaniline (1.0)		$[Rh(CO)_4]^-$	Tetraglyme-Toluene; 1542 atm; 5°

The fragmentation of $Rh_6(CO)_{16}$ by carbon monoxide (equation 4) suggests that the formation of $Rh_2(CO)_8$ followed by its condensation to $Rh_6(CO)_{16}$ may be occurring in that reaction (equation 17). In fact, preliminary results obtained by ^{103}Rh NMR (collaboration with Professor Otto Gansow) for $[Rh_{12}(CO)_{30}]^{2-}$ show a pattern of two lines in a 5:1 ratio in contrast with the pattern of three-lines in 4:1:1 ratios expected for the solid state structure.[(19)] We suggest these results could indicate that while there are five scrambling rhodium atoms (B) in the structure of $[Rh_{12}(CO)_{30}]^{2-}$ (Figure 4), the other two rhodium atoms (A) are not taking part in that scrambling. These two unique atoms could not take part in the scrambling perhaps as a consequence of the strength of the linkage between them due to a Rh-Rh bond and two edge-bridged carbonyls. These observations suggest that these atoms (A), could be then released as $Rh_2(CO)_8$ which in turn will dimerize or trimerize to give $Rh_4(CO)_{12}$ or $Rh_6(CO)_{16}$, respectively.

The participation of $Rh_2(CO)_8$ in the fragmentation-aggregation equilibrium involving $[Rh_5(CO)_{15}]^-$ (equation 18) is supported by the formation of this anion in the reaction of $Rh_4(CO)_{12}$ with $[Rh(CO)_4]^-$ (equation 1)[(20)] and the fragmentation reaction of $Rh_4(CO)_{12}$ (equation 4). We hypothesized that the involvement of $Rh_2(CO)_8$ in the fragmentation of $[Rh_5(CO)_{15}]^-$ under high pressures of carbon monoxide leads to the equilibrium below (equation 18).

$$[Rh_5(CO)_{15}]^- + 5CO \rightleftharpoons 2\,Rh_2(CO)_8 + [Rh(CO)_4]^- \qquad (18)$$

The previous hypothesis was tested by working with a solution of $[Rh_5(CO)_{15}]^-$ in 18-crown-6 prepared from $[Rh_{12}(CO)_{30}]^{2-}$ (equation 17) from which $Rh_6(CO)_{16}$ has been removed prior to use. The displacement of that equilibrium (equation 18) towards the right was achieved by reduction of the concentration of $Rh_2(CO)_8$ while the system was monitored by infrared spectroscopy. The initial reaction mixture showed only the spectrum of $[Rh_5(CO)_{15}]^-$ at 150° and 832 atm. of $CO:H_2$ (1:1) (Figure 5) but further treatment with this gas mixture under these conditions resulted in the presence of an additional band at 1900 cm^{-1} assigned to $[Rh(CO)_4]^-$; we have also observed the spectra of $Rh_2(CO)_8$ and $Rh_4(CO)_{12}$ at the same time under 200 atm. of $CO:H_2$ (1:1) and 85°. The latter cluster has been shown to be formed by the dimerization of $Rh_2(CO)_8$.

The formation of $[Rh(CO)_4]^-$ in the system above could not be ascribed to the fragmentation of $[Rh_5(CO)_{15}]^-$ caused by carbon monoxide in presence of bases, since no bases are added to the system. Neither can it result from the initial fragmentation of this cluster as shown by the spectrum taken before further treatment with $CO:H_2$. The presence of $[Rh_5(CO)_{15}]^-$ and $[Rh(CO)_4]^-$ in that solution has resulted in the formation of $[Rh_7(CO)_{16}]^{3-}$ upon venting the system to atmospheric

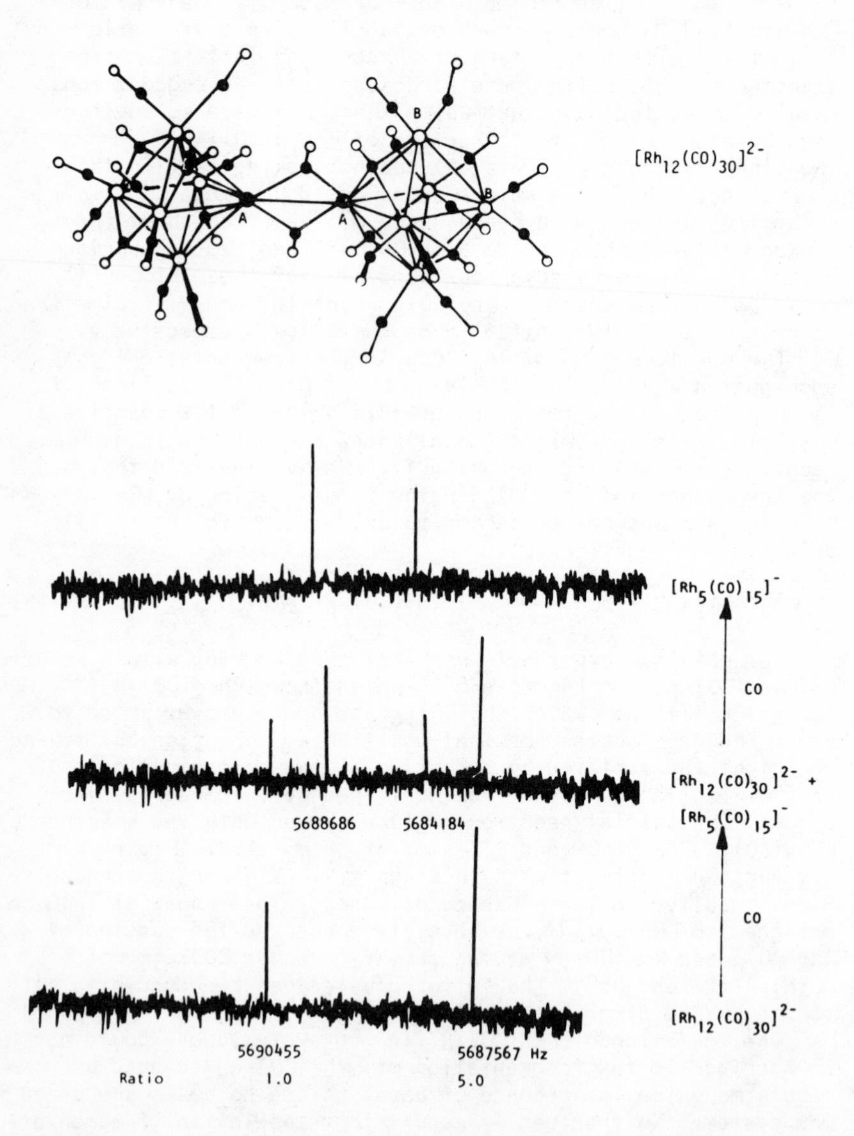

Figure 4. ^{103}*Rh NMR of* $[PhCH_2N(C_2H_5)_3]_2[Rh_{12}(CO)_{30}]$ *in* d$_6$*-acetone solution at −80°C*

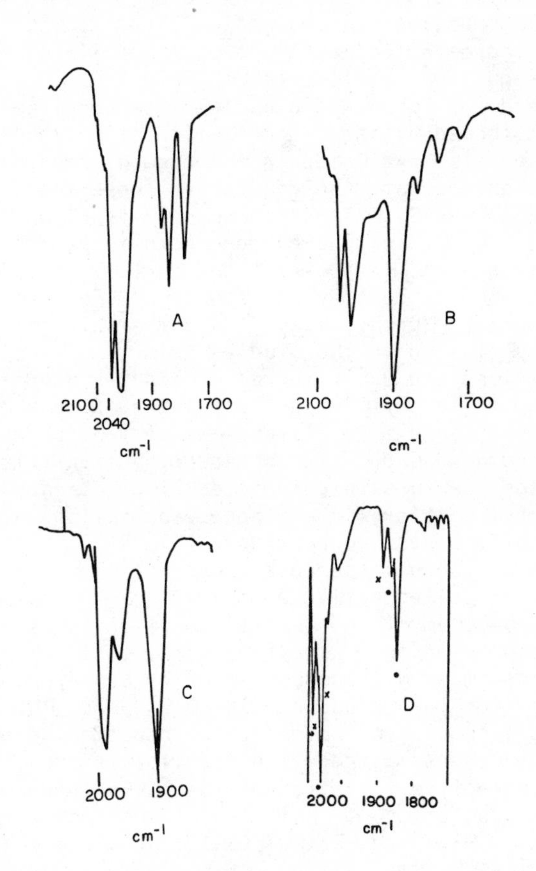

Figure 5. IR spectra in 18-crown-6 solutions of: (a) $[Rh_5(CO)_{15}]^-$ under 832 atm $CO:H_2$ (1:1) at 150°C; (b) after further treatment of solution (a) under these conditions; (c) solution in (b) under atmospheric conditions; (d) detection of $Rh_2(CO)_8$ (●), and $Rh_4(CO)_{12}$ (×), formed in (b) but observed under 200 atm of $CO:H_2$ (1:1) and 85°C

conditions (Figure 5), as expected from other results(1,3) (equation 19). Nevertheless, the mono-anionic cluster is the

$$[Rh_5(CO)_{15}]^- + 2[Rh(CO)_4]^- \rightleftharpoons [Rh_7(CO)_{16}]^{3-} + 7CO \qquad (19)$$

limiting substance in this system, while $[Rh(CO)_4]^-$ is usually the limiting substance in presence of bases e.g., Rh: N-methylmorpholine: $PhCO_2$ ratios of 2:17:1.(2) These observations indicate the magnitude of the shift of the equilibrium (equation 18) towards $[Rh(CO)_4]^-$.

The presence of mono- and binuclear- carbonyl species that could be formed by the fragmentation of polynuclear complexes has been also evaluated in the case of rhodium carbonyl clusters with encapsulated heteroatoms. These species, e.g., $[Rh_6(CO)_{15}C]^{2-}$, $[Rh_{17}S_2(CO)_{32}]^{3-}$, $[Rh_9E(CO)_{21}]^{2-}$ and $[Rh_{10}E(CO)_{22}]^{3-}$(E= P, As), have been shown to be more resistant to fragmentation into $[Rh_5(CO)_{15}]^-$ and $[Rh(CO)_4]^-$ than homo-metallic rhodium carbonyl clusters.(3) The remarkable stability shown by $[Rh_{17}S_2(CO)_{32}]^{3-}$ was the main reason for using this cluster in the studies below.

The conditions and experimental procedures used to study the fragmentation of $[Rh_{17}S_2(CO)_{32}]^{3-}$ are the same as those described above. Once more, it has been established that $Rh_2(CO)_8$ is produced under high pressure, but in this instance it has been possible to detect the presence of another rhodium carbonyl complex with infrared absorptions at 2010, 1970 cm^{-1}. This pattern has been assigned to $[Rh(CO)_2(SH)]_2$, (this complex has been prepared by reaction of H_2S and $Rh(CO)_2AcAc$ in a collaborative effort with Dr. R. A. Fiato). The infrared spectrum of the initial solution shows only the pattern corresponding to $[Rh_{17}S_2(CO)_{32}]^{3-}$, while the one taken after treatment with carbon monoxide and hydrogen shows the bonds expected for this cluster, $[Rh(CO)_4]^-$ and $[Rh(CO)_2(SH)]_2$. These results suggest the fragmentation reaction below (equation 20), while the reverse aggregation reaction is proposed on the basis of the selective formation of the bis-sulfide tri-anion upon bringing a solution of $[Rh(CO)_4]^-$ in contact with a solution of $Rh_4(CO)_{12}$ and $[Rh(CO)_2(SH)]_2$ previously placed at 150°, 400 atm.

$$[Rh_{17}S_2(CO)_{32}]^{3-} + 32CO \rightleftharpoons 6Rh_2(CO)_8 + 3[Rh(CO)_4]^- + [Rh(CO)_2(SH)]_2 \qquad (20)$$

Another simple rhodium carbonyl complex that has been shown to be involved in the aggregation and fragmentation reactions of polynuclear rhodium species is "$Rh(CO)_2^+$ ". This moiety has been shown to be involved in such reactions under ambient conditions as reported by Chini, et al.,(11) (equations 21 and 22).

$$[Rh_{15}(CO)_{27}]^{3-} + 2Br^{-} \rightleftharpoons [Rh_{14}(CO)_{25}]^{4-} + [Rh(CO)_2(Br)_2]^{-} \quad (21)$$

$$[Rh_{15}(CO)_{27}]^{3-} + 2CH_3CN \rightleftharpoons [Rh_{14}(CO)_{25}]^{4-} + [Rh(CO)_2(CH_3CN)_2]^{+} \quad (22)$$

We anticipated those results based on our observations concerning the reactions of $[Rh_{15}(CO)_{27}]^{3-}$. For instance, we found that amines, e.g., n-methylmorpholine or 1,10-orthophenantroline, are able to transform this cluster into $[Rh_{14}(CO)_{25}]^{4-}$ in acetone solution, with an equilibrium similar to those above. It is quite possible, as suggested by an infrared band at 2060 cm^{-1}, that a complex such as $[N_2Rh(CO)_2]^{+}$, a sixteen-electron species, would be generated in these cases. The coordination of the amine to the apical or capping rhodium prior to its removal from the cluster is strongly suggested by the recent report concerning [1,10-orthophenantroline. $Rh_6(CO)_{14}$].(21) The bis-amine is proposed to be coordinated to one of the rhodium atoms while the octahedral cluster cage is maintained.

We have also studied the effect of solvents on the apparent release of a "$Rh(CO)_2^{+}$" based on the appearance in the infrared spectra of such solutions of the bands of $[Rh_{14}(CO)_{25}]^{4-}$ (1960, 1810 cm^{-1}) and in some instances of $[L_2Rh(CO)_2]^{n-}$ (n = -1,0,+1) (2060, 2010 cm^{-1}). These results (Tables IV and V) indicate that it is not possible to ascribe the ability to promote this reaction to either the coordinative ability of the solvent or to its polarity. For instance, that reaction does not proceed in a polar solvent(22a) able to form complexes with transition metals such as acetone, but it proceeds in a low polarity medium, such as 18-crown-6(22b), that coordinates with alkali or transition metal ions or in a highly polar solvent such as sulfolane,(22c) that forms complexes with transition metal ions only under very specific circumstances. We do not yet have a precise explanation for the role of the solvent in the previous reactions. An acceptable hypothesis is that ion-pairing between the cluster anion and its counter-ions precludes any interaction between the apical atom(s) and the solvent. The decrease in ion-pairing that should occur in high polarity solvents could facilitate that interaction with the consequential detachment of a coordinatively unsaturated group, e.g., $[(solvent)_nRh(CO)_2]^{+}$ (n=1,2), that could react further to form a coordinatively saturated species (n=3). Obviously, the distribution between these species, (n=1, 2 or 3), should depend upon the nature of the medium and other reaction conditions. On the other hand, the driving force for the reaction with 18-crown-6 could be provided by the effective coordination of this solvent to "$Rh(CO)_2^{+}$".

TABLE IV. Gutmann Donor Number (DN),(22c) Dielectric Constant (ε),(22a) and Constant of Stability of the Cesium Complexes (log Ks)(22b,d) of the Solvent(s) Used in this Study and the Occurrence of the Reaction Below in That Medium

$$[Rh_{15}(CO)_{27}]^{3-} + nS \rightleftharpoons [Rh_{14}(CO)_{25}]^{4-} + [S_nRh(CO)_2]^+$$

Solvent(L)	DN	$\varepsilon^{25°/25°}$	log Ks(CsL^+)	Reaction Above
Tetraglyme	-	3.5	1.45	N.D.
18-Crown-6	-	4.7	4.60	Detected
Acetone	17.0	20.7	-	N.D.
Acetonitrile	14.1	36.2	-	Detected
Sulfolane	14.8	44.0	-	Detected

Our involvement in the study of the chemistry of polynuclear rhodium complexes under high pressures of $CO:H_2$ induced us to look at the reactions of $[Rh_{15}(CO)_{27}]^{3-}$ with CO or H_2. These reactions have been monitored by ^{13}C NMR and infrared spectroscopy with the results below (Table V). Those results show once more the influence of the solvent in the reactivity of this cluster. This is dramatically evident in the hydrogenation experiments. We consider that the lack of reaction with H_2 in acetone (Table V) may be a consequence of the low polarity of the solvent and the subsequent ion-pairing between the anion and the cation. By contrast, the quantitative reaction occurring in presence of $CsPhCO_2$ could be caused by the potential ability of the benzoate anion to trap the "$Rh(CO)_2^+$" group. We could not get evidence for the coordination of $PhCO_2^-$ to that moiety but it has been observed by ^2H NMR that D $PhCO_2$ is formed upon bubbling D_2 into a sulfolane solution of $[Rh_{15}(CO)_{27}]^{3-}$ containing $CsPhCO_2$. It appears that the coordination of $PhCO_2^-$ to a rhodium complex, e.g., "$Rh(CO)_2^+$", could account for these results if it is concurrent with the oxidative addition of D_2, and followed by the reductive elimination of the acid (equation 23). The solids formed in that system has been characterized as rhodium metal.

$$
\begin{aligned}
&[Rh_{15}(CO)_{27}]^{3-} + PhCO_2^- + \xrightarrow{25^o} [Rh_{14}(CO)_{25}]^{4-} + [PhCO_2.Rh(CO)_2] \\
&\quad [PhCO_2.Rh(CO)_2] + D_2 \longrightarrow [PhCO_2.Rh(CO)_2(D)_2] \\
&\quad [PhCO_2.Rh(CO)_2(D)_2] \longrightarrow PhCO_2D + D\ Rh(CO)_2" \\
&\quad\quad "D\ Rh(CO)_2" \longrightarrow Rh \ + \text{Other Products}
\end{aligned}
\qquad (23)
$$

It is also possible to establish a correlation between the results obtained under ambient and high pressure conditions in the case of "$Rh(CO)_2^+$". A solution of $[Cs(18\text{-crown-}6)_n]_3$ $[Rh_{15}(CO)_{27}]$ in 18-crown-6 containing n-methylmorpholine was placed under 10 atm. of $CO:H_2$ at 150°, and the progress of the reaction was followed by infrared spectroscopy and by atomic absorption analysis, with the results below (equation 24). We believe those results also show the detachment of "$Rh(CO)_2^+$" by either CO or/and H_2. The formation of the intermediate hydrido cluster could be an effect of the pressure of hydrogen while the deprotonation of this species by the amine results in the formation of $[Rh_{14}(CO)_{25}]^{4-}$.

TABLE V. Reactivity of $[Rh_{15}(CO)_{27}]^{3-}$ with CO or H_2 (25°, 1 atm) in Different Solvents

Gas Phase	Conditions	Sulfolane	Acetone
CO	69h	$[Rh_{15}(CO)_{27}]^{3-}$, $[Rh_{14}(CO)_{25}]^{4-}$ $Rh(CO)_2^+$	$[Rh_{15}(CO)_{27}]^{3-}$, $[Rh_{14}(CO)_{25}]^{4-}$ $Rh(CO)_2^+$
H_2	25h	As above	$[Rh_{15}(CO)_{27}]^{3-}$
	25h $CsPhCO_2$	--	$[Rh_{14}(CO)_{25}]^{4-}$, Solids $HPhCO_2$

$$[Rh_{15}(CO)_{27}]^{3-} \rightarrow [Rh_{14}(CO)_{25}H]^{3-} \rightarrow [Rh_{14}(CO)_{25}]^{4-} \quad (24)$$

$\bar{\nu}$(cm^{-1})	1990, 1840	1990, 1830	1965, 1810
time (h)	0	5	24
observation	mixture with Rh_{14}^{4-}	mixture with Rh_{15}^{3-}, Rh_{14}^{4-}	mixture with Rh_{14}^{3-}

CONCLUSIONS

These studies have indicated that simple rhodium carbonyl complexes, e.g., mono- and binuclear species are involved in the fragmentation and aggregation reactions of rhodium carbonyl clusters under high pressures of carbon monoxide and hydrogen. They indicate that it is possible to write formal equations for such reactions in the case of rhodium carbonyl anionic hydrido clusters (equation 25) and for the more particular situation when there are not hydrides present (equation 26)

$$[Rh_x(CO)_yH_z]^{m-} + (4x-y)CO \rightleftharpoons m[Rh(CO)_4]^- + z\,HRh(CO)_4 + (x-m-z/2)Rh_2(CO)_8 \quad (25)$$

$$[Rh_x(CO)_y]^{m-} + (4x-y)CO \rightleftharpoons m[Rh(CO)_4]^- + (x-m/2)Rh_2(CO)_8 \quad (26)$$

Our results also indicate that the behavior of these reactions under ambient and high pressure conditions may be similar. Further work concerning the reactivity of rhodium carbonyl clusters in related reactions will be reported elsewhere.

ACKNOWLEDGEMENT

The authorization by the management of Union Carbide Corporation for the publication of this work is appreciated, together with the continuous encouragement of Dr. G. L. O'Connor. Special recognition is due to Mssrs. B. K. Halstead and R. B. Cunningham who contributed in an essential way with the synthetic and high pressure work.

LITERATURE CITED

1. P. Chini, G. Longoni and V. G. Albano, Adv. Organometal. Chem., 14, 285 (1976).
2. W. E. Walker and R. L. Pruett, U. S. Patents, 3878290, 387214, and 3957857; J. L. Vidal, W. E. Walker and Z. C. Mester, U. S. Patent 4115428; L. A. Cosby, R. A. Fiato and J. L. Vidal, U. S. Patent 4115433; J. L. Vidal, R. A. Fiato, L. A. Cosby and R. L. Pruett, Inorg. Chem., 17, 2574 (1978); J. L. Vidal, W. E. Walker, R. L. Pruett and R. C. Schoening, Inorg. Chem., 18, 129 (1979).
3. J. L. Vidal and W. E. Walker, Inorg. Chem., 19, 864 (1980).
4. J. W. Rathke and H. A. Feder, J. Am. Chem. Soc., 100, 3623 (1978).
5. J. M. Bradley, J. Am. Chem. Soc., 101, 7419 (1979).
6. G. Henrici-Olive' and S. Olive', Angew. Chem. Intern., Ed. English, 15, 136 (1976).
7. E. L. Muetterties, Bull. Soc. Chim. Belg., 84, 959 (1975); E. Band and E. L. Muetterties, Chem. Revs., 78, 639 (1978); E. L. Muetterties, T. N. Rhodin, E. Band, C. F. Brucker and W. R. Pretzer, Chem. Revs., 79, 91 (1979).
8. L. A. Hanlan and G. A. Ozin, J. Am. Chem. Soc., 96, 6324 (1974).
9. J. L. Vidal and W. E. Walker, Inorg. Chem., 19, 0000 (1980).
10. G. Longoni and P. Chini, J. Organometal. Chem., 174, C41 (1979).
11. S. Martinengo, G. Ciani, A. Sironi and P. Chini, J. Am. Chem. Soc., 100, 7096 (1978).
12. J. L. Vidal and R. C. Schoening, Inorg. Chem., 19, 0000 (1980).
13. V. G. Albano, A. Ceriotti, P. Chini, G. Ciani, S. Martinengo and W. M. Anker, J. Chem. Soc. Chem. Comm., 1975, 859; and J. Chem. Soc. Dalton, 978 (1979).
14. R. Whymann, J. Chem. Soc. Chem. Comm., 1194 (1970).
15. R. Whymann, J. Organometal. Chem., 81, 97 (1974).
16. P. Chini and S. Martinengo, Inorg. Chim. Acta., 3, 21 (1969).
17. R. E. Dessy, R. L. Pohl and R. B. King, J. Am. Chem. Soc., 88, 5121 (1966) and D. F. Shriver, Acc. Chem. Research, 3, 231 (1970).
18. P. Chini, et al., personal communication; J. L. Vidal, no reported results.
19. O. A. Gansow, D. S. Gill, F. J. Bennis, J. R. Hutchison, J. L. Vidal and R. C. Schoening, J. Am. Chem. Soc., 102, 2449 (1980).
20. A. Fumagalli, T. F. Koetzle, F. Takusagawa, P. Chini, S. Martinengo and B. T. Heaton, J. Am. Chem. Soc., 102, 1740 (1980).
21. K. Nomiya and H. Suzuki, Bull. Chem. Soc. Japan., 52, 623 (1979).

22. (a) W. L. Jolly, The Synthesis and Characterization of Inorganic Compounds, Prentice-Hall, Inc., Englewood Cliffs, N.J., p. 101; (b) J. M. Lehn, Structure and Bonding, 16, 1, 1976); (c) Viktor Gutmann, Coordination Chemistry in Non-Aqueous Solutions, Springer-Verlag, New York, 1968, p. 19; (d) G. Chaput, G. Jeminet and J. Juillard, Can. J. Chem., 53, 2240 (1975).

Mr. Wellington E. Walker died suddenly on May 8, 1980. His death is a deep loss to his friends at Union Carbide Corporation. He was honest in his private and professional lives and never surrendered his principles; he was a good friend always ready to help; he worked every day honestly with excellence and dedication. He was an inventor of commercialized reactions for low pressure polyethylene, the synthesis of n-octanol and the rhodium catalyzed homogeneous conversion of $CO:H_2$ into polyols. He was a co-author of ca. forty-six patents and twenty-seven papers. In summary, Welly was the hero that works quietly every day.

RECEIVED December 1, 1980.

5

Photochemistry of Metal–Metal–Bonded Transition Element Complexes

MARK S. WRIGHTON , JAMES L. GRAFF, JOHN C. LUONG, CAROL L. REICHEL, and JOHN L. ROBBINS

Department of Chemistry, Massachusetts Institute of Technology, Cambridge, MA 02139

Transition metal complexes that have direct metal-metal bonds have been the objects of intense interest from the point of view of geometric and electronic structure, synthesis and catalysis (1, 2, 3, 4). Low-lying electronic excited states of metal-metal bonded complexes often involve significant changes in the electron density associated with the metal-metal bond, compared to the ground electronic state. Accordingly, study of the photochemistry of metal-metal bonded complexes not only provides potential new reaction chemistry but also provides insight into, and confirmation of, the electronic structure. This symposium volume affords us an opportunity to record the state of the field of metal-metal bond photochemistry. The aim of this article is to summarize recent research results from this laboratory and to place them in perspective in relation to results from other laboratories.

Complexes studied in this laboratory have included low-valent, organometallic clusters having two, three, or four metals. *A priori* the photoexcited complexes can be expected to undergo any reaction that is possible, but the rate constant for a given process will be different for each electronic state. Whether reaction occurs with measurable yield from a given excited state depends on the rate of the reaction relative to internal conversion of the excited state to a lower lying excited state. From an examination of products alone, therefore, it is not always possible to quantitatively assess the relative reactivity of the various electronic states. But when photochemical products differ from those obtained by thermal activation alone, profound effects from the redistribution of electron density can be inferred. In some cases the excited state may simply give the same product as from thermal activation of the ground state. However, when photoreaction occurs it should be realized that the conversion to product occurs within the lifetime of the excited state. Even the longest-lived excited metal complexes are of the order of 10^{-3} s (5, 6) in lifetime and the longest-lived metal-metal bonded complex in 298 K fluid solution is of the order of $\sim 10^{-6}$ s in lifetime (7). Thus, excited state reactions of any kind must

0097-6156/81/0155-0085$06.50/0

have large rate constants compared to those for the thermally inert ground state species. Measuring the rate constant for a given chemical reaction in two different electronic states is a procedure for directly assessing the relative chemical reactivity of the two states.

In the sections below we will describe in detail the known photochemistry of di-, tri-, and tetranuclear metal clusters. Results will be discussed in simple electronic structural terms.

Photochemistry of Dinuclear Complexes

a. Complexes Having Single M-M Bonds and π-d $\rightarrow \sigma^*$ or $\sigma_b \rightarrow \sigma^*$ Lowest Excited States. It is apparent from the description of the electronic structure of $Mn_2(CO)_{10}$, (8, 9), Scheme I, that complexes having a single M-M bond could have low-lying excited states that should be more labile than the ground state with respect to either M-ligand dissociation or M-M bond cleavage. In particular, $Mn_2(CO)_{10}$ and its third row analogue, $Re_2(CO)_{10}$, exhibit low-lying excited states arising from π-d $\rightarrow \sigma^*$ and $\sigma_b \rightarrow \sigma^*$ transitions (8, 9, 10, 11). These high symmetry ,

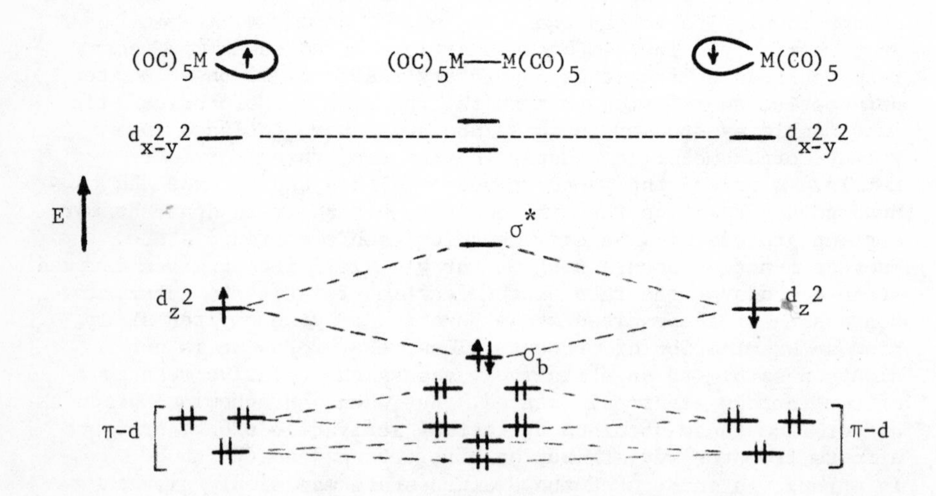

Scheme I. One-electron level diagram for $Mn_2(CO)_{10}$, $Re_2(CO)_{10}$, $W_2(CO)_{10}^{2-}$, etc.

"$d^7 - d^7$", homodinuclear dimers were the first to be subjected to a detailed photochemical investigation (10, 11). Irradiation resulting in $\pi\text{-d} \rightarrow \sigma^*$ or $\sigma_b \rightarrow \sigma^*$ transitions results in clean, quantum efficient, homolytic scission of the M-M bond to yield reactive, 17-valence electron radicals, equation (1) (10, 11).

$$M_2(CO)_{10} \xrightarrow[\text{alkane}]{h\nu} 2M(CO)_5 \tag{1}$$

The generation of 17-valence electron radicals is consistent with flash irradiation of the heterodinuclear dimer $MnRe(CO)_{10}$ that yields radical coupling products according to equation (2)

$$2MnRe(CO)_{10} \xrightarrow[\text{alkane}]{h\nu} Mn_2(CO)_{10} + Re_2(CO)_{10} \tag{2}$$

(11, 12). Photogeneration of $MnRe(CO)_{10}$ from irradiation of both $Mn_2(CO)_{10}$ and $Re_2(CO)_{10}$ also occurs, and such cross-coupling has become a definitive test of whether homolytic metal-metal bond cleavage occurs. Kinetics of recovery of ground state $Mn_2(CO)_{10}$ absorption after flash irradiations are consistent with formation of $Mn(CO)_5$ radicals from the excitation (13).

Irradiation of $M_2(CO)_{10}$ in the presence of halogen donors provides chemical evidence consistent with the quantum efficient generation of 17-valence electron radicals (6). Chemistry according to equation (3) occurs with a disappearance quantum

$$Re_2(CO)_{10} \xrightarrow[CCl_4]{313\text{ nm } (\sigma_b \rightarrow \sigma^*)} 2Re(CO)_5Cl \tag{3}$$

yield of 0.6 (10), showing that greater than half of the excited states yield the cleavage reaction. The efficiency of homolytic cleavage may be greater since cage escape of $Re(CO)_5$ radicals may be less than unity. There is a solvent viscosity effect on the disappearance quantum yield of $M_2(CO)_{10}$ in the presence of I_2, consistent with a solvent cage effect (11).

In polar solvents (pyridine, THF, alcohols, etc.) the photochemistry of simple M-M bonded systems seems to be different based on the products observed (14, 15). For example, irradiation of $Mn_2(CO)_{10}$ can give $Mn(CO)_5^-$. But this chemistry very likely originates from the 17-valence radical as the primary product. Disproportionation of the 17-valence electron species, perhaps after substitution at the radical stage, can account for the apparent heterolytic cleavage. If the excited state reaction is truly dissociative, as the evidence from cross-coupling in alkane suggests, there should be little or no influence from solvent.

Population of the σ^* orbital in $M_2(CO)_{10}$ should labilize the M-CO bond, (8, 9) and such is likely the case. However, prompt CO loss apparently does not compete with scission of the M-M bond. Irradiation of $M_2(CO)_{10}$ in the presence of potential entering ligands such as PPh_3 does lead to substitution (5, 6, 11), but the principal primary photoproduct suggests a mechanism other than dissociative loss of CO to give $M_2(CO)_9$. For example, irradiation of $Mn_2(CO)_{10}$ in the presence of PPh_3 leads to mainly $Mn_2(CO)_8(PPh_3)_2$ rather than the expected $Mn_2(CO)_9PPh_3$ (11). This led to the postulated mechanism represented by equations (4)-(6). The substitution lability of 17-valence electron radicals

$$Mn_2(CO)_{10} \underset{\Delta}{\overset{h\nu}{\rightleftarrows}} 2\ {\cdot}Mn(CO)_5 \quad (4)$$

$$2\ {\cdot}Mn(CO)_5 + 2PPh_3 \longrightarrow 2\ {\cdot}Mn(CO)_4(PPh_3) \quad (5)$$

$$2\ {\cdot}Mn(CO)_4PPh_3 \longrightarrow Mn_2(CO)_8(PPh_3)_2 \quad (6)$$

has been elegantly elaborated by T. L. Brown and his co-workers (16, 17, 18, 19).

The population of the σ^* orbital in $M_2(CO)_{10}$ obviously gives rise to considerable lability of the M-M bond. The ground state of these molecules is inert; the high quantum efficiency for photoreaction means that virtually every excited state produced yields radicals. The excited state cleavage rates can be concluded to be $>10^{10}\ s^{-1}$, since no emission has ever been detected from these species and the reactions cannot be quenched by energy transfer. The excited state cleavage rate of $>10^{10}\ s^{-1}$ is many orders of magnitude larger than from the ground state and reflects the possible consequence of a one-electron excitation. The question of whether both depopulation of σ_b and population of σ^* are necessary is seemingly answered by noting that irradiation at the low energy tail of the $\pi\text{-}d \rightarrow \sigma^*$ absorption gives a quantum yield nearly the same as that for $\sigma_b \rightarrow \sigma^*$ excitation. However, the relative quantum yields do not reflect the relative excited state rate constants for reaction. It is only safe to conclude that either $\pi\text{-}d \rightarrow \sigma^*$ or $\sigma_b \rightarrow \sigma^*$ results in dramatic labilization compared to ground state reactivity and that the rate constant is large enough in each case to compete with internal conversion to an unreactive state. Another ambiguity is that the state achieved by a given wavelength may not be the state from which reaction occurs. The reactive states, for example, could be triplet states that only give rise to weak absorptions from the ground state.

Electrochemistry and redox chemistry provide possible ways of assessing the consequence of one-electron depopulation or population of orbitals. The $M_2(CO)_{10}$ species suffer rapid M-M cleavage upon reduction (population of σ^*) or oxidation (depopulation of σ_b) (20, 21). But cyclic voltammetry only reveals that the cleavage rate from the radical anion, $M_2(CO)_{10}^{\bar{\cdot}}$, or radical cation, $M_2(CO)_{10}^{+\cdot}$, is $>\sim10^3\ s^{-1}$. This is certainly consistent but direct measurements of the rate remain to be done.

Comparison of the photochemistry and the thermal reactivity of $M_2(CO)_{10}$ is of interest. Considerable effort has been expended to show that thermolysis can lead to M-M bond cleavage. The activation energy from M-M dissociation has been examined and correlated with the position of the $\sigma_b \rightarrow \sigma^*$ absorption band from a number of derivatives of $Mn_2(CO)_{10}$. The question is: what is the relative importance of M-ligand cleavage vs. M-M cleavage in the ground state? A recent report bears on this question. Thermal reaction of $MnRe(CO)_{10}$ with PPh_3 was examined (22). The interesting finding is that the primary products do not include $Mn_2(CO)_8(PPh_3)_2$ which would be expected if $Mn(CO)_5$ were produced in the rate limiting step (see equations (4)-(6)). Thus, the ground state pathway for this complex appears to be mainly CO loss, not Mn-Re bond cleavage. The photoreactivity appears to be dominated by Mn-Re bond cleavage (11). A difference in M-ligand vs. M-M cleavage is seemingly established. The excited state results in a relatively low barrier to M-M cleavage compared to M-ligand cleavage. Since the excited state reaction is so clean and quantum efficient (*vide infra*) it is tempting to generalize the specificity found for excited state M-M' cleavage in $MnRe(CO)_{10}$. But this generalization would require that the ground state have a lower activation energy for M-ligand cleavage than for M-M cleavage. Such is apparently not the case for certain species such as $(\eta\text{-}C_5H_5)_2Cr_2(CO)_6$ that likely exist in solution in equilibrium with the $(\eta\text{-}C_5H_5)Cr(CO)_3$ radical (23). The species $[(\eta^3\text{-allyl})Fe(CO)_3]_2$ also exists in the monomeric form in solution (24). Weak M-M single bonds do exist and the ground state M-M cleavage can clearly dominate the chemistry. Light may serve only to accelerate the observed rate.

There are now known a large number of other photosensitive dinuclear complexes that can be formulated as having a 2-electron sigma bond (25-29). Designation of the electronic configuration by the d^n configuration of the 17-valence electron fragment that would be obtained by homolytic cleavage allows a convenient cataloguing of the species studied so far. The transition metal-metal bonded complexes in the summary, Table I, have all been shown to undergo efficient M-M bond homolysis upon photoexcitation (11, 25-29). Examples of d^5-d^5, d^7-d^7, d^9-d^9, d^5-d^7, d^7-d^9, and d^5-d^9 are all known. All of the complexes exhibit an optical absorption spectral feature in the near-uv or high energy visible that can be attributed to the $\sigma_b \rightarrow \sigma^*$ transition. Generally, a π-d $\rightarrow \sigma^*$ feature is also observable at lower energy than the $\sigma_b \rightarrow \sigma^*$, and irradiation at this energy results in homolysis but with a somewhat lower quantum yield. None of the complexes listed are complicated by bridging ligands (see succeeding section) and none give significant yields for CO loss as the primary photoprocess.

All of the M-M single bonded complexes can be formulated within the framework suggested by Scheme II. The 17-valence electron radicals studied all have about the same group electronegativity based on the ability to predict the position of the $\sigma_b \rightarrow \sigma^*$ absorption of heterodinuclear dimers, M-M' from the $\sigma_b \rightarrow \sigma^*$ energy in M-M and M'-M' (28). Significant ionic contribution to the M-M' bond will likely result in a situation where M-M' bond cleavage is heterolytic or where some other reaction, M-ligand cleavage for example, dominates the excited

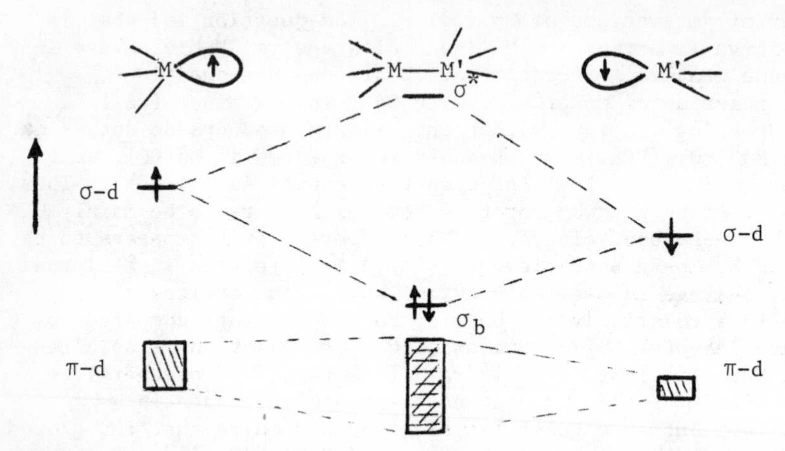

Scheme II. General one-electron level diagram for M-M' single bonded complexes.

state processes. Irradiation of the $(\eta\text{-}C_5H_5)M(CO)_3X$ (M = Mo, W) (30), $(\eta\text{-}C_5H_5)Fe(CO)_2X$ (31), or $Re(CO)_5X$ (32) (X = halide or pseudohalide) yields CO extrusion; in these cases the $\sigma_b \rightarrow \sigma^*$ transition is very high in energy.

Complexes that exhibit a formal two-electron sigma bond between a transition metal and main group metal are of interest but have received relatively little study. Species such as $(CH_3)_3SnMn(CO)_5$ are thermally quite rugged but are photosensitive (33). Irradiation of $(\eta\text{-}C_5H_5)Mo(CO)_3SnMe_3$ in the presence of $P(OPh)_3$ yields CO substitution (34). The question of course is whether the primary photoreaction is loss of CO or homolysis of the Sn-M bond. The complexes $R_3SnCo(CO)_nL_{4-n}$ represent an interesting set of Sn-M bonded complexes, because the 17-valence electron Co-fragment is of d^9 configuration and the complex has only one low-lying σ^* orbital. For example, Scheme III shows a simple, but very adequate, one-electron level diagram for $Ph_3SnCo(CO)_3L$. Irradiation of this complex does not lead to efficient formation of $Co_2(CO)_6L_2$ that would be an expected cross-coupling product from photogenerated radicals. Rather, CO substitution occurs with a quantum yield of ~0.2 at 366 nm (35). The lowest excited state is very likely a $\pi\text{-}d \rightarrow \sigma^*$ excitation and does not appear to result in sufficient Sn-M lability to compete with Co-CO cleavage within the excited state lifetime. The related system $R_3SiCo(CO)_4$ likewise only gives CO extrusion upon photoexcitation (36). In both of these systems the $\sigma_b \rightarrow \sigma^*$ is likely much higher in energy than the $\pi\text{-}d \rightarrow \sigma^*$ transitions. Even if the $\sigma_b \rightarrow \sigma^*$ excitation is achieved it is not clear the Sn-M or Si-M bond cleavage can compete with M-CO cleavage rates in the excited state. It appears that bonds between group IV elements and transition metal carbonyls are not efficiently cleaved compared to the efficiency for M-CO cleavage.

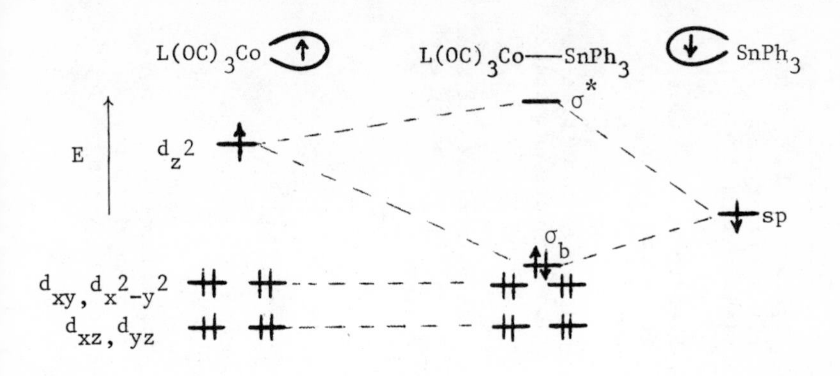

Scheme III. One-electron level diagram for a high symmetry d^9, 17-valence electron radical coupled to Ph_3Sn.

b. Photochemistry of Bridged, M-M Single Bonded Complexes. The unsupported M-M single bonds, Table I, give clean rupture of the M-M bond upon photoexcitation. Bridged M-M bonds represent a situation that could differ significantly since the bridging group could prevent the dissociation of the 17-valence electron fragments. The doubly CO-bridged $Co_2(CO)_8$, and $(\eta-C_5H_5)_2M_2(CO)_4$ (M = Fe, Ru) species have been examined. Of these, the most interesting is $(\eta-C_5H_5)_2Ru_2(CO)_4$ which exists as a non-bridged or bridged species, equation (7), in solution at 298 K. The

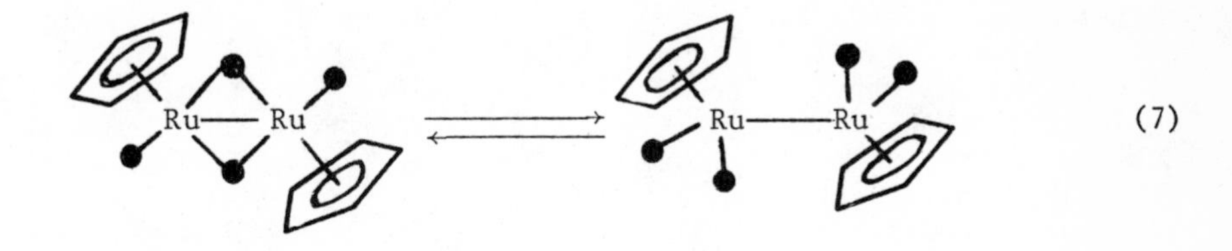

(7)

ratio of the two forms depends on the solvent with alkane solvents giving approximately a 1/1 ratio, while polar solvents such as CH_3CN give nearly 100% of the bridged form (37, 38). The 366 nm quantum yield for reaction according to equation (8) is

$$(\eta\text{-}C_5H_5)_2Ru_2(CO)_4 \xrightarrow[\text{solvent}]{h\nu,\ 0.1\ \underline{M}\ CCl_4} 2\ (\eta\text{-}C_5H_5)Ru(CO)_2Cl \quad (8)$$

Table I. Photochemistry of Dinuclear, M–M or M–M' Species Having a Two-Electron Sigma Bond.

Complex (d^n - d^n)	Primary Photoreaction (Ref.)	Comment
$(\eta\text{-}C_5H_5)_2M_2(CO)_6$ (M=Mo, W)(d^5-d^5)	M–M Bond Cleavage (25) (from cross-coupling and trapping reactions)	Population of $\sigma_b \rightarrow \sigma^*$ gives higher quantum yield than $\pi\text{-}d \rightarrow \sigma^*$.
$M_2(CO)_{10}$ (M=Mn, Re)(d^7-d^7) $MnRe(CO)_{10}$ (d^7-d^7)	M–M or Re–Mn Cleavage (11) (from cross-coupling and trapping reactions)	Electronic structure well-established (8,9).
$Co_2(CO)_6L_2$ (L=P(n-Bu)$_3$,P(OPh)$_3$(d^9-d^9)	Co–Co Bond Cleavage (29) (from cross-coupling reactions)	Halogen atom abstraction products not stable at 298 K.
$(\eta\text{-}C_5H_5)Fe(CO)_2Mn(CO)_5$ (d^7-d^7)	Fe–Mn Bond Cleavage (40) (from cross-coupling reactions)	Heterodinuclear but same d^n configuration at each center.
$(\eta\text{-}C_5H_5)Fe(CO)_2M(CO)_3(\eta\text{-}C_5H_5)$ (d^7-d^5) (M = Mo, W)	Fe–M Bond Cleavage (27,28,40) (from cross-coupling and halogen atom abstraction reactions)	
$(\eta\text{-}C_5H_5)Fe(CO)_2Co(CO)_3(P(OPh)_3)$ (d^7-d^9)	Fe–Co Bond Cleavage (29) (from cross-coupling reaction)	
$(\eta\text{-}C_5H_5)M(CO)_3M'(CO)_5$ (d^5-d^7) (M=Mo,W; M'=Mn,Re)	M–M' Bond Cleavage (26) (from cross-coupling and halogen atom reactions)	Significantly lower quantum yield for $\pi\text{-}d \rightarrow \sigma^*$ vs. $\sigma_b \rightarrow \sigma^*$.

Table I. (continued)

Complex (d^n - d^n)	Primary Photoreaction (Ref.)	Comment
$(\eta-C_5H_5)M(CO)_3Co(CO)_4$ (d^5-d^9)	M-Co Bond Cleavage (27,28,40) (from cross-coupling and halogen atom reactions)	
$(\eta-C_5H_5)Mo(CO)_3SnMe_3$ (d^5-main group)	Loss of CO (34) (from photosubstitution)	
$Mn(CO)_5SnMe_3$ (d^7-main group)	Not well-established (33)	Loss of CO or Mn-Sn cleavage would rationalize results.
$Co(CO)_3(P(OPh)_3)SnPh_3$ (d^9-main group)	Loss of CO (35) (from photosubstitution)	

essentially the same in CH_3CN and in isooctane where the fraction of incident light absorbed by the non-bridged form changes dramatically (38). The lack of a change in the quantum yield suggests that the bridged species is cleaved as efficiently as the non-bridged species by excitation corresponding to transitions that lead to population of the σ^* orbital.

The $(\eta\text{-}C_5H_5)_2Fe_2(CO)_4$ is fully bridged in any solvent. Irradiation at 298 K in the presence of CCl_4 accelerates the thermal rate of formation of $(\eta\text{-}C_5H_5)Fe(CO)_2Cl$ (38, 39). Light-induced cross-coupling of $(\eta\text{-}C_5H_5)_2Fe_2(CO)_4$ with the radical precursors $M_2(CO)_{10}$, (M = Mn, Re) $(\eta\text{-}C_5H_5)_2M_2(CO)_6$ (M = Mo, W) is observed (28, 40), consistent with clean photogeneration of the $(\eta\text{-}C_5H_5)Fe(CO)_2$ radical. Irradiation of $(\eta\text{-}C_5H_5)_2Fe_2(CO)_4$ at low temperature in solutions containing $P(OR)_3$ results in an intermediate hypothesized to be a CO-bridged species that has a ruptured Fe-Fe bond (41). This intermediate can be used to account for the reactions of photoexcited $(\eta\text{-}C_5H_5)_2Fe_2(CO)_4$ without invoking the prompt photogeneration of free radicals. However, the radical cross-coupling is likely best explained by the ultimate generation of the free radicals but this may not be required if the lifetime of the CO-bridged species not containing the Fe-Fe bond is sufficiently long. In any event, population of states where the σ^* orbital is populated leads to a significantly weakened M-M interaction and net cleavage does occur. Prompt CO ejection does not seem to be an important process for either the Fe or Ru species.

The $Co_2(CO)_8$ also exists as bridged and non-bridged forms in solution (42). Irradiation of $Co_2(CO)_8$ and $(\eta\text{-}C_5H_5)_2M_2(CO)_6$ (M = Mo, W) leads to the expected radical coupling product $(\eta\text{-}C_5H_5)M(CO)_3Co(CO)_4$ (28). But the quantum efficiency is not known. Irradiation of $Co_2(CO)_8$ at low temperature has been shown to lead to loss of CO, equation (9), and the reaction is

$$Co_2(CO)_8 \xrightarrow[\text{low temperature matrix}]{h\nu} Co_2(CO)_7 + CO \qquad (9)$$

wavelength dependent but quantum yields have not been reported (43). Reaction according to equation (9) leads to the suggestion that dissociative loss of CO can become the dominant photoreaction when the cage effect becomes severe enough to preclude separation of the 17-valence electron radicals. The importance of the bridging CO's in 298 K solutions is difficult to assess at this time. Some of the usual chemical probes are unsatisfactory here because the $Co_2(CO)_8$ is labile thermally in the presence of potential entering ligands. The $Co(CO)_4$ radical does not seem to react rapidly enough with CCl_4 to provide a reasonable measure of the efficiency of the photogeneration of $Co(CO)_4$ (27, 28). The situation is further complicated by the fact that the $Co(CO)_4X$ (X = Cl, Br, I) species are thermally labile, even if formed.

Intuitively, it would seem that bridging ligands would result in significantly lower quantum yields for M-M cleavage. Such is not found experimentally. If a retarding effect is to be quantitated it will likely come from a direct measurement of the excited state rate constants. Table II summarizes the key results to date.

c. Photochemistry of Dinuclear Complexes Having Multiple M-M Bonds. The cornerstone example of strong, multiple M-M bonds is the Re ≡ Re quadruple bonded $Re_2Cl_8^{2-}$ (44). This substance was found to suffer Re ≡ Re bond cleavage from an upper excited state produced by optical excitation in CH_3CN solution, equation (10) (45). The lowest excited state, associated with the

$$Re_2Cl_8^{2-} \xrightarrow[CH_3CN]{h\nu} 2ReCl_4(CH_3CN)_2^{-} \quad (10)$$

$\delta \rightarrow \delta^*$ transition (46), is unreactive and apparently does not disrupt the bond enough to yield cleavage. Even the upper excited state suffers cleavage via attack of the CH_3CN as determined from flash photolysis studies (47). The photoinduced cleavage places an upper limit on the Re ≡ Re bond energy, but the energetics are obscured by the fact that reaction is not dissociative homolytic cleavage. The $Re_2Cl_8^{2-}$ and $Re_2Br_8^{2-}$ were found to be luminescent from the $\delta \rightarrow \delta^*$ excited state (48, 49). In the first report (48) describing the low temperature (1.3 K) emission of $Re_2X_8^{2-}$ (X = Cl, Br) and $Mo_2Cl_8^{4-}$ the emission of ~100 ns lifetime was attributed to the triplet state, but subsequently (49) it was reported that the complexes $Re_2X_8^{2-}$ and $Mo_2Cl_4(PR_3)_4$ are emissive in fluid solution at 298 K with lifetimes in the 50-140 ns regime. The detailed study (49) led to the conclusion that the emission originates from a distorted singlet state. Related Mo≡Mo quadruple bonded complexes apparently do emit from a triplet state at low temperature (48) with a lifetime of ~2 ms for $Mo_2(O_2CCF_3)_4$ at 1.3 K. The $Mo_2Cl_4(PR_3)_4$ species have been found to undergo photooxidation in chlorocarbon solution (4,50), but like the photochemistry for $Re_2X_8^{2-}$ (45, 47), the reaction occurs from an upper excited state.

The Mo≡Mo quadruple bonded species $K_4Mo_2(SO_4)_4$ has been irradiated in aqueous sulfuric acid solution (50). The photochemistry proceeds according to equation (11). The disappearance

$$H^+_{(aq)} + Mo_2(SO_4)_4^{4-} \xrightarrow{h\nu} \tfrac{1}{2}H_2 + Mo_2(SO_4)_4^{3-} \quad (11)$$

quantum yield at 254 nm was found to be 0.17, but the reaction can also be effected with visible light. This system exemplifies the notion that if the lifetime of the excited state is not controlled by dissociative M-M bond cleavage, then bimolecular processes may be possible. The $Mo_2(SO_4)_4^{4-}$ is another example of a multiple M-M bond that cannot be efficiently cleaved in a dissociative manner. Similarly, photoexcitation of $Mo_2X_8^{4-}$ (X = Cl, Br) in aqueous acid results in photooxidation. Clean formation of $Mo_2Cl_8H^{3-}$ can be observed. Again, rupture of the M ≡ M quadruple bond is not found (50).

Table II. Photochemistry of Bridged Dinuclear Metal-Metal Bonded Complexes.

Complex	Photoreaction (Ref.)	Comment
$(\eta\text{-}C_5H_5)_2Fe_2(CO)_4$	$(\eta\text{-}C_5H_5)_2Fe_2(CO)_4 \xrightarrow[CCl_4]{h\nu} 2(\eta\text{-}C_5H_5)Fe(CO)_2Cl$ (38,39)	Quantum yield reasonable but thermal reaction is detectable suggesting relatively small degree of labilization upon photoexcitation.
	$(\eta\text{-}C_5H_5)_2Fe_2(CO)_4 + (\eta\text{-}C_5H_5)_2M_2(CO)_6 \xrightarrow[hydrocarbon]{h\nu} 2\ (\eta\text{-}C_5H_5)Fe(CO)_2\text{-}Mo(\eta\text{-}C_5H_5)(CO)_3$ (28)	Implies formation of free radicals.
	$(\eta\text{-}C_5H_5)_2Fe_2(CO)_4 + P(OR)_3 \xrightarrow{h\nu} (\eta\text{-}C_5H_5)_2Fe_2(CO)_3P(OR)_3$ (38,40)	CO-bridged, Fe-Fe cleaved intermediate observed at low temperature (40).
$(\eta\text{-}C_5H_5)_2Ru_2(CO)_4$	$(\eta\text{-}C_5H_5)_2Ru_2(CO)_4 \xrightarrow[CCl_4]{h\nu} 2(\eta\text{-}C_5H_5)Ru(CO)_2Cl$ (38)	Quantum yield independent of whether 1/1 bridged/non-bridged or 100% bridged.
$Co_2(CO)_8$	$Co_2(CO)_8 \xrightarrow[\text{low temp. matrix}]{h\nu} Co_2(CO)_7 + CO$ (43)	Low temperature form is fully bridged; room temperature form is 1/1 in hydrocarbon solution.
	$Co_2(CO)_8 + (\eta\text{-}C_5H_5)_2M_2(CO)_6 \xrightarrow[hydrocarbon]{h\nu} 2(\eta\text{-}C_5H_5)M(CO)_3Co(CO)_4$ (27,28)	

Recent studies of the hydrocarbon soluble complexes $(\eta\text{-}C_5R_5)_2Cr_2(CO)_4$ (R = H, Me) and $(\eta\text{-}C_5H_5)(\eta\text{-}C_5Me_5)Cr_2(CO)_4$, that are formulated as having a Cr ≡ Cr triple bond, (51, 52) show that the dominant photoreaction is loss of CO according to equation (12) (53). The tricarbonyl species accounts for the

$$(\eta\text{-}C_5R_5)_2Cr_2(CO)_4 \xrightarrow[\text{alkane}]{h\nu} (\eta\text{-}C_5R_5)_2Cr_2(CO)_3 + CO \qquad (12)$$

incorporation of ^{13}CO upon irradiation of $(\eta\text{-}C_5R_5)_2Cr_2(CO)_4$ under a ^{13}CO atmosphere. Irradiation of $(\eta\text{-}C_5H_5)(\eta\text{-}C_5R_5)Cr_2(CO)_4$ produces no detectable amount of $(\eta\text{-}C_5H_5)_2Cr_2(CO)_4$ or $(\eta\text{-}C_5Me_5)_2Cr_2(CO)_4$ and no measurable $(\eta\text{-}C_5H_5)(\eta\text{-}C_5Me_5)Cr_2(CO)_4$ is formed when a 1/1 mixture of the symmetrical species is irradiated. The negative results from the attempted cross-coupling experiments rule out an important role for the prompt generation of 15-valence electron fragments from photoexcitation of the Cr ≡ Cr triple bonded complexes. The quantum yield for loss of CO, equation (12), is wavelength dependent; irradiation at the lowest absorption (~600 nm) results in no reaction, but near-uv irradiation gives a quantum yield of $>10^{-2}$ (53). The wavelength dependence is reminiscent of $Co_2(CO)_8$ in low temperature matrices where the potential fragments are tethered by the bridging groups and the cage effects of the matrix (43).

Irradiation of the $(\eta\text{-}C_5H_5)_2V_2(CO)_5$, V=V double bonded species results in loss of CO, not V=V bond cleavage upon photoexcitation (54). The crucial result comes from an experiment where the $(\eta\text{-}C_5H_5)_2V_2(CO)_5$ is irradiated at -50°C in the presence of PEt_2Ph in THF solution. The only product observed is $(\eta\text{-}C_5H_5)_2V_2(CO)_4PEt_2Ph$.

From the studies of the quadruple, triple, and double bonded complexes examined thus far, it will prove difficult to photochemically cleave multiple metal-metal bonds in a dissociative fashion. Photochemical cleavage of multiple bonds is not taboo, since the light induced cleavage of O_2 and CO_2 is well known. But in metal complexes it appears that other processes such as solvent attack on the excited state, electron transfer, and ligand dissociation can lead to excited state deactivation before bond rupture can occur. As seen in the summary in Table III, dissociative cleavage of a multiple metal-metal bond remains to be accomplished. Also, upper excited state reaction is the rule in the multiple bonded systems.

d. Photochemistry of Dinuclear M-M Bonded Complexes Having Charge Transfer Lowest Excited States. Complexes such as $Re(CO)_5Re(CO)_3$(1,10-phenanthroline) (55) and $Ph_3SnRe(CO)_3$(1,10-phenanthroline) (7, 56) and related complexes can be formulated as having a M-M or M'-M single bond. Table IV summarizes the known photoprocesses for such complexes. The lowest excited state in the complexes has been identified as arising from a $(M\text{-}M)\sigma_b \rightarrow \pi^*$phen charge transfer transition (7, 55, 56). Importantly, the orbital of termination of the

Table III. Photochemistry of Dinuclear M-M Multiple-Bonded Complexes.

Complex	M-M Bond Order	Primary Photoreaction (Ref.)
$Re_2Cl_8^{2-}$	4	Solvent attack of excited state leads to Re≣Re Cleavage (45, 47)
$Mo_2Cl_4(P(n\text{-}Bu)_3)_4$	4	Emits at 298 K or lower (49); Photooxidized in chlorocarbons (4, 50)
$Mo_2X_8^{4-}$	4	Emits at 1.3 K; photooxidized in H_2O to give $Mo_2X_8H^{3-}$ (48, 50)
$(\eta\text{-}C_5R_5)_2Cr_2(CO)_4$ (R = Me, H)	3	Dissociative loss of CO from upper excited states (53)
$(\eta\text{-}C_5H_5)_2V_2(CO)_5$	2	Loss of CO (54)

Table IV. Primary Excited State Processes of $\gtrdot$MRe(CO)$_3$(1,10-phenanthroline) (7, 55, 56)

Complex	Photoprocess
$Re(CO)_5Re(CO)_3$(1,10-phenanthroline)	Emission at 77 K
	Re-Re Bond Dissociation at 298 K
$Ph_3ERe(CO)_3$(1,10-phenanthroline)	Emission at 298 or 77 K
(E = Sn, Ge)	E-Re Bond Dissociation
	Electronic Energy Transfer
	Electron Transfer

transition is not the σ^* orbital associated with the M-M or M-M' sigma bond. The orbital of termination is localized on the charge acceptor ligand and population of it is not expected to seriously disrupt the M-M or M-M' bonding. Consistent with this assertion is the fact that the complexes are reversibly reducible on the cyclic voltammetry time scale (56). However, the oxidation of the complexes is not reversible and cyclic voltammetry shows that M-M or M-M' cleavage occurs for the radical monocation at a rate of $>10^3$ s^{-1} (56).

The lowest energy electronic transition of the $Re_2(CO)_8$-(1,10-phenanthroline) and related complexes is expected to labilize the M-M bond in the sense that the M_2-core is "oxidized" by the intramolecular shift in electron density. Excitation has been shown to yield M-M bond cleavage; reaction according to equation (13) is representative (55). The quantum yield of ~0.2

$$Re_2(CO)_8(1,10\text{-phenanthroline}) \xrightarrow[CCl_4]{h\nu} ClRe(CO)_5 + ClRe(CO)_3\text{-}(1,10\text{-phenanthroline}) \quad (13)$$

was found to be independent of the excitation wavelength from 550 nm to 313 nm. This wavelength independent quantum yield is consistent with reaction that originates from the $(M\text{-}M)\sigma_b \rightarrow \pi^*$-phen CT state that is found to yield emission (but no photoreaction) at low temperature. This result shows that depopulation of an M-M core bond level labilizes the M-M bond sufficiently to allow cleavage within the lifetime of the excited state. The 77 K emission lifetime was found to be extraordinarily long (~95 μsec), but emission was not detectable at all at 298 K where photoreaction occurs with a quantum yield of ~0.20. It is possible that emission is not observable at 298 K because M-M bond cleavage occurs too fast; this would indicate that the M-M dissociation rate in the excited state significantly exceeds 10^5 s^{-1}. Given that the ground state dissociation is slow ($<10^{-6}$ s^{-1}) for these thermally inert (298 K) systems, even an excited state rate of 10^5 s^{-1} reflects an increase in dissociation rate of $>10^{11}$ compared to the ground state. An excited state rate of $>10^5$ s^{-1} is consistent with the lower limit of 10^3 s^{-1} cleavage rate of the radical monocation generated in electrochemical experiments.

Study of $Ph_3SnRe(CO)_3$(1,10-phenanthroline) resulted in the first direct determination of the rate constant for excited state cleavage of the M'-M bond (7, 56). The key is that this complex is emissive from the reactive state under the conditions where the cleavage reaction also occurs. Measurement of the emission lifetime (1.8×10^{-6} s) and the photoreaction quantum yield (~0.23) give a rate constant of 1.3×10^5 s^{-1} at 298 K for the dissociation represented by equation (14). The wavelength

$$[Ph_3Sn\text{-}Re(CO)_3L]^* \xrightarrow[\substack{L=1,10\text{-phenanthroline} \\ CH_2Cl_2/0.5\ \underline{M}\ CCl_4}]{k_{14}} Ph_3Sn + Re(CO)_3L \quad (14)$$

independence of photoreaction quantum yields (488-313 nm) and the ability to equally efficiently quench the lifetime, emission, and photoreaction with anthracene confirm that the emissive state is also the reactive state. The ~10^5 s^{-1} excited state cleavage rate is consistent with the lower limit of 10^3 s^{-1} from cyclic voltammetry for cleavage of the radical monocation.

The relatively long lifetime of the lowest excited state of $Ph_3SnRe(CO)_3$(1,10-phenanthroline) allows fast bimolecular processes to compete with the cleavage of the M'-M bond (7). For example, anthracene, having a triplet energy of ~42 kcal/mol, quenches the excited state (~50 kcal/mol) at an essentially diffusion controlled rate by electronic energy transfer. The excited state can also be quenched by electron transfer, equations (15) and (16) (56). Both of these processes are

$$[Ph_3SnRe(CO)_3L]^* + TMPD \xrightarrow{k_{15}} TMPD^{\dot{+}} + [Ph_3SnRe(CO)_3L]^{\dot{-}} \quad (15)$$

$$[Ph_3SnRe(CO)_3L]^* + MV^{2+} \xrightarrow{k_{16}} MV^{\dot{+}} + [Ph_3SnRe(CO)_3L]^{\dot{+}} \quad (16)$$

TMPD≡N,N,N',N'-tetramethyl-p-phenylenediamine; MV^{2+}≡N,N'-dimethyl-4,4'-bipyridinium

significantly downhill and the rate constants k_{15} and k_{16} are those expected for diffusion controlled reactions. The reduction to form $[Ph_3SnRe(CO)_3L]^{\dot{-}}$ results in no net chemical change, since the back electron transfer is fast and the electron added to the Re complex in equation (15) is localized on L. The excited state oxidation though results in net chemistry, since chemistry according to equation (17) competes with the back electron

$$[Ph_3SnRe(CO)_3L]^{\dot{+}} \xrightarrow[S\ =\ solvent]{k_{17}} Ph_3Sn\cdot + SRe(CO)_3L^+ \quad (17)$$

transfer. The unimolecular rate constant k_{17} is $>10^3$ s^{-1} from the electrochemistry and from the photochemistry (equation (14)) the value of k_{17} could be greater than 10^5 s^{-1}. The one-electron oxidants lead to production of 18-electron $SRe(CO)_3L^+$ products; this leads to the conclusion that the cleavage of $[Ph_3ERe(CO)_3L]^{\dot{+}}$ yields $Ph_3E\cdot$ and $Re(CO)_3L^+$, not Ph_3Sn^+ and $\cdot Re(CO)_3L$.

The experiments with the various $\geq$M-$Re(CO)_3L$ species establishes that population of the σ^* level is not required to achieve sufficient M-M bond lability to yield homolytic cleavage within the lifetime of the excited state. However, when the sigma bond order is reduced from one to approximately one-half by the depopulation of the σ_b level, the rate constant for M-M bond cleavage appears to be only ~10^5 s^{-1}. By way of contrast, M-M bond cleavage seems to occur with a rate of $>10^{10}$ s^{-1} for species such as $Re_2(CO)_{10}$ where the sigma bond order is reduced from one to zero by the $\sigma_b \rightarrow \sigma^*$ excitation.

Photochemistry of Trinuclear Complexes

Table V summarizes the key photochemistry of trinuclear metal-metal bonded complexes. The first noteworthy photochemical study of trinuclear complexes concerns $Ru_3(CO)_{12}$. This species was found to undergo declusterification to mononuclear fragments when irradiated in the presence of entering ligands such as CO, PPh_3, or ethylene. The intriguing finding is that thermal reaction of $Ru_3(CO)_{12}$ with PPh_3 results in the substitution product indicated in equation (18) whereas irradiation yields the mononuclear species given by equation (19) (57). These results

$$Ru_3(CO)_{12} \xrightarrow[PPh_3]{\Delta} Ru_3(CO)_9(PPh_3)_3 \qquad (18)$$

$$Ru_3(CO)_{12} \xrightarrow[PPh_3]{h\nu} Ru(CO)_4PPh_3 + Ru(CO)_3(PPh_3)_2 \qquad (19)$$

suggest that Ru-Ru bond cleavage occurs in the excited state whereas Ru-CO dissociation occurs in the ground state. An electronic structural study shows that the orbital of termination for the lowest energy excited states of $Ru_3(CO)_{12}$ is σ^* with respect to the Ru_3-core (58). The disappearance quantum yields for $Ru_3(CO)_{12}$, $Fe_3(CO)_{12}$ and $Ru_3(CO)_9(PPh_3)_3$ are all in the range of 10^{-2} for entering groups such as CO or PPh_3 and are independent of entering group concentration (59, 60). The electronic structure is consistent with primary photoreaction as represented by equation (20), and the overall low quantum yields are consistent

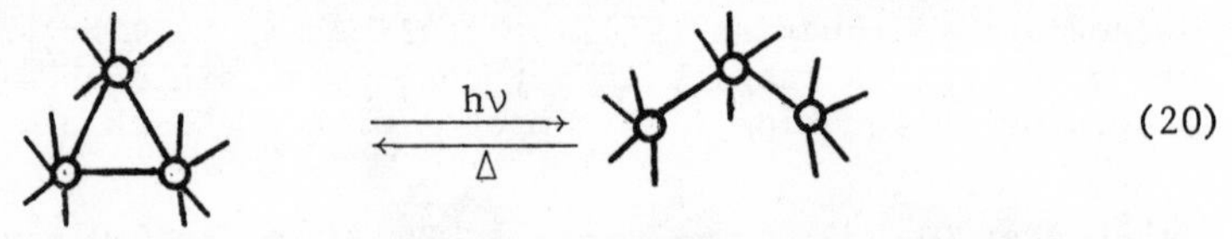

with efficient closure to regenerate the metal-metal bond.

Irradiation of $Ru_3(CO)_{12}$ (57, 60), $Fe_3(CO)_{12}$ (60), or $Ru_3(CO)_9(PPh_3)_3$ (59) under CO cleanly leads to mononuclear complexes. This fact seems to rule out dissociative loss of CO as the dominant reaction from the excited state. If loss of CO were the dominant process, the presence of CO would simply retard the decomposition of the cluster. The lack of an effect from high concentrations of entering group (1-pentene or PPh_3) on the quantum yield for photodeclusterification of $Fe_3(CO)_{12}$ or $Ru_3(CO)_{12}$ (60) is consistent with this conclusion.

The lowest excited states of triangular trinuclear complexes, like dinuclear complexes, can also be labile with respect to loss

Table V. Photochemistry of Trinuclear M-M Bonded Complexes.

Complex	Primary Photoreaction (Ref.)
$M_3(CO)_{12}$ (M = Fe, Ru)	M-M Cleavage is likely dominant (57,59,60) (low overall quantum yield).
$Ru_3(CO)_9(PPh_3)_3$	Ru-Ru Bond Cleavage (59) (low overall quantum yield).
$Os_3(CO)_{12}$	Dissociative loss of CO (60,61) (low quantum yield),
$Os_3(CO)_{12}Cl_2$	Os-Os Bond Cleavage; Linear Structure; High Quantum Yields (61)
$H_3M_3(CO)_{12}$ (M = Mn, Re)	CO loss or M-M Bond Cleavage (62)
$RCCo_3(CO)_9$ (R = CH_3, H)	CO loss Co-Co Bond Cleavage (63)

of CO as well as metal-metal bond cleavage. Irradiation of $Os_3(CO)_{12}$ under conditions where $Ru_3(CO)_{12}$ is declusterified leads to substitution of CO, equation (21) (60, 61). This

$$Os_3(CO)_{12} \xrightarrow[PPh_3]{h\nu} Os_3(CO)_n(PPh_3)_{12-n} \qquad n = 11, 10, 9 \tag{21}$$

change in photoreactivity may be due to the fact that the lowest excited state is $\sigma \rightarrow \sigma^*$ in the $Ru_3(CO)_{12}$ case whereas it is $\sigma^{*'} \rightarrow \sigma^*$ in the case of $Os_3(CO)_{12}$. An alternative explanation may be simply that the stronger Os-Os bonds have lower dissociation rates while Os-CO and Ru-CO cleavage rates are similar. In any event, there is a striking difference in the qualitative features of $M_3(CO)_{12}$ (M = Fe, Ru) vs. $Os_3(CO)_{12}$. Absolute photoreaction quantum yields for any photoreaction are small, but it does appear that M-M bond cleavage dominates for M = Ru or Fe while M-CO cleavage dominates for M = Os.

The clean photodeclusterification represented by equation (22)

$$2\ H_3Re_3(CO)_{12} \xrightarrow{h\nu} 3\ H_2Re_2(CO)_8 \tag{22}$$

is a case where photoexcitation leading to either Re-Re cleavage or dissociative loss of CO could allow rationalization of the observed chemistry (62). The observed quantum yield of ~0.1 is the highest observed for a triangular metal-metal bonded system. This high quantum yield and the fact that the Re-Re bonds are bridged by H atoms suggest that the dissociative loss of CO is the likely result of photoexcitation in this case.

Photolysis of the complex $RCCo_3(CO)_9$ (R = H, CH_3) under an H_2 atmosphere has been reported to yield declusterification with low quantum efficiency (63). As in the case above it is not clear whether reaction begins with M-M or M-CO bond cleavage; the excited state should be more labile than the ground state with respect to either process.

In contrast to the generally low quantum yields for the triangular-M_3 systems, $Os_3(CO)_{12}Cl_2$, that has only two Os-Os bonds, undergoes photoinduced M-M bond cleavage with a high quantum yield (61). This result lends support to the assertion that the tethered diradical center, equation (20), may be important in giving net quantum yields that are low compared to species such as $Mn_2(CO)_{10}$. A possible contribution to low quantum yields for the triangular-M_3 core systems is the fact that the one-electron excitations promoted by optical absorption are not simply localized labilization of two of the three bonds.

Photochemistry of Tetranuclear Complexes

Complexes having the tetrahedrane-M_4 core have six direct M-M bonds and it is not likely that one-electron excitation will result in enough bonding disruption to extrude mononuclear fragments. *If* an M-M bond does cleave, equation (23), cleavage of

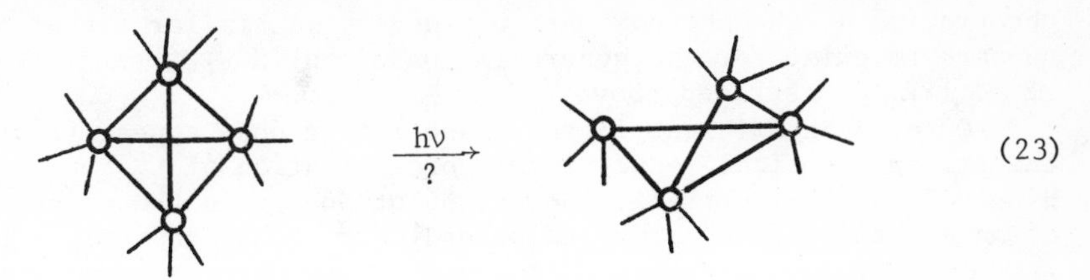

(23)

two other M-M bonds would be required in order to generate the mononuclear fragment. Cleavage in the sense suggested by equation (23) can seemingly occur, though, since photoreaction according to equation (24) has been observed (63). If

$$Co_4(CO)_{12} \xrightarrow{h\nu} 2\ Co_2(CO)_8 \qquad (24)$$

dissociative loss of CO occurs the presence of added CO would seemingly only lead to back reaction with no net chemical change. Likewise, $HFeCo(CO)_{12}$ yields $Co_2(CO)_8$ when irradiated under CO (63). The quantum yields for these reactions are low and it is not clear where the inefficiency lies: is the M-M bond cleavage (equation (23)) a low quantum yield process or is the reformation of the M-M so fast that it is inefficiently trapped by CO?

Irradiation of $Ir_4(CO)_{12}$ in the presence of $(MeCO_2)_2C_2$ results in the retention of an Ir_4 complex but the Ir_4-core in the $Ir_4(CO)_8\{(MeCO_2)_2C_2\}_4$ product is a rectangle. Such a product could arise from either CO loss or from trapping of a photogenerated diradical. As in the trinuclear $M_3(CO)_{12}$ complexes, the result for the Ir_4 species (Ir_4 retention) compared to the Co_4 species (fragmentation (63)) may signal a trend in M-M bond retention for the third row systems where M-M bonds are expected to be stronger.

Not surprisingly, the tetranuclear $Fe_4(CO)_4(\eta-C_5H_4)_4$, that has a tetrahedrane-Fe_4 core with the CO's triply face bridging, is photoinert in solution with respect to Fe-Fe or Fe-CO bond rupture (65). In the presence of halocarbons such as CCl_4 there is a clean photooxidation reaction, equation (25), resulting from

$$Fe_4(CO)_4(\eta-C_5H_5)_4 \xrightarrow[CCl_4]{h\nu,\ CTTS} [Fe_4(CO)_4(\eta-C_5H_5)_4]^+ \qquad (25)$$

irradiation corresponding to absorption due to a charge transfer to solvent transition. Ferrocene exhibits a similar photoreactivity (66); to extend the comparison it is noteworthy that the potential for ferricenium/ferrocene and $[Fe_4(CO)_4(\eta-C_5H_5)]^{+/0}$ is nearly the same (67), the CTTS is at about the same energy and intensity for a given halocarbon, and the quantum yields for photooxidation (after correction for intramolecular absorption) are quite similar (65). The resilience of the $Fe_4(CO)_4(\eta-C_5H_5)_4$ with respect to light induced bond cleavage reactions allows the observation of the photooxidation process. A similar situation appears to exist for the quadruple bonded $Mo_2(SO_4)_4^{2-}$ and $Mo_2Cl_4(PR_3)_4$ described above (50).

Quantum inefficient ligand dissociation does appear to be the primary chemical result from photoexcitation of $H_4Ru_4(CO)_{12}$ (68). Irradiation at 436 or 366 nm in the presence of an entering ligand proceeds according to equation (26). The

$$H_4Ru_4(CO)_{12} \xrightarrow[L=P(OMe)_3,\ PPh_3]{h\nu} H_4Ru_4(CO)_{11}L \qquad (26)$$

reaction has the same quantum efficiency ($5 \pm 1 \times 10^{-3}$) for L = $P(OMe)_3$ or PPh_3 and for a variation in concentration of L from 0.01 to 0.1 M. These observations support the prompt generation of $H_4Ru_4(CO)_{11}$ from photoexcitation. The complex undergoes substitution thermally but photoexcitation accelerates the rate dramatically. Photoexcitation of $H_4Os_4(CO)_{12}$ does give chemistry and photoreaction may begin with dissociative loss of CO subsequent to photoexcitation (69).

Irradiation of $H_4Re_4(CO)_{12}$ does not result in any significant photoreaction (70). The lack of any Re-Re bond cleavage may be associated with the fact that the Re-Re bonds have multiple bond character (71). The $H_4Re_4(CO)_{12}$ exhibits emission from the lowest excited state. This finding prompted a comparison of the excited state decay of the $D_4Re_4(CO)_{12}$. Generally, the highest energy vibrational modes are important in non-radiative decay (72), and for metal complexes the highest energy M-L vibrations may be most important. The hydrogen atoms in $H_4Re_4(CO)_{12}$ are believed to be triply face bridging (71) with a Re_3-H stretching frequency of ~1023 cm^{-1} (73). The $H_4Re_4(CO)_{12}$ complex emits in hydrocarbon solution or as the pure solid at 298 or 77 K. The emission (~14,300 cm^{-1}) onset overlaps the absorption and thus a large distortion of the complex upon excitation does not appear to occur. The emission can be quenched by the triplet quencher anthracene, having a triplet energy of ~42 kcal/mol (74). The lifetime is in the range 0.1 - 16 μsec depending on conditions and is of the order of 20-30% longer for the ^2H substituted complex. Likewise the emission quantum yields for the $D_4Re_4(CO)_{12}$ are 20-30% longer than for the ^1H species. Thus, replacing ^1H by ^2H in $H_4Re_4(CO)_{12}$ has the expected effect of reducing the rate of non-radiative decay. But the effect does not lead to a situation where the excited decay is dominated by the radiative decay rate.

The studies of the tetrahedrane-M_4 clusters, Table VI, show quite generally that the importance of dissociative processes within the excited state lifetime is low. Clean, but low quantum yield, photoreactions are detectable in some cases. The resistance of the complexes to declusterification may be exploited to study and utilize bimolecular processes, and in the case of the $H_4Re_4(CO)_{12}$ the reactions and non-chemical, non-radiative decay are sufficiently low that the excited state lifetime allows

Table VI. Photochemistry of Tetranuclear Complexes.

Complex	Photoprocess (Ref.)
$Co_4(CO)_{12}$	Likely Co-Co Bond Cleavage; $Co_2(CO)_8$ formed under CO (63)
$HFeCo_2(CO)_{12}$	Primary process likely M-M Bond Cleavage (63)
$Fe_4(CO)_4(\eta\text{-}C_5H_5)_4$	Inert in Hydrocarbon; Photooxidation via CTTS in Halocarbon (65)
$Ir_4(CO)_{12}$	Derivative with Ir_4 unit retained can be formed (64)
$H_4Ru_4(CO)_{12}$	Loss of CO; Low Quantum yield (68)
$H_4Os_4(CO)_{12}$	Loss of CO (69)
$H_4Re_4(CO)_{12}$	Photoinert; Emits at 298 or 77 K (70)

radiative decay to be observed. Such long-lived excited states may allow the development of a bimolecular excited state chemistry of such species.

Acknowledgements. We thank the National Science Foundation and the Office of Naval Research for partial support of the research described herein. Support from GTE Laboratories, Inc., is also gratefully acknowledged. M.S.W. acknowledges support as a Dreyfus Teacher-Scholar grant recipient, 1975-1980.

Literature Cited

1. Cotton, F.A. Acc. Chem. Res., 1969, 2, 240.
2. Cotton, F.A. Acc. Chem. Res., 1978, 11, 225.
3. Cotton, F.A. and Wilkinson, G. "Advanced Inorganic Chemistry" 4th ed., John Wiley and Sons: New York, 1980, Ch. 26, pp. 1080-1112, and references cited on p. 1112.
4. Trogler, W.C.; Gray, H.B. Acc. Chem. Res., 1978, 11, 232.
5. Balzani, V.; Carassiti, V. "Photochemistry of Coordination Compounds", Academic Press: New York, 1970.
6. Geoffroy, G.L.; Wrighton, M.S. "Organometallic Photochemistry", Academic Press: New York, 1979.
7. Luong, J.C.; Faltynek, R.A.; Wrighton, M.S. J. Am. Chem. Soc., 1979, 101, 1597.
8. Levenson, R.A.; Gray, H.B. J. Am. Chem. Soc., 1975, 97, 6042.
9. Levenson, R.A.; Gray, H.B.; Caesar, G.P. J. Am. Chem. Soc., 1970, 92, 3653.
10. Wrighton, M.; Bredesen, D. J. Organometal. Chem., 1973, 50, C35.
11. Wrighton, M.S.; Ginley, D.S. J. Am. Chem. Soc., 1975, 97, 2065.
12. Evans, G.O.; Sheline, R.K. J. Inorg. Nucl. Chem., 1968, 30, 2862.
13. Hughey, IV, J.L.; Anderson, C.P.; Meyer, T.J. J. Organometal. Chem., 1977, 125, C49.
14. Allen, D.M.; Cox, A.; Kemp, T.J.; Sultann, Q.; Pitts, R.B. J.C.S. Dalton, 1976, 1189.
15. Burkett, A.R.; Meyer, T.J.; Whitten, D.G. J. Organomet. Chem., 1974, 67, 67.
16. Byers, B.H.; Brown, T.L. J. Am. Chem. Soc., 1975, 97, 947 and 3260; 1977, 99, 2527.
17. Absi-Halabi, M.; Brown, T.L. J. Am. Chem. Soc., 1977, 99, 2982.
18. Kidd, D.R.; Brown, T.L. J. Am. Chem. Soc., 1978, 100, 4103.
19. Brown, T.L. Ann. New York Acad. Sci., 1980, 333, 80.
20. Lemoine, P.; Giraudeau, A.; Gross, M. Electrochim. Acta, 1976, 21, 1.
21. Lemoine, P.; Gross, M. J. Organomet. Chem., 1977, 133, 193.
22. Sonnenberger, D.; Atwood, J.D. J. Am. Chem. Soc., 1980, 102, 3484.

23. Adams, R.D.; Collins, D.E.; Cotton, F.A. J. Am. Chem. Soc., 1974, 96, 749.
24. Murdoch, H.D.; Lucken, E.A.C. Helv. Chim. Acta, 1964, 47, 1517.
25. Wrighton, M.S.; Ginley, D.S. J. Am. Chem. Soc., 1975, 97, 4246.
26. Ginley, D.S.; Wrighton, M.S. J. Am. Chem. Soc., 1975, 97, 4908.
27. Abrahamson, H.B.; Wrighton, M.S. J. Am. Chem. Soc., 1977, 99, 5510.
28. Abrahamson, H.B.; Wrighton, M.S. Inorg. Chem., 1978, 17, 1003.
29. Reichel, C.L.; Wrighton, M.S. J. Am. Chem. Soc., 1979, 101, 6769.
30. Alway, D.G.; Barnett, K.W. Inorg. Chem., 1980, 19, 1533.
31. Alway, D.G.; Barnett, K.W. Inorg. Chem., 1978, 17, 2826.
32. Wrighton, M.S.; Morse, D.L.; Gray, H.B.; Ottesen, D.K. J. Am. Chem. Soc., 1976, 98, 1111.
33. Clark, H.C.; Tsai, J.H. Inorg. Chem., 1966, 5, 1407.
34. King, R.B.; Pannell, K.H. Inorg. Chem., 1968, 7, 2356.
35. Reichel, C.L.; Wrighton, M.S. to be submitted.
36. Reichel, C.L.; Wrighton, M.S. Inorg. Chem., 1980, 19, 0000.
37. Fischer, R.D.; Vogler, A.; Noack, K. J. Organomet. Chem., 1967, 7, 135.
38. Abrahamson, H.B.; Palazzotto, M.C.; Reichel, C.L.; Wrighton, M.S. J. Am. Chem. Soc., 1979, 101, 4123.
39. Giannotti, C.; Merle, G. J. Organomet. Chem., 1976, 105, 97.
40. Ginley, D.S. Ph.D. Thesis, M.I.T., 1976.
41. Tyler, D.R.; Schmidt, M.A.; Gray, H.B. J. Am. Chem. Soc., 1979, 101, 2753.
42. Noack, A. Helv. Chim. Acta, 1964, 47, 1055 and 1064.
43. Sweany, R.L.; Brown, T.L. Inorg. Chem., 1977, 16, 421.
44. Cotton, F.A.; Harris, C.B. Inorg. Chem., 1965, 4, 330.
45. Geoffroy, G.L.; Gray, H.B.; Hammond, G.S. J. Am. Chem. Soc., 1974, 96, 5565.
46. Trogler, W.C.; Cowman, C.D.; Gray, H.B.; Cotton, F.A. J. Am. Chem. Soc., 1977, 99, 2993.
47. Flemming, R.H.; Geoffroy, G.L.; Gray, H.B.; Gupta, A.; Hammond, G.S.; Kliger, D.; Miskowski, V.M. J. Am. Chem. Soc., 1976, 98, 48.
48. Trogler, W.C.; Solomon, E.I.; Gray, H.B.; Inorg. Chem., 1977, 16, 3031 and Trogler, W.C.; Solomon, E.I.; Trajberg, I.; Ballhausen, C.J.; Gray, H.B. Inorg. Chem., 1977, 16, 828.
49. Miskowski, V.M.; Goldbeck, R.A.; Kliger, D.S.; Gray, H.B. Inorg. Chem., 1979, 18, 86.
50. Erwin, D.K.; Geoffroy, G.L.; Gray, H.B.; Hammond, G.S.; Solomon, E.I.; Trogler, W.C.; Zagars, A.A. J. Am. Chem. Soc., 1977, 99, 3620; Trogler, W.C.; Gray, H.B. Nouv. J. Chim., 1977, 1, 475 and Trogler, W.C.; Erwin, D.K.; Geoffroy, G.L.; Gray, H.B. J. Am. Chem. Soc., 1978, 100, 1160.

51. King, R.B.; Efraty, A. J. Am. Chem. Soc., 1971, 93, 4950.
52. Hackett, P.; O'Neill, P.S.; Manning, A.R. J.C.S. Dalton, 1974, 1625.
53. Robbins, J.L.; Wrighton, M.S. submitted for publication.
54. Lewis, L.N.; Caulton, K.G. Inorg. Chem., 1980, 19, 1840.
55. Morse, D.L.; Wrighton, M.S. J. Am. Chem. Soc., 1976, 98, 3931.
56. Luong, J.C.; Faltynek, R.A.; Wrighton, M.S. J. Am. Chem. Soc., 1980, 102, 0000.
57. Johnson, B.F.G.; Lewis, J.; Twigg, M.V. J. Organomet. Chem., 1974, 67, C75 and J.C.S. Dalton, 1976, 1876.
58. Tyler, D.R.; Levenson, R.A.; Gray, H.B. J. Am. Chem. Soc., 1978, 100, 7888.
59. Graff, J.L.; Sanner, R.D.; Wrighton, M.S. J. Am. Chem. Soc., 1979, 101, 273.
60. Austin, R.G.; Paonessa, R.S.; Giordano, P.J.; Wrighton, M.S. Adv. Chem. Ser., 1978, 168, 189.
61. Tyler, D.R.; Altobelli, M.; Gray, H.B. J. Am. Chem. Soc., 1978, 102, 3022.
62. Epstein, R.A.; Gaffney, T.R.; Geoffroy, G.L.; Gladfelter, W.L.; Henderson, R.S. J. Am. Chem. Soc., 1979, 101, 3847.
63. Geoffroy, G.L.; Epstein, R.A. Inorg. Chem., 1977, 16, 2795 and Adv. Chem. Ser., 1978, 168, 132.
64. Heveldt, P.F.; Johnson, B.F.G.; Lewis, J.; Raithby, P.R.; Sheldrick, G.M. J.C.S. Chem. Comm., 1978, 340.
65. Bock, C.R.; Wrighton, M.S. Inorg. Chem., 1977, 16, 1309.
66. Brand, J.C.D.; Snedden, W. Trans. Faraday Soc., 1957, 53, 894.
67. Ferguson, J.A.; Meyer, T.J. J. Am. Chem. Soc., 1972, 94, 3409.
68. Graff, J.L.; Wrighton, M.S. J. Am. Chem. Soc., 1980, 102, 2123.
69. Bhaduri, S.; Johnson, B.F.G.; Kelland, J.W.; Lewis, J.; Raithby, P.R.; Rehani, S.; Sheldrick, G.M.; Wong, K.; McPartlin, M. J.C.S. Dalton, 1979, 562.
70. Graff, J.L.; Wrighton, M.S. to be submitted.
71. Saillant, R.B.; Barcelo, G.; Kaesz, H.D. J. Am. Chem. Soc., 1970, 92, 5739.
72. Turro, N.J. "Modern Molecular Photochemistry", Benjamin Cummings: Menlo Park, California, 1978.
73. Lewis, J.; Johnson, B.F.G. Gazz. Chim. Ital., 1979, 109, 271.
74. Calvert, J.G.; Pitts, Jr., J.N, "Photochemistry", John Wiley & Sons: New York, 1966.

RECEIVED December 1, 1980.

6

Thermal and Photochemical Reactivity of $H_2FeRu_3(CO)_{13}$ and Related Mixed-Metal Clusters

GREGORY L. GEOFFROY, HENRY C. FOLEY, JOSEPH R. FOX, and WAYNE L. GLADFELTER

Department of Chemistry, Pennsylvania State University, University Park, PA 16802

For the past few years metal carbonyl clusters have been under study in this laboratory from several viewpoints. First, we have a continuing interest in developing better methods for the directed synthesis of mixed-metal clusters and considerable progress has been made in this area (1-6). We have more recently been evaluating the reactivity features of mixed-metal clusters with a variety of substrates. We have chosen to concentrate our efforts on one particular cluster, $H_2FeRu_3(CO)_{13}$, 1, and to examine its reactivity in as much detail as possible.

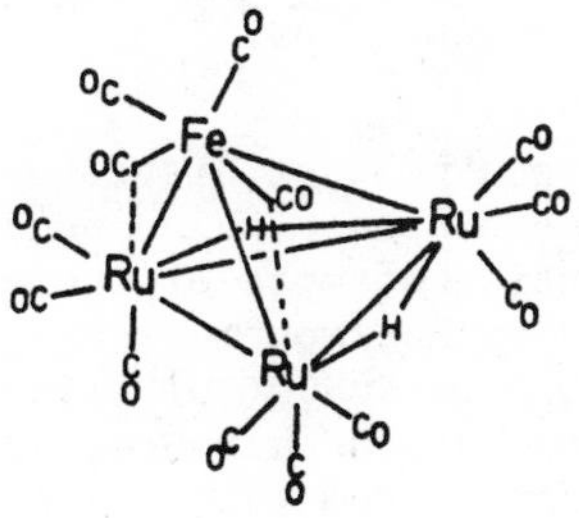

1

It has been our aim to try to fully understand the kind of chemical reactions which this cluster undergoes as well as their mechanistic course, with the expectation that this knowledge will prove applicable to other cluster systems. Complete details of these various studies will be provided in future, separate publications, but it is the purpose of this article to summarize the more important results and to present the overall reactivity picture of $H_2FeRu_3(CO)_{13}$ as we now perceive it.

0097-6156/81/0155-0111$06.00/0

Results and Discussion

Reactivity of $H_2FeRu_3(CO)_{13}$ and Related Mixed-Metal Clusters With Carbon Monoxide. Transition metal cluster compounds are currently under intense scrutiny as catalysts for a variety of reactions involving carbon monoxide. These include the reduction of CO to produce hydrocarbons (7,8), alcohols (9), and ethylene glycol (10), hydroformylation and related olefin reactions (11,12,13), and the water-gas shift reaction (13,14,15, 16). With the interest in chemistry of this type so intense, it seems essential that the basic reactivity patterns of clusters with CO be understood. We have accordingly examined the reactions of a series of mixed-metal clusters with CO with the aim of understanding the types of reactions that can occur as well as their mechanistic course. It was our finding that most of the clusters examined in this work undergo fragmentation, under rather mild conditions (1 atm CO, 25°-50°C), generally to produce monomeric and trimeric products, Table II and eqs. 1-3 (17).

$$H_2FeRu_3(CO)_{13} + CO \xrightarrow[42\ h,\ 10\%\ conv.]{50°C} Ru_3(CO)_{12} + Fe(CO)_5 + H_2 \quad (1)$$

$$[HFeRu_3(CO)_{13}]^- + CO \xrightarrow[THF,\ 90\%\ conv.]{25°C,\ 96\ h} [HRu_3(CO)_{11}]^- + Fe(CO)_5 \quad (2)$$

$$[CoRu_3(CO)_{13}]^- + CO \xrightarrow[CH_2Cl_2,\ 100\%\ conv.]{25°C,\ 96\ h} Ru_3(CO)_{12} + [Co(CO)_4]^- \quad (3)$$

Despite the fact that numerous product distributions are possible, especially for the clusters which possess three different metals, fragmentation of most of the compounds proceeds to give a specific trimer/monomer pair. $H_2FeRu_2Os(CO)_{13}$, for example, reacts to give only $Ru_2Os(CO)_{12}$ and none of the $RuOs_2$, $FeRu_2$, or FeRuOs trimers; furthermore, $Fe(CO)_5$ is the only monomeric product observed from this reaction.

The reactivity of a particular cluster towards CO is greatly dependent on its metal composition, Table I. $HCoRu_3(CO)_{13}$, for example, shows complete fragmentation in <1 h when stirred at 25°C, whereas only 10% fragmentation is observed for $H_2FeRu_3(CO)_{13}$ when maintained at 50°C for 42 h, both under 1 atm CO pressure. As expected, the clusters become more resistant to fragmentation with increasing third-row metal content, as illustrated by the data given in Table I for $H_2FeRu_3(CO)_{13}$, $H_2FeRu_2Os(CO)_{13}$, and $H_2FeRuOs_2(CO)_{13}$.

For comparative purposes, the polyhydride clusters $H_4FeRu_3(CO)_{12}$, $H_4Ru_4(CO)_{12}$, and $[H_3FeRu_3(CO)_{12}]^-$ were also studied (17). As illustrated in eqs. 4-6, these first undergo substitution of CO for H_2 to form the corresponding di- and

Table I. Fragmentation of Mixed-Metal Tetranuclear Clusters with Carbon Monoxide at 1 Atmosphere Pressure (17)

Initial Cluster	Fragmentation Products	Typical Conditions
$HCoRu_3(CO)_{13}$	$Ru_3(CO)_{12}$, $Ru(CO)_5$, $Co_2(CO)_8$[a]	25°C, 1 h, hexane, 100% conv.
$H_2Ru_4(CO)_{13}$	$Ru_3(CO)_{12}$, $Ru(CO)_5$, H_2	25°C, 1 h, hexane, 30% conv.
$[CoRu_3(CO)_{13}]^-$	$Ru_3(CO)_{12}$, $[Co(CO)_4]^-$	25°C, 96 h, CH_2Cl_2, 100% conv.
$[HFeRu_3(CO)_{13}]^-$	$[HRu_3(CO)_{11}]^-$, $Fe(CO)_5$	25°C, 96 h, THF, 90% conv.
$H_2FeRu_3(CO)_{12}(PMe_2Ph)$	$FeRu_2(CO)_{11}(PMe_2Ph)$, $Ru(CO)_5$	25°C, 185 h, CH_2Cl_2, 40% conv.
	$Ru_3(CO)_{11}(PMe_2Ph)$ (trace), H_2	
$H_2FeRu_3(CO)_{13}$	$Ru_3(CO)_{12}$, $Fe(CO)_5$, H_2	50°C, 42 h, hexane, 10% conv.
$H_2FeRu_2Os(CO)_{13}$	$Ru_2Os(CO)_{12}$, $Fe(CO)_5$, H_2	25°C, 16 d, hexane, 5% conv.
$H_2FeRuOs_2(CO)_{13}$	$RuOs_2(CO)_{12}$, $Fe(CO)_5$, H_2	50°C, 24 d, hexane, 5% conv.

[a] Plus other unidentified Co containing products.

monohydride clusters, which then undergo their characteristic fragmentation.

$$H_4Ru_4(CO)_{12} + CO \rightarrow H_2Ru_4(CO)_{13} + H_2 \quad (4)$$

$$H_4FeRu_3(CO)_{12} + CO \rightarrow H_2FeRu_3(CO)_{13} + H_2 \quad (5)$$

$$[H_3FeRu_3(CO)_{12}]^- + CO \xrightarrow[THF]{66°C,\ 5\ min} [HFeRu_3(CO)_{13}]^- + H_2 \quad (6)$$

The reactivity of the two tetrahydride clusters is vastly different: after 68 h at 25°C in hexane solution, >90% of $H_4FeRu_3(CO)_{12}$ is converted to products whereas after 100 h at 50°C only ~30% of $H_4Ru_4(CO)_{12}$ has disappeared. In the latter case, the initially formed $H_2Ru_4(CO)_{13}$ was detected by IR spectroscopy and by analytical liquid chromatography, but its concentration did not build up since it rapidly reacts with CO to produce $Ru_3(CO)_{12}$, $Ru(CO)_5$, and H_2.

In order to define the mechanistic course of these reactions, we undertook a series of kinetic studies on $H_2FeRu_3(CO)_{13}$ and for comparison, $H_2Ru_4(CO)_{13}$. The derived rate law which fit the kinetic data for fragmentation of both $H_2Ru_4(CO)_{13}$ and $H_2FeRu_3(CO)_{13}$ is that shown in eq. 7 (17).

$$-\frac{d[cluster]}{dt} = \{k_1 + k_2[CO]\}[cluster] \quad (7)$$

Specific rate constants and activation parameters are given in Table II. Under the conditions of our experiments the first order term is negligible and the reactions are essentially second order overall.

The mechanism which we believe is most consistent with this rate law and with the activation parameters obtained, Table II, involves association of CO with the intact cluster concomitant with cleavage of one of the metal-metal bonds to give an $H_2M_4(CO)_{14}$ butterfly cluster as the first intermediate. The activation profile envisaged for this process is shown in Scheme 1.

This mechanism implies that attack of CO on the cluster is concerted with metal-metal bond breakage. The latter is necessary because the initial cluster is coordinatively and sterically saturated. A comparison of the activation parameters indicates that the relative degree of CO association and M-M bond cleavage in the transition state varies from $H_2Ru_4(CO)_{13}$ to $H_2FeRu_3(CO)_{13}$. The larger values of $\Delta H^{\ddagger}$ (20.0 kcal/mole) and the less negative values of $\Delta S^{\ddagger}$ (-25.4 cal/mole-K) imply that in the transition state, M-CO bond formation occurs to a lesser extent for $H_2FeRu_3(CO)_{13}$ than for $H_2Ru_4(CO)_{13}$ ($\Delta H^{\ddagger}$ = 12.5 kcal/mole; $\Delta S^{\ddagger}$ = -36.6 cal/mole-K). This is consistent with the

Table II. Kinetic Parameters for the Reaction of $H_2FeRu_3(CO)_{13}$ and $H_2Ru_4(CO)_{13}$ with CO[a] (17)

Cluster	Temp	k_1 (sec^{-1})	k_2 $(M^{-1}sec^{-1})$	$\Delta H^{\circ\ddagger}$ (kcal/mole)	$\Delta S^{\circ\ddagger}$ (cal/mole-K)
$H_2FeRu_3(CO)_{13}$	70°C	6×10^{-7}	3.0×10^{-4}	20.0	-25.4
$H_2Ru_4(CO)_{13}$	30°C	0	5.7×10^{-3}	12.5	-36.6

[a] [cluster] = 6×10^{-4} M, n-heptane solution; $[CO]_{70°} = 1.42 \times 10^{-2}$ M, $[CO]_{30°} = 1.25 \times 10^{-2}$ M.

Scheme 1

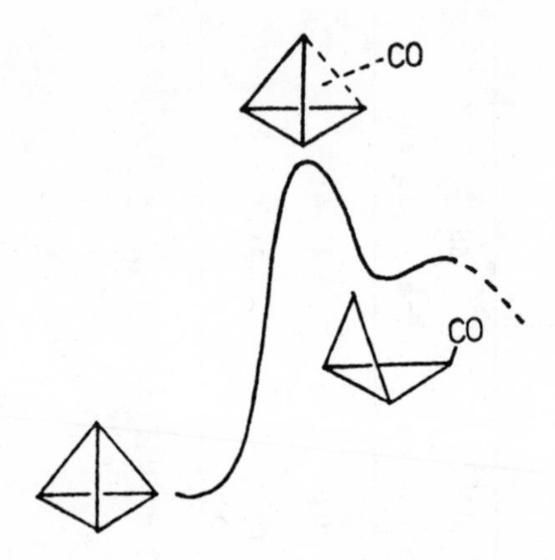

notion that Ru is larger than Fe and there is more room on the surface of intact $H_2Ru_4(CO)_{13}$ to accomodate the incoming CO than in $H_2FeRu_3(CO)_{13}$. The relative degree of CO association and M-M bond cleavage in the transition state will surely vary with the metal composition of the cluster and may account for the greatly different reactivity observed for the compounds listed in Table I.

Reactions of $H_2FeRu_3(CO)_{13}$ With Tertiary Phosphines and Phosphites. $H_2FeRu_3(CO)_{13}$ reacts with a series of tertiary phosphine and phosphite ligands to yield mono- and disubstituted clusters in 20-30% yield, eq. 8 (18).

$$H_2FeRu_3(CO)_{13} + PR_3 \rightarrow H_2FeRu_3(CO)_{12}(PR_3) + H_2FeRu_3(CO)_{11}(PR_3)_2 \quad (8)$$

$(PR_3 = PMe_3, PMe_2Ph, PMePh_2, PEt_2Ph, PPh_3, P(i\text{-}Pr)_3, P(OMe)_3, P(OEt)_3, P(OEt)_2Ph)$

These substitution reactions proceed cleanly and give only trace amounts of side products unless the reactions are carried out under forcing conditions. With the aid of ^1H and ^{31}P NMR spectroscopy, we have been able to take advantage of the low symmetry of these derivatives and determine the specific sites at which substitution occurs in $H_2FeRu_3(CO)_{13}$. Figure 1, for example, shows the ^1H NMR spectrum of $H_2FeRu_3(CO)_{12}(PMe_2Ph)$ at -50°C which indicates the presence of two substitutional isomers for the compound. The intense doublet at τ 28.07 ppm

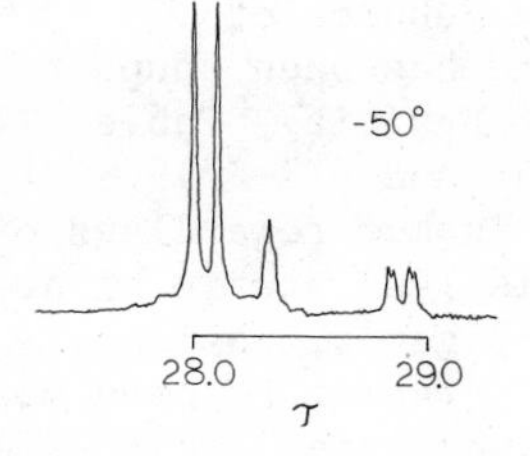

Figure 1. ^1H NMR spectrum at −50°C of $H_2FeRu_3(CO)_{12}(PMe_2Ph)$ in $CDCl_3$ solution

(J_{P-H} = 10.3 hz) is attributed to the two equivalent hydrogens of the C_s isomer shown below while the pseudo triplet at τ 28.35 ppm and the doublet of doublets at τ 28.93 ppm (J_{P-H} = 9.7 hz; J_{H-H} = 2.6 hz) are assigned to H_A and H_B respectively of the C_1 isomer.

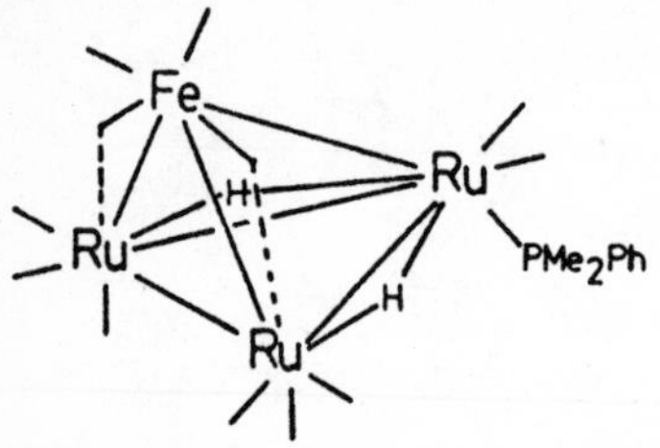

C_s 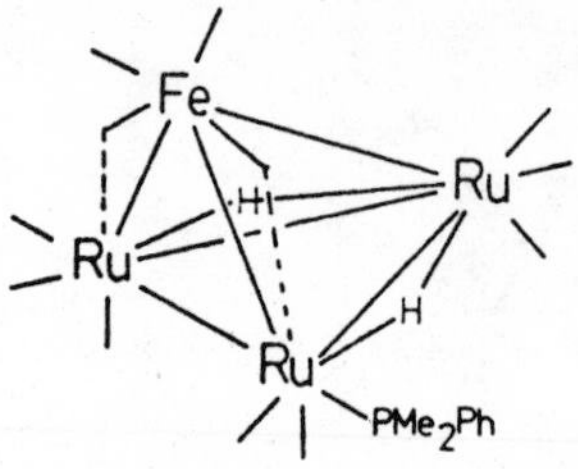 C_1

Similar substitutional isomers result for the mono-substituted derivatives of the other PR_3 ligands but the relative ratios of these two isomers is greatly dependent on the nature of the particular ligand. The choice of substitution site for a particular ligand has been found to depend on both the ligand's size and its basicity, Table III (18,19). With large ligands such as PPh_3 and $P(i\text{-}Pr)_3$, substitution occurs to give only the C_s isomer regardless of the ligand basicity. Presumably there is less steric hindrance in this substitution site. For smaller ligands, basicity becomes the controlling factor and the C_1 isomer becomes more abundant as the basicity of the ligand increases. This effect is best illustrated by comparing the ligands PMe_3 and $P(OMe)_3$, Table II. The former is highly basic and it gives a $C_1 \rightleftarrows C_s$ equilibrium constant of 0.4 whereas the smaller but less basic $P(OMe)_3$ gives $K_{C_1 \rightleftarrows C_s}$ = 10.

The increased abundance of the C_1 isomer with increasing ligand basicity presumably occurs because the semi-bridging carbonyl attached to the substituted metal can become more full-bridging and remove the excess electron density released by the basic phosphine. Consistent with this hypothesis is the

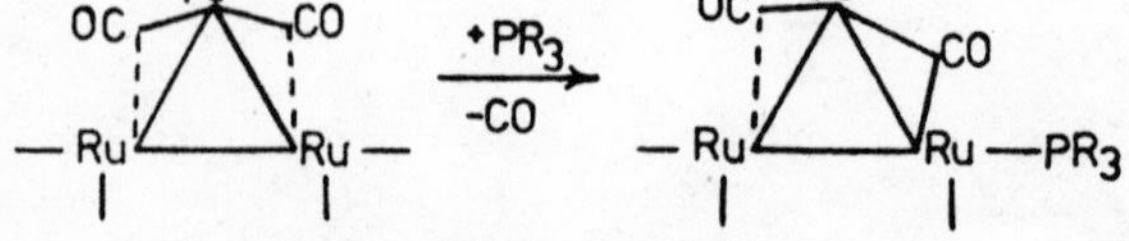

$$\text{Fe(OC)(CO)Ru}_2 \xrightarrow[-CO]{+PR_3} \text{Fe(OC)(CO)Ru}_2\text{–PR}_3$$

Table III. Effect of Ligand Size and Basicity on the $C_1 \rightleftarrows C_s$ Equilibrium of $H_2FeRu_3(CO)_{12}L$ (18)

L	Cone Angle[a] (degrees)	Basicity[a] (cm^{-1})	$K_{C_1 \rightleftarrows C_s}$
$P(i\text{-}Pr)_3$	160	2059.2	>100
PPh_3	145	2068.9	>100
$PMePh_2$	136	2067.0	11
PEt_2Ph	136	2063.7	11
$P(OMe)_3$	107	2079.5	10
$P(OEt)_2Ph$	116	2074.2	5.5
PMe_2Ph	122	2065.3	1.8
PMe_3	118	2064.1	0.4

[a]Ref. 19.

increased downfield chemical shift of one of the bridging carbonyls in the ^{13}C NMR spectrum of $H_2FeRu_3(CO)_{12}(PMe_2Ph)$ (δ 249 ppm) compared to $H_2FeRu_3(CO)_{13}$ (δ 229 ppm) and also the shift to lower energy of the bridging carbonyl infrared absorption in the C_1 isomer (~1806 cm^{-1} vs. ~1850 cm^{-1}).

In an attempt to understand the mechanism by which these substitution reactions occur, a kinetic study of the reaction of PPh_3 with $H_2FeRu_3(CO)_{13}$ in hexane solution was undertaken. The kinetic data obtained implied the rate equation shown below, eq. 9, with $k_1 = 6.96 \times 10^{-4}$ sec^{-1} at 50°C (18).

$$-\frac{d[H_2FeRu_3(CO)_{13}]}{dt} = k_1[H_2FeRu_3(CO)_{13}] \qquad (9)$$

The rate of substitution is first order with respect to cluster but zero order with respect to PPh_3 even up to very high PPh_3 concentrations. The activation parameters obtained for this substitution are $\Delta H^{\circ\ddagger} = 25.7$ kcal/mole and $\Delta S^{\circ\ddagger} = 4.8$ cal/mole-K. The zero-order dependence on $[PPh_3]$, the positive value of $\Delta S^{\circ\ddagger}$, and the decrease in the reaction rate under a CO atmosphere all argue for a dissociative mechanism in which the rate-determining step is loss of CO ligand from $H_2FeRu_3(CO)_{13}$. This would generate an unsaturated cluster that could rapidly add PPh_3 to give the monosubstituted derivative, Scheme 2.

Scheme 2

$$H_2FeRu_3(CO)_{13} \underset{k_2}{\overset{k_1\ \text{(slow)}}{\rightleftharpoons}} H_2FeRu_3(CO)_{12} + CO$$

$$H_2FeRu_3(CO)_{12} + PPh_3 \xrightarrow[\text{fast}]{k_3} H_2FeRu_3(CO)_{12}PPh_3$$

Reaction of $H_2FeRu_3(CO)_{13}$ With Alkynes. Isomeric products also result from the reaction of $H_2FeRu_3(CO)_{13}$ with a series of internal alkynes, eq. 10 (20).

$$H_2FeRu_3(CO)_{13} + RC{\equiv}CR' \rightarrow H_2 + CO + \underset{\text{(isomers)}}{FeRu_3(CO)_{12}(RC{\equiv}CR')} \qquad (10)$$

H_2 is evolved during the course of these reactions and the isomeric $FeRu_3(CO)_{12}(RC{\equiv}CR')$ products are conveniently separated by liquid chromatography on silica gel. With PhC≡CPh, two isomers of $FeRu_3(CO)_{12}(PhC{\equiv}CPh)$ are obtained in an overall yield of 65%. Reaction with MeC≡CMe gives two analogous isomers of $FeRu_3(CO)_{12}(MeC{\equiv}CMe)$, and three isomers of $FeRu_3(CO)_{12}(MeC{\equiv}CPh)$ result from the reaction with MeC≡CPh.

The structures of the isomers of $FeRu_3(CO)_{12}(PhC{\equiv}CPh)$ have been determined by x-ray diffraction (21, 22). They are isomorphous and differ only in the positioning of the Fe atom as indicated by the "axial" and "equatorial" labels given in the drawings below.

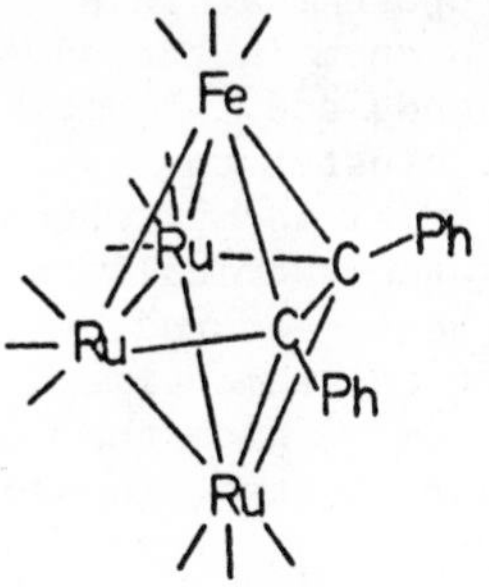

"axial" isomer

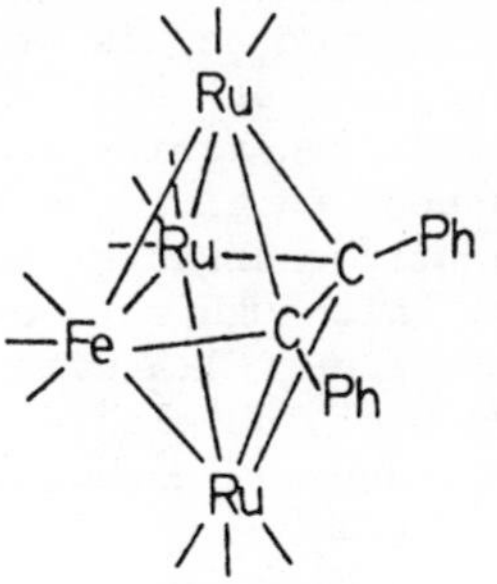

"equatorial" isomer

In essence the alkyne has inserted across a metal-metal bond to give a closo-$FeRu_3C_2$ cluster. Such a structure is fully consistent with Wade's skeletal electron counting rules for a closo structure with 6 vertices (23,24). The three isomers of $FeRu_3(CO)_{12}(MeC{\equiv}CPh)$ arise because in the equatorial isomer the methyl substituent can occupy a position cis or trans to the Fe atom.

Interestingly, we observed that the isomers of these $FeRu_3(CO)_{12}(RC{\equiv}CR')$ clusters readily interconvert upon heating to yield an equilibrium mixture, e.g., eq. 11 (20).

$$\underset{\text{equatorial}}{FeRu_3(CO)_{12}(PhC{\equiv}CPh)} \underset{\text{hexane}}{\overset{70°C}{\rightleftharpoons}} \underset{\text{axial}}{FeRu_3(CO)_{12}(PhC{\equiv}CPh)} \qquad K_{eq} \simeq 9.0 \tag{11}$$

For $FeRu_3(CO)_{12}(PhC{\equiv}CPh)$ the equilibrium mixture contains about 90% of the axial isomer. Although detailed mechanistic experiments were not conducted for the isomerization process, it was observed that the rate of isomerization occurs more slowly in concentrated solutions than in dilute solutions and more slowly under an atmosphere of CO than under an N_2 atmosphere. Furthermore, no exchange of coordinated alkyne with free alkyne occurs during the isomerization process. These various experiments indicate that the isomerization must be an intramolecular process and likely proceeds via initial CO dissociation.

It is significant to note that although the axial isomer of $FeRu_3(CO)_{12}(PhC{\equiv}CPh)$ is the more thermodynamically stable, eq. 11, the equatorial isomer is formed in greatest yield in the synthesis of the compound. This observation suggests that the $FeRu_3(CO)_{12}(RC{\equiv}CR')$ derivatives may form via the sequence of reactions outlined in Scheme 3. The first step presumably involves dissociation of CO from $H_2FeRu_3(CO)_{13}$ to generate $H_2FeRu_3(CO)_{12}$ which rapidly adds the alkyne to give a substituted derivative. We specifically propose that this initial substitution occurs at the unique Ru atom in the same substitution site that large phosphorus ligands add. From this site, the alkyne can easily swing under the cluster to insert into a Ru-Ru bond to give the kinetic equatorial isomer which then subsequently rearranges to the more thermodynamically stable axial isomer. Although rate data have not been measured, the reaction of $H_2FeRu_3(CO)_{13}$ with alkynes is of the same time scale as the reaction of the cluster with phosphines to give the substituted products, in accord with the initial step proposed in Scheme 3.

Scheme 3

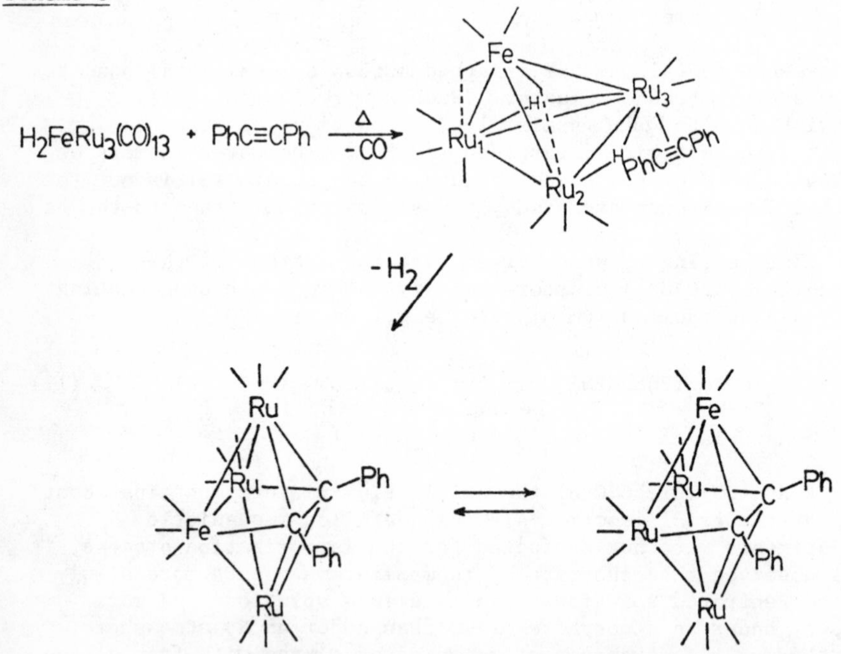

$H_2FeRu_3(CO)_{13}$ + PhC≡CPh $\xrightarrow[-CO]{\Delta}$ [Fe, Ru_1, Ru_2, Ru_3 cluster with H, H, PhC≡CPh]

$-H_2$

[Ru, Ru, Fe, Ru cluster with C–Ph, C–Ph] ⇄ [Fe, Ru, Ru, Ru cluster with C–Ph, C–Ph]

Molecular Dynamics of $H_2FeRu_3(CO)_{13}$ and Related Mixed-Metal Clusters. Metal clusters have been shown to undergo a wide variety of fluxional processes in which carbonyls, hydrides, and even the metals themselves undergo rearrangement (25). Mixed-metal clusters are ideally suited for studies of fluxional processes because of the low symmetry which is inherent within their metal framework. In such clusters, the majority of ligands are in chemically non-equivalent positions and thus are distinguishable by NMR spectroscopy. Furthermore, a homologous series of mixed-metal clusters allows one to study the effects of metal substitution on the fluxional processes and their activation parameters. In this regard, the series of clusters $H_2FeRu_3(CO)_{13}$, $H_2FeRu_2Os(CO)_{13}$, and $H_2FeRuOs_2(CO)_{13}$ in which Ru is progressively replaced by Os has been studied in our laboratory (6).

The trimetallic $FeRu_2Os$ and $FeRuOs_2$ clusters exist in the two isomeric forms shown in Figure 2 which also gives their corresponding symmetry labels and carbonyl labeling schemes. The low-temperature limiting ^{13}C NMR spectra of these derivatives are shown in Figure 3. The limiting ^{13}C NMR spectrum of $H_2FeRuOs_2(CO)_{13}$ is particularly illustrative. This cluster exists in C_1 and C_s isomeric forms in solution and examination of the structures shown in Figure 2 illustrates that of the 26 carbonyls in these two isomers, 21 are in distinctly different chemical environments. Each should thus show its own separate, characteristic ^{13}C NMR resonance. Indeed, 19 separate resonances are resolvable in the static ^{13}C NMR spectrum of this derivative, Figure 3, and explicit carbonyl assignments were derived for each (5). Similar assignments were made for $H_2FeRu_3(CO)_{13}$ and $H_2FeRu_2Os(CO)_{13}$. Significantly, the terminal carbonyls bound to the different metal atoms in this series group together in characteristic chemical shift regions. The chemical shift decreases relative to TMS upon descending the triad: Fe (204-211 ppm) > Ru (184-180 ppm) > Os (168-177 ppm).

Having made the assignments of specific carbonyls to the individual resonances in the static spectrum, it is basically a simple matter to analyze the changes in the NMR spectra as the temperature is raised. Three distinctly different fluxional processes have been found to occur in each of these clusters from such analysis (5). As the temperature is raised from the low-temperature limiting spectrum, the first process to occur is bridge-terminal interchange localized on iron. The mechanism which we have proposed for this exchange is shown in Scheme 4 and involves opening one of the carbonyl bridges, a subsequent trigonal twist of the resultant $Fe(CO)_3$ unit and finally reformation of the CO bridge (5).

The next exchange process to occur at slightly higher temperatures in each of these clusters involves migration of the carbonyls around the Fe-M-M triangle which possesses the bridging carbonyls. It seems reasonable to propose that the intermediates in this cyclic movement are the tautomers which have the semi-bridging carbonyls bound mainly to Ru or Os, instead of Fe, as indicated in Scheme 5.

The third and final process which occurs is unique, involving a subtle shift in the metal framework. This process and its implications are best illustrated by consideration of the drawings shown in Scheme 6 which depict the metal framework of

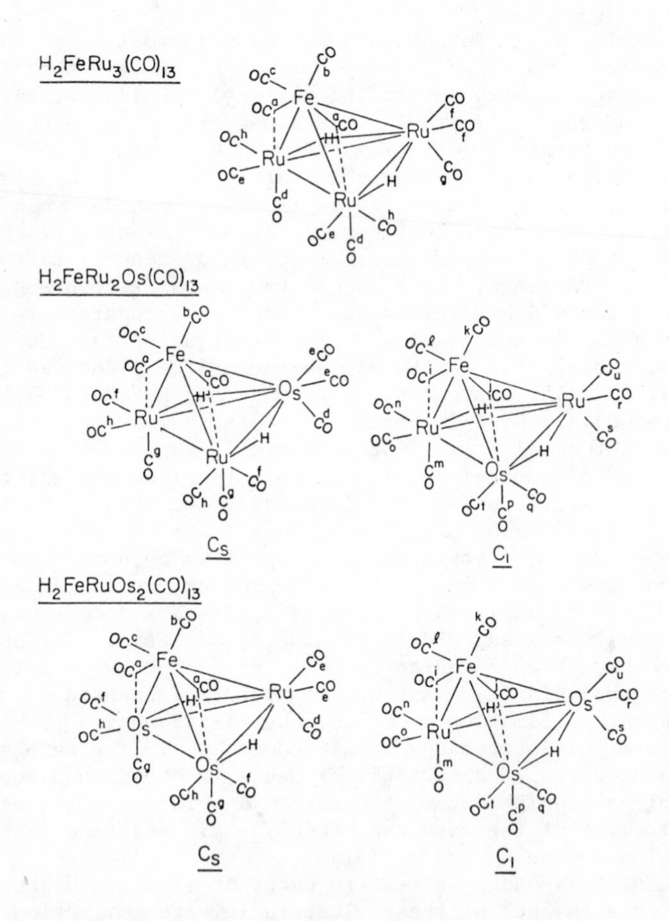

Figure 2. Carbonyl labeling schemes for $H_2FeRu_3(CO)_{13}$ and the C_s and C_1 isomers of $H_2FeRu_2Os(CO)_{13}$ and $H_2FeRuOs_2(CO)_{13}$

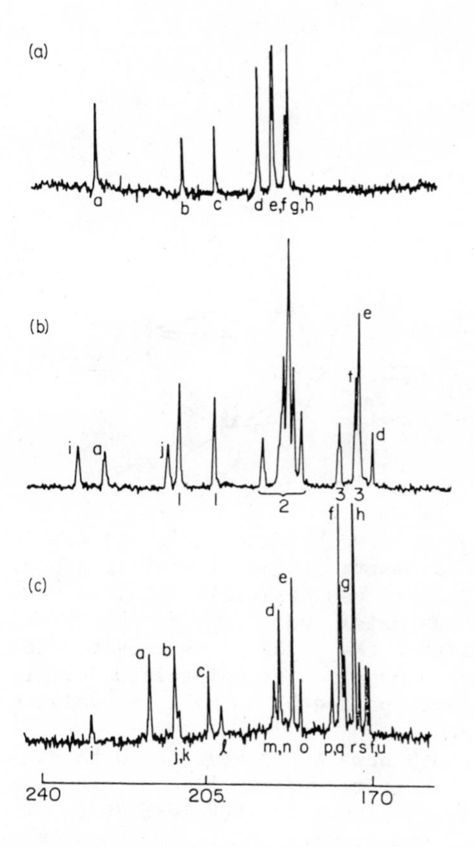

Figure 3. Low-temperature limiting ^{13}C NMR spectra of (a) $H_2FeRu_3(CO)_{13}$ ($-95°C$); (b) $H_2FeRu_2Os(CO)_{13}$ ($-90°C$); and (c) $H_2FeRuOs_2(CO)_{13}$ ($-60°C$)

Scheme 4

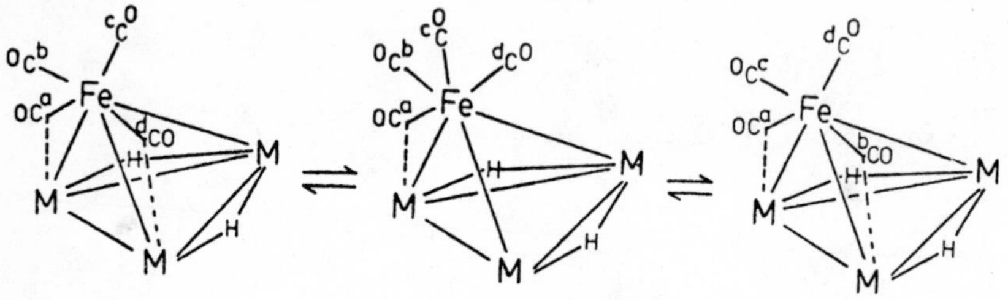

Scheme 5

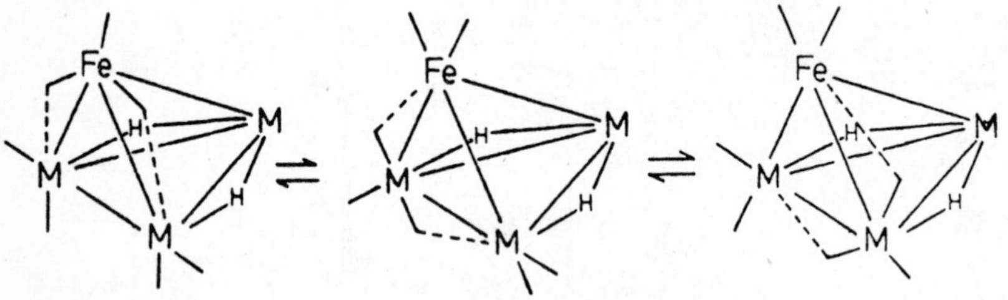

$H_2FeRuOs_2(CO)_{13}$, the two bridging CO's and the two bridging hydrides. The asymmetry of the cluster is grossly exaggerated for clarity. The process basically involves movement of the Fe atom away from one metal and closer to another with a concomitant shift of the bridging carbonyls. It must also involve a slight elongation or compression of all the M-M bonds and it must be accompanied by a shift in position of one of the bridging hydrides.

Referring to Scheme 6 and starting with the C_{1a} enantiomer, if the Fe moves away from Os_1 towards Os_2, it generates the C_{1b} enantiomer. Movement of Fe away from Ru in either of the C_1 enantiomers and toward both Os atoms gives the C_s isomer. Each time the cluster rearranges, the carbonyls execute the cyclic process about a different Fe-M-M face and hence involve different carbonyl ligands in that process. The cyclic processes coupled with the framework rearrangement lead to total exchange of all the carbonyl ligands of the cluster in this final fluxional process.

The actual magnitude of the shifts within the metal framework must be relatively small. Although the crystal structure of $H_2FeRuOs_2(CO)_{13}$ has not yet been determined, that published for the isostructural $H_2FeRu_3(CO)_{13}$ cluster shows that the greatest change expected in any one bond length during the

Scheme 6

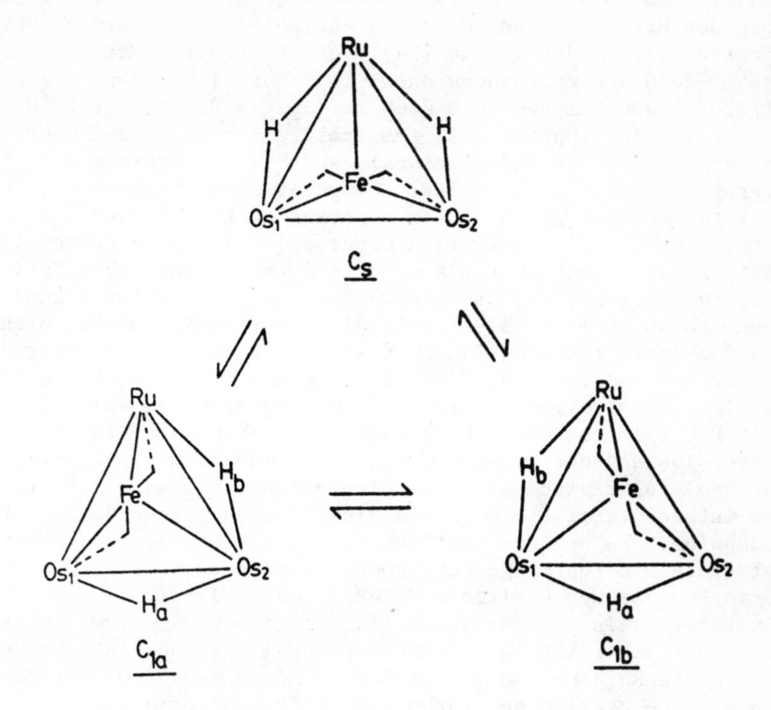

rearrangement is 0.11 Å (26). We tend to view the rearrangement process as more of a breathing motion of the metal framework, but one which has coupled to it motions of the carbonyls and hydride ligands.

Although within the series $H_2FeRu_3(CO)_{13}$, $H_2FeRu_2Os(CO)_{13}$, and $H_2FeRuOs_2(CO)_{13}$ the exchange processes are identical, the activation barrier for each process increases as the Os content of the cluster increases. It is unlikely that the activation energy increase can be accounted for solely on the basis of the size increase in the metal involved in the fluxional process since Ru and Os probably have similar atomic radii in these clusters. In $Ru_3(CO)_{12}$ and $Os_3(CO)_{12}$, for example, the metal atomic radii are 1.43 Å and 1.44 Å, respectively (27). On the other hand, it has been demonstrated that Os forms stronger metal-metal bonds and M-CO bonds than does Ru (28) and this greater bond strength could increase the barriers for the various fluxional processes.

The intrametallic rearrangement process has profound consequences in the fluxionality of the phosphine and phosphite substituted $H_2FeRu_3(CO)_{13}$ derivatives which were discussed above. Each of the monosubstituted derivatives can exist in C_1

and C_s isomeric forms and for $H_2FeRu_3(CO)_{12}(PMe_2Ph)$ the $C_1 \rightleftarrows C_s$ equilibrium constant is ~1.8. However, these two isomers rapidly interconvert as evidenced by the variable temperature ^1H NMR spectra of this derivative (29). As the temperature is raised above −50°C the resonances shown in Figure 1 broaden, coalesce at ~20°C, and sharpen to a doublet (J_{P-H} = 9.6 hz) at 70°C. The latter implies that at this temperature the $C_1 \rightleftarrows C_s$ isomerization is rapid on the ^1H NMR time scale and the two hydrogens see an average chemical environment. Computer simulation of the ^1H NMR spectral changes gives a rate constant of k = 500 sec^{-1} at 20°C for the $C_1 \rightleftarrows C_s$ isomerization process. This is an extremely fast rate of isomerization for a process in which two substitutional isomers interconvert, especially since the phosphine ligand is bound to Ru atoms in distinctly different environments in the two isomers. The question then arises as to how the interconversion occurs. The first clue is the presence of the doublet in the high-temperature limiting spectrum which implies that P-H spin correlation is maintained throughout the interconversion and hence the isomerization may occur by a purely intramolecular process. This eliminates the possibility that the interconversion occurs via dissociation-reassociation of the phosphine. We believe that the isomerization occurs by an intrametallic rearrangement process analogous to that discussed above for the unsubstituted $H_2FeRu_{3-x}Os_x(CO)_{13}$ (x = 0–2) clusters. This process is outlined in Scheme 7 below. Consider the C_s isomer first in which the phosphine ligand is attached to the unique Ru atom (Ru_3). If the Fe moves away from Ru_1 towards Ru_3 and the hydride and carbonyls shift appropriately, the C_1 isomer is generated. In order to bring the phosphine into the required equatorial position, rotation of the $Ru(CO)_2(PMe_2Ph)$ unit must accompany the rearrangement. Note, however, that the phosphine has remained attached to the same Ru atom but that this Ru has altered its environment via the intrametallic rearrangement process. In essence, the phosphine appears to exchange positions but *the exchange occurs by the phosphine staying attached to Ru_3 with the cluster rearranging around the ligand*.

Scheme 7

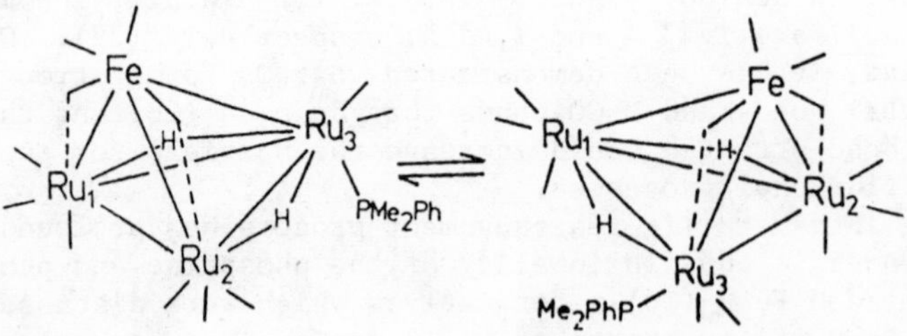

Photochemical Reactivity of $H_2FeRu_3(CO)_{13}$, $H_2FeOs_3(CO)_{13}$, and $H_2Ru_4(CO)_{13}$. Monomeric metal carbonyl complexes generally lose CO when irradiated whereas for dinuclear carbonyls, such as $Mn_2(CO)_{10}$, the dominant photoreaction involves cleavage of the M-M bond and fragmentation (30). The question then arises as to what are the primary photochemical properties of metal carbonyl clusters containing three or more metal atoms; i.e., will CO loss or M-M bond cleavage preferentially obtain? With trinuclear clusters, M-M bond cleavage appears to dominate the photochemistry although truly definitive mechanistic studies are lacking. Certainly $Fe_3(CO)_{12}$ (31), $Ru_3(CO)_{12}$ (32), and $H_3Re_3(CO)_{12}$ (33) fragment upon photolysis, although in the latter case a detailed mechanistic study which we conducted was not able to resolve whether Re-Re bond cleavage or CO loss occurs in the primary photochemical event. $Os_3(CO)_{12}$ does not fragment upon photolysis but rather clean photosubstitution chemistry obtains (34). Even here, however, such substitution could occur via a pathway involving initial cleavage of an Os-Os bond.

Few studies of tetranuclear clusters have been reported. Johnson, Lewis and coworkers (35,36) have shown that photolysis of $H_4Os_4(CO)_{12}$ gives clean substitution chemistry and analogous results were obtained by Wrighton and Graff (37) for $H_4Ru_4(CO)_{12}$. The tentative conclusion that may be drawn is that for clusters with nuclearity $\geq$ 4, photoinduced fragmentation does not obtain and the net photochemistry evolves from CO loss. Whether this is because the metal-metal excited states, which clearly lie lowest in energy (30,34), are too delocalized to give cleavage of a single M-M bond, or whether M-M bond cleavage does indeed obtain but that net fragmentation is not observed because the remaining M-M bonds hold the cluster together is still an unanswered question.

In order to test the hypothesis that CO dissociation is the dominant photoreaction for tetranuclear clusters we have undertaken a study of the photochemistry of $H_2FeRu_3(CO)_{13}$, a cluster for which we know the efficiency of thermal CO dissociation and fragmentation as detailed above. For comparison, the clusters $H_2Ru_4(CO)_{13}$ and $H_2FeOs_3(CO)_{13}$ have also been examined. With each of these clusters CO loss is clearly the dominant photoreaction (38). In the presence of PPh_3, each cluster gives clean photosubstitution chemistry to yield primarily the monosubstituted derivatives, eq. 12.

$$H_2M_4(CO)_{13} + PPh_3 \xrightarrow{h\nu} H_2M_4(CO)_{12}(PR_3) + CO \quad (12)$$

$$M_4 = FeRu_3,\ FeOs_3,\ Ru_4$$

Prolonged photolysis also leads to formation of di- and trisubstituted products. Photolysis under H_2 atmospheres leads to the $H_4M_4(CO)_{12}$ clusters in each case, eq. 13.

$$H_2M_4(CO)_{13} + H_2 \xrightarrow{h\nu} H_4M_4(CO)_{12} + CO \quad (13)$$

However, these reactions are not as clean as the PPh_3-photo-substitution reactions since $H_4FeRu_3(CO)_{12}$ and $H_4FeOs_3(CO)_{12}$ are themselves photosensitive and readily degrade to other products. $H_2Ru_4(CO)_{13}$, however, can be quantitatively converted to $H_4Ru_4(CO)_{12}$, which is photochemically inactive under H_2.

These various experiments indicate that CO loss occurs in the primary photochemical event for each cluster to produce coordinatively unsaturated $H_2M_4(CO)_{12}$ which subsequently adds PPh_3 or H_2, eqs. 14-15.

$$H_2M_4(CO)_{13} \xrightarrow{h\nu} CO + H_2M_4(CO)_{12} \quad (14)$$

$$H_2M_4(CO)_{12} + L \longrightarrow H_2M_4(CO)_{12}L \quad (L = H_2,\ PPh_3) \quad (15)$$

Quantum yields have been measured for PPh_3 substitution and these are summarized in Table IV. The quantum yields vary slightly with the nature of the cluster but for $H_2FeRu_3(CO)_{13}$ are independent of PPh_3 concentration, consistent with the formation of an $H_2M_4(CO)_{12}$ intermediate which subsequently scavenges PPh_3.

Table IV. Quantum Yield Data for PPh_3 Substitution in Isooctane Solution

Cluster	[Cluster]	[PPh_3]	ϕ_{366}
$H_2FeRu_3(CO)_{13}$	3.65×10^{-4}	3.65×10^{-4}	0.025 ± 0.006
	3.65×10^{-4}	1.83×10^{-3}	0.029 ± 0.006
	3.65×10^{-4}	3.65×10^{-3}	0.030 ± 0.006
$H_2FeOs_3(CO)_{13}$	5.28×10^{-4}	2.65×10^{-3}	0.057 ± 0.012
$H_2Ru_4(CO)_{13}$	3.00×10^{-4}	1.5×10^{-3}	0.016 ± 0.002

We have observed that prolonged photolysis of these clusters under CO atmospheres does induce fragmentation, but the quantum yields are far too low to be measured on our apparatus. For $H_2FeRu_3(CO)_{13}$ an upper limit quantum yield of 2.4×10^{-6} can be estimated based on our detection limits. The very low quantum yield for fragmentation of a cluster which we know can fragment thermally under CO (see above) provides strong evidence that the primary photochemical reaction is not metal-metal bond cleavage, which should readily lead to fragmentation in the presence of CO, but rather involves CO dissociation.

Summary and Outlook

One can begin to assemble a composite picture of the reactivity of $H_2FeRu_3(CO)_{13}$ with the reactivity information discussed above now available. By far the most rapid process that this cluster undergoes involves intramolecular exchange of the carbonyl ligands and the hydrides and the rearrangement of the metal framework. These processes occur with rate constants ranging from 10-50 sec^{-1} at 50°C. On a far slower time scale, $H_2FeRu_3(CO)_{13}$ undergoes thermal dissociation of CO to yield $H_2FeRu_3(CO)_{12}$. This species can then subsequently add phosphines and phosphites to give substituted derivatives, alkynes to first yield substituted clusters followed by reaction to give the closo-$FeRu_3(CO)_{12}$(RC≡CR') products, ^{13}CO to effect $^{13}CO/^{12}CO$ exchange, and H_2 to give $H_4FeRu_3(CO)_{12}$. The first step, dissociation of CO, occurs with a rate constant of 6.96×10^{-4} sec^{-1} at 50°C. Thus, $H_2FeRu_3(CO)_{13}$ undergoes the various fluxional processes 1,000-10,000 times between each CO dissociative event. Finally, on a much slower time scale, $H_2FeRu_3(CO)_{13}$ undergoes fragmentation with CO to produce $Ru_3(CO)_{12}$, $Fe(CO)_5$, and H_2 by a mechanism which we believe involves association of CO concerted with breakage of one of the metal-metal bonds. Recent photochemical experiments indicate that CO dissociation, but not fragmentation, can be photoinduced, and this reaction appears to offer potential for the synthesis of unusual substituted derivatives (38).

Obviously there is still much to be learned concerning the chemistry of $H_2FeRu_3(CO)_{13}$ and related mixed-metal clusters. An important feature for which good information is not at all available concerns the relative strength of the metal-metal bonds in clusters of this type. Is an Fe-Ru bond stronger or weaker than Fe-Fe and Ru-Ru bonds? This question could presumably be addressed directly by a series of microcalorimetric studies or indirectly by a detailed kinetic examination of the reactions of a series of iso-structural mixed-metal clusters with CO. We know nothing concerning the redox properties of these particular clusters and that should be a subject for further study. Will they retain their integrity on oxidation and reduction or might they open to yield "butterfly" clusters upon 2-electron reduction? An interesting question concerns exactly which CO is labilized in the CO dissociation step: *i.e.*, a CO bound to Fe or Ru; axial Ru or equatorial Ru? Even though we know which sites ligands prefer to add to give the substituted products, this implies nothing about the specific CO initially lost.

Acknowledgments

The research described herein has been generously supported by the Office of Naval Research, the National Science Foundation and the Department of Energy, Office of Basic Energy Sciences. GLG gratefully acknowledges the Camille and Henry Dreyfus Foundation for a Teacher-Scholar Grant, the Alfred P. Sloan Foundation for a research fellowship.

Literature Cited

1. Gladfelter, W. L.; Geoffroy, G. L., Adv. Organomet. Chem., 1980,

2. Geoffroy, G. L.; Gladfelter, W. L., J. Am. Chem. Soc., 1977, 99, 7565.

3. Steinhardt, P. C.; Gladfelter, W. L.; Harley, A. D.; Fox, J. R.; Geoffroy, G. L., Inorg. Chem., 1980, 19, 332.

4. Burkhardt, E. W.; Geoffroy, G. L., J. Organomet. Chem., in press.

5. Gladfelter, W. L.; Geoffroy, G. L., Inorg. Chem.,

6. Epstein, R. A.; Withers, H. W.; Geoffroy, G. L., Inorg. Chem., 1979, 18, 942.

7. Demitras, G. C.; Muetterties, E. L., J. Am. Chem. Soc., 1977, 99, 2796.

8. Thomas, M. G.; Beier, B. F.; Muetterties, E. L., J. Am. Chem. Soc., 1976, 98, 1296.

9. Bradley, J., J. Am. Chem. Soc., 1979, 101, 7419.

10. Pruett, R. L.; Walker, W. E., U.S. Patent 2,262,318 (1973); 3,944,588 (1976); Ger. Offen. 2,531,103 (1976).

11. Laine, R. M., J. Am. Chem. Soc., 1978, 100, 6451.

12. Laine, R. M.; Thomas, D. W.; Cary, L. W.; Buttrill, S. E., J. Am. Chem. Soc., 1978, 100, 6527.

13. Kang, H. C.; Mauldin, C. H.; Cole, T.; Slegeir, W.; Cann, K.; Pettit, R., J. Am. Chem. Soc., 1977, 99, 8323.

14. Laine, R. M.; Rinker, R. G.; Ford, P. C., J. Am. Chem. Soc., 1977, 99, 252.

15. Ford, P. C.; Rinker, R. G.; Ungermann, C.; Laine, R. M.; Landis, V.; Moya, S. A., J. Am. Chem. Soc., 1978, 100, 4595.

16. Ungermann, C.; Landis, V.; Moya, S. A.; Cohen, H.; Walker, H.; Pearson, R. G.; Rinker, R. G.; Ford, P. C., J. Am. Chem. Soc., 1979, 101, 5922.

17. Fox, J. R.; Gladfelter, W. L.; Geoffroy, G. L., Inorg. Chem., 1980, 19, 2574.

18. Fox, J. R.; Gladfelter, W. L.; Wood, T. G.; Smegal, J. A.; Foreman, T. K.; Geoffroy, G. L., J. Am. Chem. Soc., submitted for publication.

19. Tolman, C. A., Chem. Rev., 1977, 77, 313.

20. Fox, J. R.; Gladfelter, W. L.; Geoffroy, G. L., J. Am. Chem. Soc., submitted for publication.

21. Fox, J. R.; Gladfelter, W. L.; Geoffroy, G. L.; Day, V. W.; Meguid, S.-A.; Tavanaiepour, I., J. Am. Chem. Soc., submitted for publication.

22. Tavanaiepour, I., Ph.D. Dissertation, University of Nebraska, Lincoln, Nebraska, 1980.

23. Wade, K., Chem. Britain, 1975, 11, 1977.

24. Wade, K., Adv. Inorg. Chem. Radiochem., 1976, 18, 1.

25. Band, E.; Muetterties, E. L., Chem. Rev., 1978, 78, 639.

26. Gilmore, C. J.; Woodward, P., J. Chem. Soc. A, 1971, 3453.

27. Churchill, M. R.; Hollander, F. J.; Hutchinson, J. P., Inorg. Chem., 1977, 16, 2655.

28. Conner, J. A., Top. Curr. Chem., 1977, 71, 71.

29. Gladfelter, W. L.; Fox, J. R.; Smegal, J. A.; Wood, T. G.; Geoffroy, G. L., J. Am. Chem. Soc., submitted for publication.

30. Geoffroy, G. L.; Wrighton, M. S.; "Organometallic Photochemistry," Academic Press, New York, 1979.

31. Austin, R. G.; Pavnessa, R. S.; Giordano, P. J.; Wrighton, M. S., Adv. Chem. Ser., 1978, 168, 189.

32. Johnson, B. F. G.; Lewis, J.; Twigg, M. U., J. Organomet. Chem., 1974, 67, C75.

33. Epstein, R. A.; Gaffney, T. R.; Geoffroy, G. L.; Gladfelter, W. L.; Henderson, R. S., J. Am. Chem. Soc., 1979, 101, 3847.

34. Tyler, D. R.; Altobelli, M.; Gray, H. B., J. Am. Chem. Soc., 1980, 102, 3022.

35. Johnson, B. F. G.; Kelland, J. W.; Lewis, J.; Rehani, S. K., J. Organomet. Chem., 1976, 113, C42.

36. Bhoduri, S.; Johnson, B. F. G.; Kelland, J. W.; Lewis, J.; Raithby, P. R.; Rehani, S.; Sheldrick, G. M.; Wong, K.; McPartlin, M., J. Chem. Soc. Dalton Trans., 1979, 562.

37. Graff, J. L.; Wrighton, M. S., J. Am. Chem. Soc., 1980, 102, 2123.

38. Foley, H. C.; Geoffroy, G. L., to be submitted.

RECEIVED December 1, 1980.

Kinetic Studies of Thermal Reactivities of Metal–Metal–Bonded Carbonyls

ANTHONY POË

J. Tuzo Wilson Laboratories, Erindale College, University of Toronto, Mississauga, Ontario, Canada L5L 1C6

The strengths of the metal-metal bonds in dimetal and metal-cluster carbonyls have been a matter of some interest ever since $Mn_2(CO)_{10}$ was shown to have a "pure" Mn-Mn bond (1), i.e. one unsupported by the sort of bridging carbonyls found previously in $Fe_2(CO)_9$ (2) and, soon after, in $Co_2(CO)_8$ (3). Whereas the lengths of the Fe-Fe and Co-Co bonds in these latter complexes were close to those found in the pure metals the length of the Mn-Mn bond was ca. 0.5 Å longer than that in manganese metal and this "excessive" length was thought to imply some intrinsic weakness in unsupported metal-metal bonds. An early attempt (4) to obtain a thermochemical measure of the strength of the Mn-Mn bond resulted in a value of 142 ± 54 kJ mol^{-1} and illustrated the basic difficulty of obtaining precise values.

A few estimates of metal-metal bond energies have been made on the basis of mass spectrometric measurements. These involve measurement of appearance potentials as, for instance (5), in eq 1-3. The energy required for eq 3 is the difference between the

$$Mn_2(CO)_{10} + e \longrightarrow Mn(CO)_5^+ + \cdot Mn(CO)_5 + 2e \quad (1)$$

$$\cdot Mn(CO)_5 + e \longrightarrow Mn(CO)_5^+ + 2e \quad (2)$$

$$Mn_2(CO)_{10} \longrightarrow 2 \cdot Mn(CO)_5 \quad (3)$$

appearance potentials of $Mn(CO)_5^+$ through processes shown in eq 1 and 2. It has been argued (6) that there may be a difference between the $\cdot Mn(CO)_5$ radicals involved in these two processes and that some ambiguity remains as to the meaning of the value of 80 kJ mol^{-1} found for eq 3. Alternatively, direct measurement (7) of the value of ΔH° for reaction 4, for example, was obtained, the

$$[Mn_2(CO)_{10}]_g \rightleftharpoons 2[Mn(CO)_5]_g \quad (4)$$

relative concentrations being estimated from ion currents due to $Mn_2(CO)_{10}^+$ and $Mn(CO)_5^+$ over the temperature range 210-300°C.

0097-6156/81/0155-0135$08.00/0

This estimation depends on the assumption that the cross-sections for formation of the ions are not temperature dependent and led to a value of $\Delta H° = 104 \pm 8$ kJ mol^{-1}. Similar measurements on $Co_2(CO)_8$ led to a value of $\Delta H° = 61 \pm 8$ kJ mol^{-1} for the equilibrium.

More recently a reasonably large set of thermochemical measurements has been used to obtain metal-metal bond energies for a number of dimetal and metal cluster carbonyls (8). These make use of M-CO bond energies determined mass spectrometrically and rely on the assumption that M-CO bond energies are the same in mononuclear and all polynuclear carbonyls of the same metal.

Several directly measured values of $\Delta H°$ for homolytic dissociation of a metal-metal-bonded carbonyl in solution have been obtained (9). This was for the complexes $[(\eta^3\text{-}C_3H_5)Fe(CO)_2L]_2$ where L = CO or a number of different P-donor ligands. The low value $\Delta H° = 56.5$ kJ mol^{-1} when L = CO was not unexpected for such a sterically crowded molecule. The P-donor substituents increased the steric crowding and displaced the equilibria in favor of the monomers but the effect seemed to be controlled more by $\Delta S°$ than $\Delta H°$. In general metal-metal bond energies, however they may have been estimated, are too large to allow for direct measurement of equilibrium constants in solution in this way.

Although all of these measurements, however uncertain their precision, have their own intrinsic interest, none of them has any direct bearing on the reactivity of the complexes concerned. This applies not only to the reactivity of the metal-metal bonds that they contain but also to their reactivity towards dissociation of the CO ligands. The latter point is illustrated by the relative values of the average M-CO bond energies in binary mononuclear carbonyls such as $M(CO)_6$ (M = Cr, Mo, and W) (8). Even after allowance for the effect of different valence state promotion energies the average energies increase monotonically by over 25% along the series (8, 10) whereas the activation energies for CO dissociation are 168, 133, and 167 kJ mol^{-1}, respectively (11). The metal-metal-bonded carbonyls $M_2(CO)_{10}$ (M = Mn, Tc, and Re) would have a similarly large increase in the average M-CO bond dissociation energies and the M-M bond strength is predicted from thermochemical data almost to double along the series (8). This contrasts with an increase of reactivity (as measured by the activation enthalpies for substitution or thermal decomposition) of only 8% (6, 12). This discrepancy is independent of any uncertainty in the mechanism since it exists whether CO dissociation or homolytic fission of the M-M bonds is rate-determining. Not only do the relative values of the thermochemically estimated bond strengths fail to predict the scale of the relative reactivities but some of the absolute values are grossly misleading as well. Thus the thermochemically derived value for the Mn-Mn bond strength in $Mn_2(CO)_{10}$ is 67 kJ mol^{-1} (8). Since this represents the strength of the interaction between the two halves of the molecule in its undisturbed ground state it must represent an upper limit for the enthalpy of activation for homolytic fission of the Mn-Mn bond. (Any adjustment of the coordination around each metal atom as the Mn-Mn bond stretches to form the transition state must decrease the energy required to form it.) The activation energy for substitution or decomposition is, however, close to 153 kJ mol^{-1} (12). If this value were to be assigned (incorrectly; see below) to the CO dissociative process and the value of 67 kJ mol^{-1} were assigned to the activation enthalpy of homolytic fission of the Mn-Mn bond a value of $\Delta S^{\ddagger}$ of <u>less than</u> - ca. 160 J K^{-1} mol^{-1} would have to be assigned to the homolytic fission process in order for CO dissociation to be the preferred path. This is a totally unreasonable value for a metal-metal bond breaking process that must certainly involve a substantially positive value of $\Delta S^{\ddagger}$. It therefore seems most probable that the effective strength of the Mn-Mn bond is over twice as great as that estimated thermochemically. The relationship of the thermochemical estimates to the reactivity of metal-metal-bonded carbonyls is therefore very tenuous. The results of using them to make absolute predictions, or even correlations, are so uncertain as to be almost positively misleading. Clearly the only way of obtaining reliable, precise, quantitative measurements of the reactivity of metal-metal-bonded carbonyls is to carry out careful kinetic studies. Relating these to the reactivity specifically of the metal-metal bonds in the complexes requires even more extensive and detailed mechanistic studies if the role of the metal-metal bonds in determining the energetics of the reactions is to be elucidated.

Reactivities and Reaction Mechanisms of $M_2(CO)_{10}$ (M_2 = Mn_2, Tc_2, MnRe, and Re_2) and some Substituted Derivatives

For studies of reactivities of such metal-metal-bonded carbonyls to be directly related to the strengths of the metal-metal bonds reactions have to be found for which homolytic fission of the metal-metal bond is the rate-determining step. If the fission of the metal-metal bond is reversible then a reaction scheme as shown in eq 5-7 can be envisaged. The radical species M is

$$M_2 \underset{k_{-1}}{\overset{k_1}{\rightleftharpoons}} 2M \tag{5}$$

$$M \xrightarrow{k_2} \text{product} \tag{6}$$

$$M + X \xrightarrow{k_3} \text{product} \tag{7}$$

allowed to form product by a path that is either zero or first order in the concentration of some added reagent X. The rate equation, in its non-integrated form, for this scheme (13) is shown in eq 8 and 9. R_{obsd} is the rate observed at a given concentration, $[M_2]$, of complex and R_1 is the limiting rate that is,

$$R_{obsd} = R_1 - 4R^2_{obsd}k_{-1}/(k_2+k_3[X])^2 \quad (8)$$

$$k_{obsd} = k_1 - 4k^2_{obsd}[M_2]k_{-1}/(k_2+k_3[X])^2 \quad (9)$$

or would be, observed at sufficiently high concentrations of X for the reverse of reaction 5 not to occur. k_{obsd} is an apparent pseudo-first-order rate constant defined by $R_{obsd} = k_{obsd}[M_2]$. At very low values of $[M_2]$ and/or high values of $k_2 + k_3[X]$ the reactions will be first order in $[M_2]$ and governed by the rate constant k_1. However, as $[M_2]$ increases and/or $k_2 + k_3[X]$ decreases the reaction will approach half order in $[M_2]$ and be governed by a half-order rate constant $k_{1/2} = 0.5(k_1/k_{-1})^{1/2}$-$(k_2+k_3[X])$. For a given value of $[M_2]$, k_{obsd} should increase with increasing [X] to a limiting value but the rate of increase will depend on $[M_2]$ and the reactions proceeding at less than the limiting rate should be less than first order in $[M_2]$. It is important to note that it is only when rates less than the limiting values are observable that positive kinetic evidence becomes available in favor of this mechanism. If only limiting rates are observed then the first-order rate constants obtained do not distinguish whether the rate-determining process involves fission, formation of some more energetic isomer of M_2, or even CO dissociation. Equally well, of course, observations of limiting rates cannot disprove the homolytic fission mechanism. Even if reactions are found to be inhibited by CO this does not disprove homolytic fission nor does it prove CO dissociation. A scheme such as that shown in eq 10–13 for a substitution reaction allows for

$$[M(CO)]_2 \rightleftharpoons 2MCO \quad (10)$$

$$M(CO) + L \rightleftharpoons ML + CO \quad (11)$$

$$ML + M(CO) \longrightarrow M_2(CO)L \quad (12)$$

and/or

$$2ML \longrightarrow [ML]_2 \quad (13)$$

retardation of the rates in the presence of free CO and only careful study of the reactions that proceed at less than the limiting rates can possibly distinguish between the homolytic fission and CO-dissociative mechanisms (14).

The first full study that showed positive evidence for homolytic fission was made with $Mn_2(CO)_{10}$ (15). Reaction with $X = O_2$ in decalin led to decomposition but the kinetics were quite clean. Reaction at 125°C under 100% O_2 occurred at the limiting rate over a 500-fold range of $[Mn_2(CO)_{10}]$ but reaction under partial pressures of oxygen of 0.21 and 0.053 showed a clear change from almost first-order dependence on $[Mn_2(CO)_{10}]$ at the lowest concentrations to half-order dependence at the highest concentrations. Decomposition under 50% O_2 at 155°C was unaffected by CO. Decomposition under Ar at 155°C or under CO at 170°C showed half-order dependence over a 100-fold range of $[Mn_2(CO)_{10}]$. Decomposition was unaffected by CO from 80-155°C although an additional reaction path, inhibited by CO, became detectable above 155°C. All the extensive data were in excellent quantitative agreement with the rate behavior predicted by eq 8-9. Reactions with I_2 under CO at 125°C led to formation of $Mn(CO)_5I$ in virtually quantitative yields (6) according to the rate equation $k_{obsd} = k_a + k_b[I_2]$. The value of k_a was unaffected by the presence or absence of CO and was equal to the limiting rate of reaction under O_2 from 105 to 125°C. There can be little doubt that the reaction scheme shown in eq 5-7 is being followed and that the activation enthalpy of 153.8 ± 1.6 kJ mol^{-1} found for reaction under 100% O_2 corresponds to that for homolytic fission.

Reaction with PPh_3 proceeds under Ar at a limiting rate that is 35% faster than that with O_2 whereas reaction under 100% CO proceeds at the same rate as that with O_2 (6). Reaction with PPh_3 under O_2 proceeds at exactly the same rate as under Ar. No carbonyl product is observed. This shows that the reactions with PPh_3 and O_2 do not proceed via two completely independent paths in which case the rate of reaction with PPh_3 under O_2 should proceed at the sum of the rates observed with each separately. It appears, therefore, that ca. 25% of the substitution reaction occurs via a CO-inhibited path (probably CO dissociation) but that the remainder occurs via homolytic fission. The fact that no $Mn_2(CO)_9PPh_3$ or $Mn_2(CO)_8(PPh_3)_2$ is formed during reaction with PPh_3 under O_2 is explicable because they are both quite unstable to O_2 under these conditions (12, 16). Presumably they are formed in ca. 25% yield via the CO-inhibited path but decompose rapidly when formed.

Very similar, though rather less extensive, studies have been made of decomposition reactions of $MnRe(CO)_{10}$ (15), $Tc_2(CO)_{10}$ (17), and $Re_2(CO)_{10}$ (13). All the kinetics are consistent with the rather stringent predictions of rate eq 9 for the homolytic fission mechanism and are totally inconsistent with a simple CO-dissociative mechanism. Other workers have studied substitution reactions of $Mn_2(CO)_{10}$ (18) and $MnRe(CO)_{10}$ (19) and have commented, correctly, that their kinetic results are consistent with the dissociative reaction. This seems to have led to some doubts about the correctness of our conclusion that the reactions of all the decacarbonyls proceed mainly or totally via homolytic fission. These doubts do not arise from the kinetics because the kinetics observed by these workers (18, 19) are equally consistent with the homolytic fission mechanism or, indeed, virtually any first-order activation of the complexes. The conditions under which the reactions were followed were simply not those capable of leading to a kinetic distinction between the various mechanisms proposed. The absence of any homonuclear products of the reactions of $MnRe(CO)_{10}$ was offered (19) as evidence against the homolytic fission mechanism. This point had been raised before (15, 17) when it was countered by emphasizing the absence of sufficient data on dimerization of such radical species for the point to be sustained. Initial studies of reaction 14 showed (18) that the rate increased

$$Mn_2(CO)_8(PPh_3)_2 + P(OPh)_3 \longrightarrow Mn_2(CO)_8(PPh_3)P(OPh)_3 + PPh_3 \qquad (14)$$

to a limiting value with increasing $[P(OPh)_3]$. A dissociative mechanism was proposed although it was acknowledged that some ambiguities existed. Later studies of this reaction (14, 20) showed that reactions at less-than-limiting rates were in fact half-order in $[Mn_2(CO)_8(PPh_3)_2]$ and a very extensive set of data was fully in accord with the quite complicated rate equation appropriate to initial homolytic fission and totally incompatible with a simple dissociative process.

Other kinetic studies have shown that $Mn_2(CO)_8\{P(OPh)_3\}_2$ (21) and $Re_2(CO)_8(PPh_3)_2$ (22) undergo reaction in a way characteristic of initial reversible homolytic fission. $Mn_2(CO)_8L_2$ ($L = PPh_3$, P-n-Bu_3, and $P(C_6H_{11})_3$) all react with $C_2H_2Cl_4$ to form cis-$Mn(CO)_4ClL$ in essentially quantitative yields at exactly the same rates as those for reaction with O_2 (12). Substitution reactions of such complexes usually proceed at rates similar to, though slightly different from, those for reaction with O_2 (12) but the activation parameters are very close. Thus, although these reactions with O_2 seem to go via some path in addition to homolytic fission, they do go in large part by the homolytic fission path and the activation parameters are likely to be generally very close to those for homolytic fission. It has therefore been concluded (12, 23) that activation parameters equal, or very close, to those for homolytic fission can be assigned to reactions of the decacarbonyls and many of their axially disubstituted derivatives with CO. These activation parameters can be taken as an excellent quantitative measure of the reactivity of the metal-metal bonds and the activation enthalpies are a good measure of the strengths of the metal-metal bonds.

The position of the monosubstituted complexes $Mn_2(CO)_9L$ is not so clear. If they reacted only via homolytic fission then reactions with O_2, L, and CO should all proceed at the same rate. However, for $Mn_2(CO)_9(PPh_3)$ (16) the rates are not the same, nor are the activation enthalpies very close. The assignment (12) to homolytic fission of the activation parameters for reaction of the monosubstituted complexes with O_2 is therefore less conclusive than in the other cases.

In total there are 23 complexes of these group 7 metals for which activation parameters can be assigned to homolytic fission (12, 23). Eight of these are based on good kinetic evidence, 9 are based on analogy to very closely related complexes in the first group (e.g. $Mn_2(CO)_8(PEt_3)_2$ vs. $Mn_2(CO)_8(P$-n-$Bu_3)$, and $Mn_2(CO)_8\{PPh_2(OMe)\}_2$ vs. $Mn_2(CO)_8(PPh_3)_2$) or on indirect kinetic evidence, and the remaining 6 are based on more distant analogy or must be considered tentative. These values are listed in Table I and enable the following trends to be discernible.

Table I

Activation Enthalpies (in kJ mol^{-1}) for Homolytic Fission of Some Metal-Metal Bonds

Complex	$\Delta H^{\ddagger}_{hf}$	Complex	$\Delta H^{\ddagger}_{hf}$
Group A[a]			
$Mn_2(CO)_{10}$	153.8 ± 1.6	$Mn_2(CO)_8\{P(OPh)_3\}_2$	151.0 ± 2.1
$Tc_2(CO)_{10}$	160.1 ± 0.8	$Mn_2(CO)_8(PPh_3)_2$	117.2 ± 0.8
$MnRe(CO)_{10}$	162.8 ± 0.8	$Mn_2(CO)_8(P\text{-}n\text{-}Bu_3)_2$	132.2 ± 1.6
$Re_2(CO)_{10}$	165.5 ± 0.8	$Mn_2(CO)_8\{P(C_6H_{11})_3\}_2$	99.1 ± 1.3
Group B[b]			
$Re_2(CO)_8(PPh_3)_2$	162.2 ± 2.1	$Mn_2(CO)_8(PEt_2Ph)_2$	134.0 ± 0.4
$Mn_2(CO)_8(PEt_3)_2$	136.0 ± 1.6	$Mn_2(CO)_8\{P(p\text{-}MeOC_6H_4)_3\}_2$	123.6 ± 1.1
$Mn_2(CO)_8\{P(OMe)_3\}_2$	152.5 ± 0.4	$Mn_2(CO)_8\{PPh(OMe)_2\}_2$	138.7 ± 0.4
$Mn_2(CO)_8(PEtPh_2)_2$	131.9 ± 0.2	$Mn_2(CO)_8\{PPh_2(OMe)\}_2$	152.1 ± 0.6
$Mn_2(CO)_8(AsPh_3)_2$	133.4 ± 2.4		
Group C[c]			
$MnRe(CO)_8(PPh_3)_2$	155.2 ± 1.3	$Mn_2(CO)_9(PBu_3)$	140.9 ± 3.3
$Tc_2(CO)_8(PPh_3)_2$	138.0 ± 2.2	$Mn_2(CO)_9P(C_6H_{11})_3$	141.0 ± 1.5
$Mn_2(CO)_9(PPh_3)$	133.1 ± 2.1	$Mn_2(CO)_9\{P(OPh)_3\}$	151.2 ± 0.9

[a] Positive kinetic evidence for reversible homolytic fission. [b] Assignment of $\Delta H^{\ddagger}_{hf}$ made by analogy with very closely related complexes. [c] Assignment of $\Delta H^{\ddagger}_{hf}$ uncertain.

(i) Dependence on the Metal. The order of $\Delta H^{\ddagger}_{hf}$ is $Mn_2(CO)_{10}$ < $Tc_2(CO)_{10}$ < $MnRe(CO)_{10}$ < $Re_2(CO)_{10}$. This trend is in agreement with the generally accepted increase of metal-metal bond strengths with increasing atomic number (8, 24) and there is a good correlation (25) with the force constants for the metal-metal stretching vibration (26, 27) and also with the values of $h\nu$ ($\sigma \rightarrow \sigma^*$), the energy of electronic transition from the metal-metal bonding molecular orbital to the corresponding anti-bonding orbital (28, 29). The overall change along the series is only about 12 kJ mol^{-1}, however.

(ii) The Effect of the Presence of Two Axial PPh_3 or $AsPh_3$ Ligands. The presence of two axial PPh_3 substituents results in a decrease in $\Delta H^{\ddagger}_{hf}$, the decrease being in the order Mn_2 > Tc_2 > MnRe > Re_2. This is suggestive of a steric effect that decreases with increasing size of the metal. The introduction of two $AsPh_3$ ligands into $Mn_2(CO)_{10}$ lowers $\Delta H^{\ddagger}_{hf}$, but to a lesser extent than introduction of PPh_3. This is compatible with the smaller size of $AsPh_3$ (23).

(iii) The Effect of Successive Substitution of P-donor Ligands into $Mn_2(CO)_{10}$. In spite of the uncertainty of assignment of values of $\Delta H^{\ddagger}_{hf}$ for the monosubstituted complexes the successive introduction of PPh_3 or PCy_3 ligands results in a progressive decrease in $\Delta H^{\ddagger}_{hf}$. The same may be true for P-n-Bu_3 but is not true for $P(OPh)_3$ for which no overall decrease is observed.

(iv) Correlation of $\Delta H^{\ddagger}_{hf}$ with Values of $h\nu$ ($\sigma \rightarrow \sigma^*$). The correlation found for the decacarbonyls (see above) is also found to include some complexes substituted with one or even two smaller P-donor ligands as shown in Figure 1. The electronic effects, either of changing the metal or of the presence of small P-donor ligands, is therefore quite small whereas steric effects seem to be much greater. This is confirmed as follows.

(v) Correlation of $\Delta H^{\ddagger}_{hf}$ with Cone Angles of L in $Mn_2(CO)_8L_2$. An excellent correlation with the size of the substituent is found when $\Delta H^{\ddagger}_{hf}$ is plotted against the cone angles of L (12, 23) as shown in Figure 2. Introduction of small ligands has very little effect but when the cone angle is $\gtrsim 125°$ $\Delta H^{\ddagger}_{hf}$ begins to decrease and when L = $P(C_6H_{11})_3$ (cone angle = ca. 180°) the value of $\Delta H^{\ddagger}_{hf}$ has been reduced from ca. 154 kJ mol^{-1} to 100 kJ mol^{-1}.

A rather detailed analysis of the steric effects of ligands $PPh_n(OMe)_{3-n}$ and PPh_nEt_{3-n} suggests (23) that values of $\Delta H^{\ddagger}_{hf}$ can even reflect the more subtle problems caused by the detailed meshing of an axial ligand, with essentially a 3-fold axis of symmetry, with the manganese atom, surrounded by four equatorial CO ligands.

The absence of such strong steric effects on the values of $h\nu$ ($\sigma \rightarrow \sigma^*$) (29) suggests that the strong steric effects on the ease

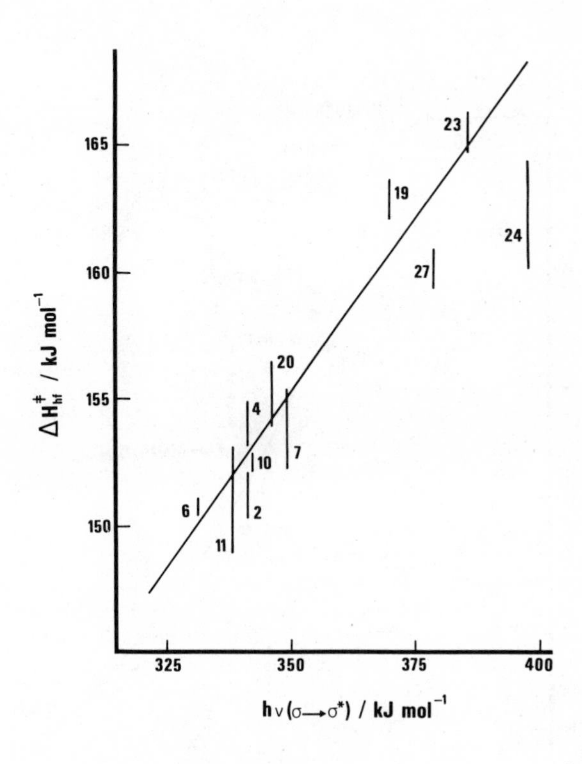

Inorganic Chemistry

Figure 1. Dependence of enthalpies of activation for homolytic fission, $\Delta H_{hf}^{\ddagger}$, *on the energies of the corresponding* $\sigma \rightarrow \sigma^*$ *transitions* (12).

Complexes are: (1) $Mn_2(CO)_{10}$*; (2)* $Mn_2(CO)_9P(OPh)_3$*; (4)* $Mn_2(CO)_9PBu_3$*; (6)* $Mn_2(CO)_9PPh_3$*; (10)* $Mn_2(CO)_8\{P(OMe)_3\}_2$*; (11)* $Mn_2(CO)_8\{P(OPh)_3\}_2$*; (19)* $MnRe(CO)_{10}$*; (20)* $MnRe(CO)_8(PPh_3)_2$*; (21)* $Tc_2(CO)_{10}$*; (23)* $Re_2(CO)_{10}$*; (24)* $Re_2(CO)_8(PPh_3)_2$.

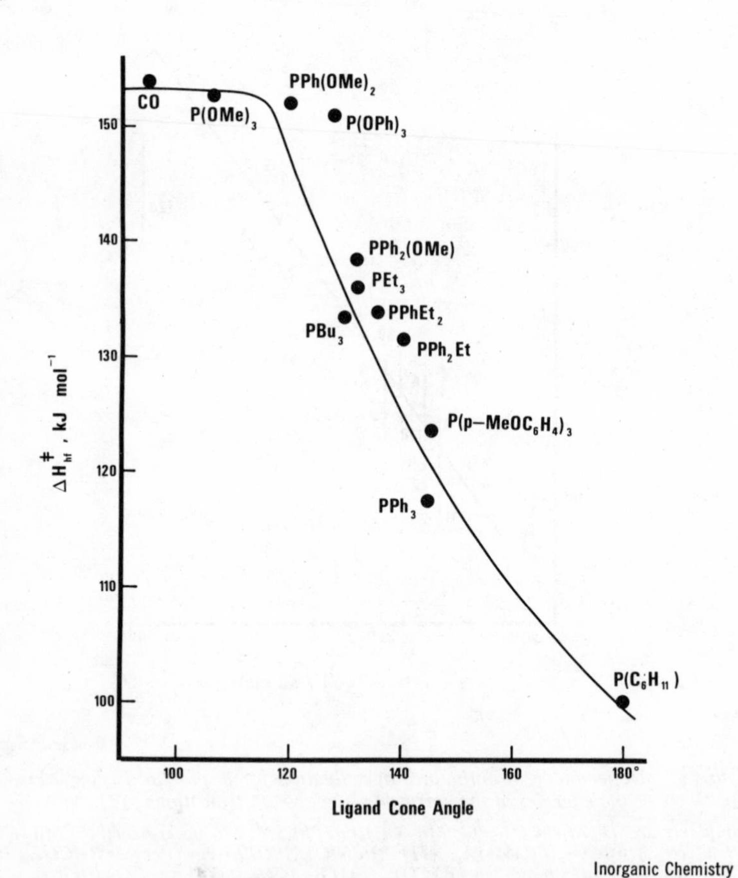

Figure 2. Dependence of $\Delta H_{hf}^{\ddagger}$ for $Mn_2(CO)_8L_2$ on the cone angle of L (23).

of homolytic fission of the Mn_2 complexes arises mainly from the release of steric strain that occurs on stretching the Mn-Mn bond, rather than from a weakening of the Mn-Mn bond strength in the undisturbed molecule.

These steric effects show up not only in the values of $\Delta H^{\ddagger}_{hf}$ but in the rates themselves. Thus, while $Mn_2(CO)_{10}$ and its derivatives containing smaller ligands react at convenient rates (convenient, that is, for kinetic or synthetic purposes) only at temperatures significantly above 100°C, $Mn_2(CO)_8(PPh_3)_2$ reacts smoothly at ca. 50°C (16), while $Mn_2(CO)_8\{P(C_6H_{11})_3\}_2$ reacts quite rapidly at room temperature (12).

Reactions with Iodine and Bromine. All the reactions discussed above proceed at limiting rates that are independent of the concentration and nature of the reactant. However, reactions of iodine with $Mn_2(CO)_{10}$ (6) and $Re_2(CO)_{10}$ (30) proceed by paths that are first order in $[I_2]$ as well as by the $[I_2]$-independent homolytic fission paths. P-donor substituents increase the rates of reaction with I_2 by several orders of magnitude (31) so that they proceed rapidly even at room temperature. Thus reaction of $Mn_2(CO)_8\{P(C_6H_{11})_3\}_2$ is estimated to occur over 10^8 times faster than $Mn_2(CO)_{10}$ by a path first order in $[I_2]$ at 25°C in cyclohexane. In all cases the reactions proceed with fission of the metal-metal bonds to form the mononuclear iodo complexes.

Detailed kinetic studies of these reactions have shown that they frequently proceed according to very complicated rate equations (31). Thus $Mn_2(CO)_8\{P(OPh)_3\}_2$ reacts according to eq 15.

$$k_{obsd} = \{k_3[I_2]^2 + k_4[I_2]^3\}/\{1 + \beta_2[I_2]^2 + \beta_3[I_2]^3\} \quad (15)$$

Thirty-five values of k_{obsd} over the range $[I_2] = (1\text{-}40) \times 10^{-3}$M fit this equation with a standard deviation of 4.6%. Setting $\beta_3 = 0$ increases this to 5.9%. Reaction of $Re_2(CO)_9(PPh_3)$ with I_2 includes a term in $[I_2]^4$ which appears definitely to be finite although it is rather imprecisely determined.

This kinetic behavior is ascribed to the rapid preformation of adducts containing up to four I_2 molecules followed by slower fission of the metal-metal bonds and subsequent formation of the mononuclear iodo complexes. The formation of adducts can in some cases be complete and is quite dependent on the nature of the solvent. A term $k_3[I_2]^2$ is common to most of the reactions and k_3 is found to increase dramatically with increasing basicity of L in $Mn_2(CO)_8L_2$. Thus it varies from 6.7×10^{-4} M^{-2} s^{-1} when L = $P(OPh)_3$ to ca. 10 M^{-2} s^{-1} when L = $PPhEt_2$ for reaction in cyclohexane at 25°C. Quite a good LFER is shown by a plot of log k_3 against Δ(hnp), the relative half-neutralization potential for titration of pure substituent ligand against perchloric acid in nitromethane. This increase in rate is also closely related to a decrease in ν_{CO} when L = phosphine. The value of log k_3 for $Mn_2(CO)_8\{P(C_6H_{11})_3\}_2$ deviates by at least 3 orders of magnitude from these linear correlations but no such pronounced, and presumably steric, deviations are shown by other ligands. It is concluded that the I_2 molecules in the adduct are attached close to the O atoms of the CO ligands and that their role is to weaken the metal-metal bonds by inductively withdrawing electron density from the metal centers through the CO ligands (31).

Reactions of Br_2 show essentially the same type of behavior (32) with substituted carbonyls although the extent of adduct formation is significantly lower. Reaction of $Mn_2(CO)_{10}$ with Br_2 is, however, unique (33) in proceeding via a chain reaction, the propagation steps presumably being as in eq 16 and 17.

$$\cdot Mn(CO)_5 + Br_2 \longrightarrow Mn(CO)_5Br + \cdot Br \quad (16)$$

$$\cdot Br + Mn_2(CO)_{10} \longrightarrow Mn(CO)_5Br + \cdot Mn(CO)_5 \quad (17)$$

Reactivities and Reaction Mechanisms of $Co_2(CO)_8$ and Some Substituted Derivatives

Relatively few kinetic studies of such complexes have been undertaken and none has suggested that spontaneous homolytic fission of the Co-Co bond is implicated as a rate-determining step although the importance of metal-centered radicals has been shown (34, 35). Reaction of $Co_2(CO)_8$ with alkynes to form the alkyne-bridged $Co_2(CO)_6(C_2R_2)$ has been shown to involve rate-determining dissociation of one CO ligand and this is quite rapid even at room temperature (36). Reaction of the axially disubstituted complex $Co_2(CO)_6(P\text{-}n\text{-}Bu_3)_2$ with C_2Ph_2 is very much slower (37) and leads to formation of $Co_2(CO)_5(P\text{-}n\text{-}Bu_3)(C_2Ph_2)$ in decalin at 100°C. A major path involves dissociative loss first of a CO ligand and then of a $P\text{-}n\text{-}Bu_3$ ligand and this is followed by attack of the C_2Ph_2 on the $Co_2(CO)_5(P\text{-}n\text{-}Bu_3)$ intermediate. This intermediate may well involve a Co≡Co triple bond by analogy with the well-characterized complexes of the type $Cp_2Mo_2(CO)_4$ (Cp = η^5-C_5H_5) (38). The dissociative mechanism is therefore probably favored by the stabilization, due to increased Co-Co bonding, of what would otherwise be a coordinatively unsaturated intermediate. In this sense the metal-metal bonding is increasing the reactivity of the complex towards ligand dissociation. The formation of this intermediate proceeds via the first formed $Co_2(CO)_5(P\text{-}n\text{-}Bu_3)_2$ and the vacant coordination site on one Co atom may also be stabilized by sideways-on bonding of a CO ligand on the other Co atom in the way found in $Mn_2(CO)_5(dppm)_2$ (dppm = $Ph_2PCH_2PPh_2$) (39). Again, the presence of the neighboring Co atom, i.e. of the metal-metal bond, must be helping determine the reactivity of the complex towards dissociation of one CO ligand. The $Co_2(CO)_5(P\text{-}n\text{-}Bu_3)$ can also be formed by dissociation first of a $P\text{-}n\text{-}Bu_3$ ligand followed by loss of a CO ligand. This sequence of dissociations is considerably less important than loss of CO first and then $P\text{-}n\text{-}Bu_3$. An almost identical pattern is found for reaction of $Co_2(CO)_6\{P(C_6H_{11})_3\}_2$ with C_2Ph_2 (40).

A third path followed in this reaction involves highly reversible formation of an energetically excited isomer of the complex which can be attacked directly by the C_2Ph_2 before subsequent reversible loss of a CO and a P-n-Bu_3 ligand. No direct evidence for the nature of this intermediate is available but it has been speculated (37) to be a CO-bridged complex as shown in I, the reaction involving a metal migration analogous to the well-known methyl migration process (41). The vacant coordination position on one of the Co atoms provides a site for direct attack by the C_2Ph_2. A very close analogue, II, has been implicated in the photochemical reaction of $Cp_2Fe_2(CO)_4$ with P(O-i-Pr)$_3$ to form III at -78°C in ethyl chloride or THF (42). I was also postulated as the intermediate involved in the formation of IV by the insertion of $SnCl_2$ into $Co_2(CO)_6$(P-n-Bu_3)$_2$ in THF (43). This reaction proceeds at a limiting rate at high [$SnCl_2$] and is much more rapid than reaction with C_2Ph_2. This may well be due to the greater nucleophilicity of $SnCl_2$ which effectively prevents reformation of the original $Co_2(CO)_6$(P-n-Bu_3)$_2$ from I when [$SnCl_2$] is high enough. The kinetics show that attack by C_2Ph_2 on I is slow compared to reversion to the reactant complex. This mechanism is an example of the wider variety of mechanisms available to metal-metal-bonded complexes, this one being possible because of the special ability of a carbonyl ligand to bridge two metal atoms. This type of mechanism has also been proposed for insertion of stannous halides into other metal-metal-bonded complexes (44).

A very similar intermediate, V, has been implicated in the reaction of VI with alkynes (45), and with CO or PPh_3 to form VII (46). In this case the reactivity of VI is controlled by the ease of breaking the Co-Co bond, the conversion of a bridging CO into a terminal one involving little or no expenditure of energy. Since the reactions with CO or PPh_3 are strictly first order in the concentration of entering ligand (46) the overall activation enthalpy is made up of $\Delta H°$ for the highly reversible formation of V from VI plus the value of $\Delta H^{\ddagger}_L$ for attack by CO or PPh_3 on V. The overall values of $\Delta H^{\ddagger} = \Delta H° + \Delta H^{\ddagger}_L$ are 92.8 ± 1.7 and 72.4 ± 1.4 kJ mol^{-1} for L = CO and PPh_3, respectively. The difference between these two values is rather larger than one might expect for simple attack of L on a vacant coordination site and this may be an indication that V is stabilized to some extent by sideways-on bridging of a CO group as in VIII. This would stabilize the intermediate and make it more selective towards nucleophilic attack.

Reactivities and Reaction Mechanisms of Some Metal-Carbonyl Clusters

Reactions of $Ru_3(CO)_{12-n}L_n$ (L = PPh_3, PBu_3, and $P(OPh)_3$; n = 0-3). $Ru_3(CO)_{12}$ reacts thermally with CO to form $Ru(CO)_5$ only at relatively high temperatures under high pressures of CO, and full

O
C
$(P\text{-}n\text{-}Bu_3)(OC)_3Co$ $Co(CO)_2(P\text{-}n\text{-}Bu_3)$

I

O
C
$Cp(CO)_2Fe$ $Fe(CO)Cp^*$

II

O
C
$Cp(OC)_2Fe$ $Fe(CO)\{P(O\text{-}i\text{-}Pr)_3\}Cp$

III

Cl Cl
Sn
$(P\text{-}n\text{-}Bu_3)(OC)_3Co$ $Co(CO)_3(P\text{-}n\text{-}Bu_3)$

IV

Ph Ph
Ge
$(OC)_4Co$ $Co(CO)_3$

V

Ph Ph
Ge
$(OC)_3Co$ — $Co(CO)_3$
C
O

VI

Ph Ph
Ge
$(OC)_4Co$ $Co(CO)_3L$

VII

Ph Ph
Ge
$(OC)_3Co$ $Co(CO)_3$
C
O

VIII

$^*Cp = \eta^5\text{-}C_5H_5$

kinetic data are not available (47). However, the complexes $Ru_3(CO)_{12-n}(PPh_3)_n$ (n = 1-3) can undergo thermal fragmentation reactions in decalin at 130-170°C to form the well-defined mononuclear products $Ru(CO)_4(PPh_3)$ and/or $Ru(CO)_3(PPh_3)_2$ (48, 49). When [CO] and $[PPh_3]$ are high enough the rates are found to be independent of [CO] and $[PPh_3]$ so that spontaneous fragmentation of the complexes is occurring without preliminary dissociation of CO or PPh_3 and without assistance from bimolecular attack by these ligands. The kinetic measurements are therefore a good measure of the intrinsic kinetic stability of the Ru_3 clusters. The activation enthalpies increase dramatically with n, being 86.4 ± 0.9, 124.2 ± 3.6, and 147.7 ± 5.0 kJ mol^{-1}, respectively, when n = 1, 2, or 3. The activation parameters fall on a good isokinetic plot of $\Delta H^{\ddagger}$ against $\Delta S^{\ddagger}$ with an isokinetic temperature of ca. 200°C. The relative reactivities at the temperatures used are therefore enthalpy-controlled and the rates would decrease by a factor of 100 at 100°C when n changes from 1 to 3. The kinetics themselves do not distinguish whether one, two, or even three Ru-Ru bonds are in the process of being broken when the transition state is reached. Concerted fragmentation into three $Ru(CO)_3PPh_3$ moieties is unlikely since reactions such as these are reversible and the aggregation reaction would then have to be trimolecular. Simply breaking one Ru-Ru bond would be unlikely to lead to the trends in activation parameters observed, a decrease of $\Delta H^{\ddagger}$ with n being more probable due to steric effects of the sort discussed above for the group 7 metal carbonyls. The activation parameters can be rationalized (49) if the complexes break into two fragments, one of which is always $Ru(CO)_3(PPh_3)$, the other being $Ru_2(CO)_8$, $Ru_2(CO)_7(PPh_3)$, or $Ru_2(CO)_6(PPh_3)_2$, respectively, as n changes from 1 to 3. In order to attain an 18-electron configuration the Ru_2 intermediates would have to contain a Ru=Ru double bond of the type which is now quite well known for somewhat analogous complexes (50). The enthalpy of these dinuclear intermediates, however, would be expected to increase with increasing numbers of substituent PPh_3 ligands because of ligand-ligand repulsions and so the activation enthalpy to form these intermediates would be expected to increase in the direction observed. The increase in $\Delta S^{\ddagger}$ along the series probably implies that the Ru-Ru bond strength actually decreases along the series to an extent that vibrational entropy is much increased. Whether this mechanistic interpretation is correct or not, the fact remains that the activation enthalpies, and to some extent the reactivities, do show a clear dependence on the number of substituents, a dependence that is opposite to that in $Mn_2(CO)_{10-n}(PPh_3)_n$ (n = 0-2) (12). Generalization of the effects of ligand substitution on the reactivity of metal-metal bonded carbonyls is therefore premature at this stage.

In addition to spontaneous fragmentation there is a more rapid path that is inhibited by CO and that therefore involves loss of CO in the rate-determining step. This leads to fragmentation for $Ru_3(CO)_9L_3$ (L = PPh_3, P-n-Bu_3, and $P(OPh)_3$) (51) and simple substitution for $Ru_3(CO)_{12-n}(PPh_3)_n$ (n = 1 or 2) (49). For $Ru_3(CO)_9(P\text{-}n\text{-}Bu_3)_3$ the product is simply $Ru(CO)_3(P\text{-}n\text{-}Bu_3)_2$ but for $Ru_3(CO)_9\{P(OPh)_3\}_3$ the mononuclear product was the orthometallated $Ru(CO)_2\{P(OC_6H_4)(OPh)_2\}_2$. $H_4Ru_4(CO)_9\{P(OPh)_3\}_3$ and $Ru_2H(CO)_3\{P(OC_6H_4)(OPh)_2\}_2OP(OPh)_2$ were also formed so aggregation and orthometallation of intermediate fragments can also occur in this case. This did not appear to have any effect on the form of the kinetics. However, for both these two substituted trinuclear clusters the kinetics are more complicated than would be expected from attack by pure ligand on initially generated $Ru_3(CO)_8L_3$ (51). Further, the activation parameters for formation of the intermediates $Ru_3(CO)_8L_3$ depend greatly on the nature of L varying from a low $\Delta H^\ddagger$ (81.6 ± 4.6 kJ mol^{-1}) and rather negative $\Delta S^\ddagger$ (-88.7 ± 4.2 JK^{-1} mol^{-1}) for L = P-n-Bu_3 to a much higher $\Delta H^\ddagger$ (138.1 ± 4.6 kJ mol^{-1}) and a positive $\Delta S^\ddagger$ (27.2 ± 10.9 J K^{-1} mol^{-1}) when L = PPh_3. This implies that the structure of the $Ru_3(CO)_8L_3$ intermediates varies considerably with the nature of L. The nature of these intermediates could vary from IX to XII as $\Delta H^\ddagger$ decreases in energy. Formation of IX might occur if the thermal energy cannot simply be concentrated on breaking just an Ru-CO bond. Not only do the activation parameters require major differences in the structures of the intermediates but the fact that the kinetics themselves (for L = P-n-Bu_3 and $P(OPh)_3$) require the existence of two kinetically distinct isomeric forms of $Ru_3(CO)_8L_3$ to be produced means that the need to speculate on the possible structures is imposed by the form of the kinetics as well as by the energetics (51). Such a range of intermediates is only conceivable by virtue of the cluster nature of the complexes so the wide range of reactivities can be directly attributed to the polynuclear nature of the complex, i.e. the reactivity towards an essentially dissociative process is determined by energetic properties of the metal cluster.

In addition to the CO-dissociative path the complexes can also undergo fragmentation via rate-determining P-donor-ligand dissociation and this is much faster than either the spontaneous fragmentation or the CO-dissociative paths (48, 49, 51). In addition to mononuclear products, $Ru_3(C_6H_4)(CO)_7(PPh_2)_2$, $Ru_3(CO)_9H\{P(C_6H_4)Ph_2\}(PPh_3)$, and $Ru_2(CO)_6\{P(C_6H_4)Ph_2\}_2$ appeared to be formed from $Ru_3(CO)_9(PPh_3)_3$. Quantitative kinetic studies were not made for the P-n-Bu_3 or $P(OPh)_3$ dissociative reactions of $Ru_3(CO)_9(P\text{-}n\text{-}Bu_3)_3$ or $Ru_3(CO)_9\{P(OPh)_3\}_3$ and no data are available for comparison with the results for PPh_3 dissociation from $Ru_3(CO)_9(PPh_3)_3$.

Another, and quite unique, type of kinetic behavior is shown by $Ru_3(CO)_9(PPh_3)_3$ in its reactions with O_2 in the presence of sufficient free PPh_3 to suppress the dissociation of the complexed PPh_3 (52). This reaction in decalin at 55-75°C leads to a yellow insoluble product that has not yet been fully characterized. One mole of PPh_3 is oxidized to $OPPh_3$ for every mole of Ru

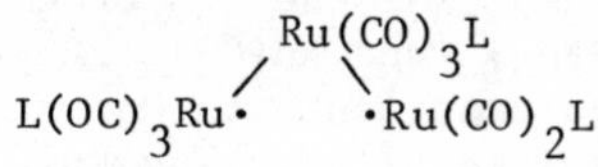

$$\mathrm{L(OC)_3Ru\cdot \quad \overset{Ru(CO)_3L}{/ \quad \backslash} \quad \cdot Ru(CO)_2L}$$

IX

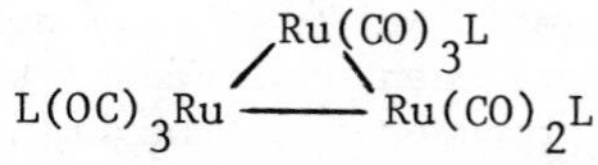

$$\mathrm{L(OC)_3Ru \overset{Ru(CO)_3L}{/ \quad \backslash} — Ru(CO)_2L}$$

X

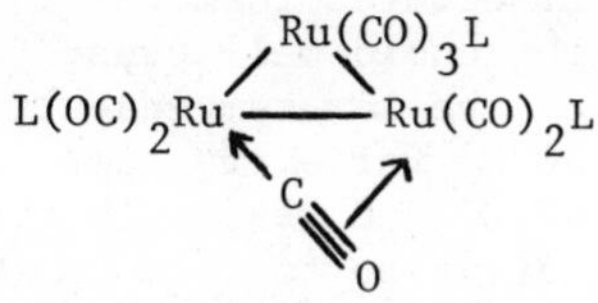

$$\mathrm{L(OC)_2Ru \overset{Ru(CO)_3L}{/ \quad \backslash} — Ru(CO)_2L \; (\leftarrow C \equiv O \rightarrow)}$$

XI

$$\mathrm{L(OC)_2Ru \overset{Ru(CO)_3L}{/ \quad \backslash} = Ru(CO)_2L \; (\mu\text{-}C{=}O)}$$

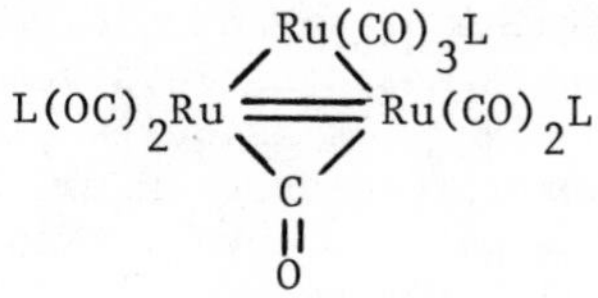

XII

in the original complex. Although the detailed natures of the overall reaction products were not elucidated the kinetic data were clear enough to be susceptible to detailed analysis. Of particular importance in this was the precise dependence of the rates on the concentrations of $Ru_3(CO)_9(PPh_3)_3$ and O_2, there being no dependence on $[PPh_3]$ provided this was sufficient to suppress the PPh_3-dissociative path. The analysis showed quite conclusively that the complex was undergoing <u>reversible fragmentation into two species, only one of which reacted with O_2 at a rate proportional to $[O_2]$</u>. It was concluded that the two fragments were most likely to be $Ru_2(CO)_6(PPh_3)_2$ and $Ru(CO)_3(PPh_3)$. The former was envisaged not to undergo reaction with O_2, some form of further decomposition being required first. The $Ru(CO)_3(PPh_3)$ could not be in its spin-paired, planar d^8 form. That intermediate must certainly be involved in the CO-dissociative reaction of $Ru(CO)_4(PPh_3)$ with PPh_3 to form $Ru(CO)_3(PPh_3)_2$, the yield of which, and its rate of formation, are quite unaffected by the presence of oxygen (53). If diamagnetic $Ru(CO)_3(PPh_3)$ were an intermediate $Ru(CO)_3(PPh_3)_2$ should have been observed as a product but it was not. It was therefore proposed that the $Ru(CO)_3(PPh_3)$ was in the high-spin, pseudo-tetrahedral configuration, a form that would be expected to be readily susceptible to attack by O_2. This conclusion requires the $Ru(CO)_3(PPh_3)$ formed in the high-temperature fragmentation in the presence of CO and PPh_3 to be in the low-spin, planar form. Intermediates formed in the reaction of O_2 with high-spin $Ru(CO)_3PPh_3$ could well cause the formation of $OPPh_3$.

The limiting rate of the low-temperature fragmentation process is such that, at 100°C, for every time spontaneous reaction occurs in the presence of CO and PPh_3 with eventual formation of $Ru(CO)_3(PPh_3)_2$ the fragmentation to form $Ru_2(CO)_6(PPh_3)_2$ and high-spin $Ru(CO)_3(PPh_3)$ must occur over 10^4 times. Because of the absence of O_2 this fragmentation leads nowhere and is always followed by recombination. Similarly, for every time CO dissociation occurs this fragmentation and its reverse must occur 10^3 times. It must be emphasized that this conclusion is quite unaffected by the correctness or otherwise of any of the speculations about the detailed nature of the intermediates involved and is a direct consequence of the kinetic data. It is quite clear that the susceptibility of this Ru_3 cluster towards eventual fragmentation depends greatly on the particular path followed. There are four of these available, the one actually followed being determined by the particular conditions.

A surprising qualitative observation relating to the stability of Ru_3 clusters was made when the kinetics of reaction of $Ru_3(CO)_{12}$ with P-n-Bu_3 were studied (54). $Ru_3(CO)_9(P\text{-}n\text{-}Bu_3)_3$ can easily be prepared by this reaction but the products in the kinetic study were $Ru(CO)_4(P\text{-}n\text{-}Bu_3)$ and $Ru(CO)_3(P\text{-}n\text{-}Bu_3)_2$ in the mole ratio 2:1. The fragmentation cannot have occurred after formation of $Ru_3(CO)_9(P\text{-}n\text{-}Bu_3)_3$ which is quite stable under the conditions pertaining (51, 54). It was concluded that the sequence of reactions 18-22 best explained the results. At the

$$Ru_3(CO)_{12} + P\text{-}n\text{-}Bu_3 \longrightarrow Ru_3(CO)_{11}(P\text{-}n\text{-}Bu_3) + CO \quad (18)$$

$$Ru_3(CO)_{11}(P\text{-}n\text{-}Bu_3) \longrightarrow 2Ru(CO)_4 + Ru(CO)_3(P\text{-}n\text{-}Bu_3) \quad (19)$$

$$2Ru(CO)_4 + P\text{-}n\text{-}Bu_3 \longrightarrow 2Ru(CO)_4(P\text{-}n\text{-}Bu_3) \quad (20)$$

$$Ru(CO)_3(P\text{-}n\text{-}Bu_3) + P\text{-}n\text{-}Bu_3 \longrightarrow Ru(CO)_3(P\text{-}n\text{-}Bu_3)_2 \quad (21)$$

$$Ru(CO)_3(P\text{-}n\text{-}Bu_3) \longrightarrow 1/3\ Ru_3(CO)_9(P\text{-}n\text{-}Bu_3)_3 \quad (22)$$

low concentrations of complex and relatively high values of [P-n-Bu_3] used in the kinetic study reactions 20-21 predominate over 22 and the latter only becomes effective when the concentration of complex is high as in preparative work. This shows qualitatively that $Ru_3(CO)_{11}(P\text{-}n\text{-}Bu_3)$ is peculiarly susceptible to fragmentation under much milder conditions (ca. 50°C) compared with all the other fragmentation reactions studied (48, 49, 51, 52). It also shows that an apparently simple substitution reaction can in fact be going via successive fragmentation and aggregation, just as is observed for reaction 14 (14).

Finally, it is observed that $Ru_3(CO)_{12}$ undergoes bimolecular attack by P-donor ligands in addition to simple CO dissociation (55, 56). (The absence of second-order paths is asserted (57), incorrectly, in the latest edition of a standard text.) The dependence of the rates on the basicity of the nucleophiles and the magnitude of steric effects shown by PPh_3 and, more clearly, by $P(C_6H_{11})_3$ shows that the cluster is quite selective and that the degree of bond-making in the transition state is quite high. Bimolecular reactions of this sort are very rare with simple binary carbonyls (11) and it appears that some degree of ambiguity in the assignment of oxidation state, i.e. of electron distribution, is a prerequisite for nucleophilic attack to be possible (58). Thus the presence of ligands such as NO or η^5-C_5H_5 enhances susceptibility to nucleophilic attack in mononuclear complexes and various bridging groups do this in binuclear complexes (59, 60, 61). It seems likely that the metal-metal bonding in clusters is of a sufficiently polarizable nature that it can adjust relatively easily to the approach of a nucleophile in such a way as to lower the activation energy. Indeed, some reactions can be envisaged (61) as involving rapid adduct formation by donation of electrons into the LUMO in the metal-metal bonding of the cluster before subsequent displacementof CO. The nature of the metal-metal bonding in such clusters is therefore controlling their reactivity towards bimolecular ligand substitution.

Kinetics of Reactions of Some Tetrahedral Metal Carbonyl Clusters. $Ir_4(CO)_{12}$ has been shown to undergo CO substitution with PPh_3 by a predominantly bimolecular reaction (62) and this is the case also with other P-donor ligands (63). When one PPh_3 is introduced further substitution is very much more rapid and occurs via a simple CO-dissociative path (62). This was ascribed to the presence of bridging carbonyls, not present in $Ir_4(CO)_{12}$, which somehow enhance the ease of CO dissociation. It is not clear that associative reaction is actually disfavored in an absolute sense since even if bimolecular reaction occurred at the same rate as with $Ir_4(CO)_{12}$ it would not have been observed in competition with the very much enhanced rate of CO dissociation. It has, however, been shown that the introduction of a P-donor ligand onto one CO atom in $Co_2(CO)_6(C_2Ph_2)$ strongly inhibits bimolecular attack at the other Co atom compared with that on the initial unsubstituted complex (59). $Ir_4(CO)_{12}$ and its derivatives only undergo fragmentation into Ir_2 complexes under extreme conditions and no kinetic data are available for such processes (64).

$Co_4(CO)_{12}$, on the other hand, undergoes reaction with CO under only moderately severe conditions to form $Co_2(CO)_8$ (65). A detailed kinetic study of this reaction has shown that it occurs by paths first and second order in [CO] (66). This was interpreted to mean that initial, highly reversible formation of $Co_4(CO)_{13}$ occurred and that this could then undergo Co-Co bond breaking, either spontaneously or with the assistance of attack by an additional CO molecule. No difficulty exists in postulating structures allowing for formal 18-electron configurations around the CO atoms. Stepwise formation of intermediates containing fewer and fewer Co-Co bonds can be envisaged. This study provides a clear example of cluster fragmentation being induced by nucleophilic attack.

$Co_4(CO)_{12}$ is also known to undergo straightforward substitution reactions. The rate of the displacement of the first CO ligand is extremely rapid with P-, As-, and Sb-donor ligands and no kinetic data are available (see below). However, it has been shown that with $P(OMe)_3$ at least one of the subsequent stages of reaction is bimolecular but that it leads to substitution and not fragmentation (67).

Studies of thermal reactions of other tetranuclear metal carbonyl clusters seem to be rather rare. Reports on some interesting reactions of complexes such as $H_4Ru_4(CO)_{12}$ (68, 69) have recently appeared as have descriptions of fragmentation kinetics of complexes such as $H_2Ru_4(CO)_{13}$ (70).

Reactivities and the Strengths of Metal-Metal Bonds in Organometallic Compounds

Estimates of metal-metal bond energies of organometallic complexes in solution have been made by measurements of $\Delta H°$ for homolytic fission of $(\eta^3\text{-}C_3H_5)_2Fe_2(CO)_4L_2$ (see above) (9) and a value of $\Delta H^‡ = 96$ kJ mol^- for homolytic fission of $Cp_2Fe_2(CO)_4$ in benzene has been reported (71). The latter was derived from nmr monitoring of the scrambling reactions between $(\eta^5\text{-}C_5H_5)_2Fe_2(CO)_4$

and $(\eta^5\text{-}C_5H_4COMe)_2Fe_2(CO)_4$, making the assumption that the acetyl substituent in the cyclopentadienyl ligands did not affect the Fe-Fe bond strength. Mass spectrometric measurements have led to the value $\Delta H° = 234.4 \pm 3.8$ kJ mol^{-1} for the homolytic fission of $Cp_2W_2(CO)_6$ in the gas phase over the temperature range 200-300°C (72). This is the strongest single metal-metal bond quantitatively characterized to date (8).

More needs to be known quantitatively of the strengths of metal-metal bonds in organometallic compounds and of the factors that affect them. We have recently made studies of some reactions of $Cp_2Mo_2(CO)_6$ in order to see what their mechanisms are and to see whether a value can be obtained for the strength of the Mo-Mo bond (73).

Reactivity and Mechanisms of Reactions of $Cp_2Mo_2(CO)_6$. Thermal decomposition of $Cp_2Mo_2(CO)_6$ under Ar is very slow but proceeds with quite clean kinetics under O_2 at 5-100% partial pressures, the first order rate constants being independent of $[O_2]$ and unaffected when the proportion of CO in $CO\text{-}O_2$ mixtures above the solution is varied from 0 to 95%. The activation parameters obtained from rate constants measured from 70-135°C are $\Delta H^{\ddagger} = 135.9 \pm 2.2$ kJ mol^{-1} and $\Delta S^{\ddagger} = 56.2 \pm 5.6$ J K^{-1} mol^{-1}.

Reaction with $C_{16}H_{33}I$ occurs very rapidly at 135°C in thoroughly deoxygenated decalin to form $CpMo(CO)_3I$ in high yield. The reaction is, however, inhibited by traces of O_2 when the rate is the same as with O_2 alone although the product is still $CpMo(CO)_3I$. This is a good indication that the rate-determining step in both the reactions is homolytic fission of the Mo-Mo bond and that the value 135.9 ± 2.2 kJ mol^{-1} can be taken as a precise kinetic measure of the strength of the Mo-Mo bond. This is considerably higher than the kinetic strength of the Fe-Fe bond in $Cp_2Fe_2(CO)_4$ in spite of the larger coordination number of the Mo atoms. However, the strength of the Cr-Cr bond in $Cp_2Cr_2(CO)_6$ is apparently much lower since finite amounts of the monomer are detectable by esr (74). On the other hand, the value $\Delta H° = 234.4$ kJ mol^{-1} for the homolytic fission of $Cp_2W_2(CO)_6$ in the gas phase (72) shows that the strength of the metal-metal bonds increases greatly with increase of atomic weight of the metal in these complexes.

Reaction of $Cp_2Mo_2(CO)_6$ with alkynes leads to the alkyne-bridged complexes $Cp_2Mo_2(CO)_4(C_2R_2)$ and it has been suggested that both the photochemical (75) and thermal (76) reactions go via initial homolytic fission. The kinetic results do not support this. The thermal reaction with C_2Ph_2 proceeds cleanly at 145°C in thoroughly deoxygenated solutions in decalin to form $Cp_2Mo_2(CO)_4(C_2Ph_2)$ in high yield. The reaction proceeds via two paths, one of which is first order in $[C_2Ph_2]$ up to high concentrations while the other increases to a limiting rate. When $[C_2Ph_2] \gtrsim 0.1M$ the observed pseudo-first-order rate constant follows eq 23.

$$k_{obsd} = k_a(\text{lim}) + k_b[C_2Ph_2] \tag{23}$$

The value of k_a(lim) is reduced by CO in such a way as to suggest the mechanism shown in eq 24-26 for which the predicted rate behavior is described in eq 27. At high values of $[C_2Ph_2]$ this would

$$Cp_2Mo_2(CO)_6 \underset{k_{-1}}{\overset{k_1}{\rightleftharpoons}} Cp_2Mo_2(CO)_5 + CO \quad (24)$$

$$Cp_2Mo_2(CO)_5 \underset{k_{-2}}{\overset{k_2}{\rightleftharpoons}} Cp_2Mo_2(CO)_4 + CO \quad (25)$$

$$Cp_2Mo_2(CO)_4 + C_2Ph_2 \xrightarrow{k_3} Cp_2Mo_2(CO)_4(C_2Ph_2) \quad (26)$$

$$k_a = \frac{k_1k_2k_3[C_2Ph_2]}{k_{-1}k_{-2}[CO]^2 + k_{-1}k_3[CO][C_2Ph_2] + k_2k_3[C_2Ph_2]} \quad (27)$$

lead to values of k_a(lim) that should follow eq 28. A plot of

$$1/k_a(\text{lim}) = 1/k_1 + (k_{-1}/k_1k_2)[CO] \quad (28)$$

$1/k_a$(lim) against [CO] is indeed a good straight line (Figure 3) of positive gradient with a finite intercept so the data are entirely consistent with the mechanism proposed.

This reaction scheme clearly parallels that for one path for reaction of C_2Ph_2 with $Co_2(CO)_6(P\text{-}n\text{-}Bu_3)_2$ in which two ligands have to be lost by reversible dissociation before attack by the alkyne can occur (37). In the Mo_2 system the intermediate $Cp_2Mo_2(CO)_4$ is, of course, expected to be identical with the fully characterized Mo≡Mo complex (38) and the energy required for successive dissociation of two CO ligands must be lowered by the ability of the Mo atoms to form multiple bonds. The same factor has been proposed to be operative in the reaction of $Cp_2(CO)_6$-$(P\text{-}n\text{-}Bu_3)_2$ with C_2Ph_2 (see above). $Cp_2Mo_2(CO)_4$ is known to undergo rapid addition of alkynes (77).

An important feature of this reaction is that the value of k_1 derived from the analysis is ca. 10 × lower than the rate constant for homolytic fission obtained by extrapolation. Indeed, during the formation of $Cp_2Mo_2(CO)_4(C_2Ph_2)$ in high yield under an atmosphere of CO homolytic fission <u>and its reverse</u> must be occurring ca. 10^3 times for every event leading to $Cp_2Mo_2(CO)_4(C_2Ph_2)$. This is another example, therefore, of a reaction which proceeds by one path at the same time as another much more facile one is occurring, but occurring reversibly and without leading anywhere.

Another important feature of this reaction has consequences for the synthesis of the alkyne-bridged products. As the reaction temperature is lowered the yield of the product decreases so that at 100°C it is negligible. The product is thermally stable under these conditions and the observations can be explained if it is

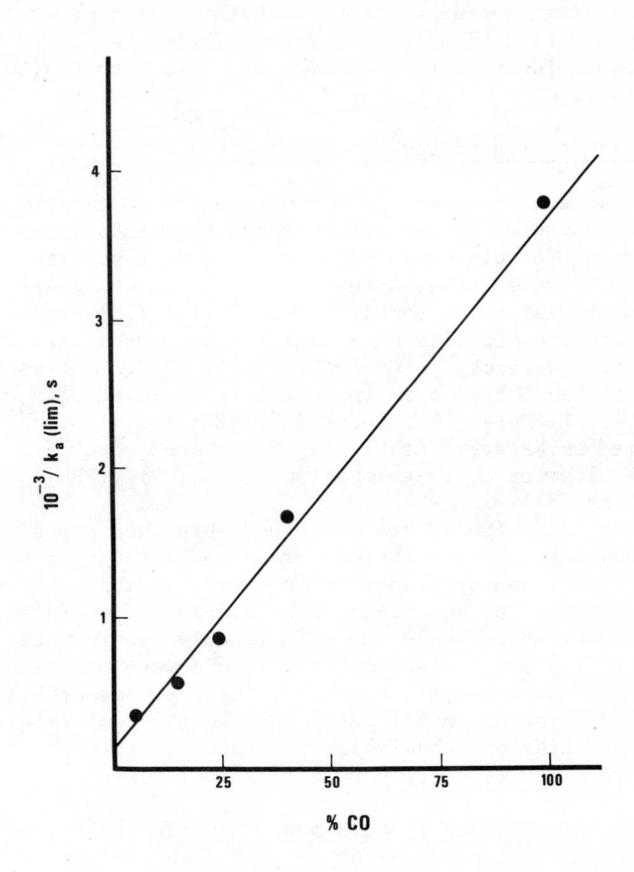

Figure 3. Plot of $1/k_a$(lim) against partial pressure of CO for the reaction of $Cp_2Mo_2(CO)_6$ with C_2Ph_2 in decalin under CO—N_2 mixtures

assumed that the intermediate $Cp_2Mo_2(CO)_5$ can be attacked by C_2Ph_2 in a reaction, leading eventually to decomposition, that has a lower activation enthalpy than dissociative loss of the second CO ligand to form $Cp_2Mo_2(CO)_4$. This proposal is not unreasonable since $Cp_2Mo_2(CO)_5(C_2Ph_2)$ would contain a monodentate, non-bridging C_2Ph_2 ligand and could well be thermally unstable towards decomposition, probably for steric reasons.

The mechanism governing the path that is first order in $[C_2Ph_2]$ is not well defined but it seems likely that it is similar to the corresponding path for reaction of C_2Ph_2 with $Co_2(CO)_6$-$(P\text{-}n\text{-}Bu_3)_2$ (37).

Recent Kinetic Studies of Some Co_4 Clusters

Apart from the reaction of $Co_4(CO)_{12}$ under CO to form Co_2-$(CO)_8$ mentioned above (66) no quantitative kinetic studies of fragmentation of Co_4 clusters appear to have been reported. Reaction of $Co_4(CO)_{12}$ with PPh_3, $AsPh_3$, or $SbPh_3$ at room temperature led to the monosubstituted complexes $Co_4(CO)_{11}L$ (78) whereas attempts to prepare clusters more highly substituted with PPh_3 appeared to lead directly to $Co_2(CO)_6(PPh_3)_2$ (79). Tetrasubstituted clusters $Co_4(CO)_8L_4$ seem to have been isolated only with L = $P(OPh)_3$ (80), $P(OMe)_3$ (79), and $P(OCH_2)_3CEt$ (79). Since these results seemed to be attributable to the greater size of PPh_3 a study of the kinetics of fragmentation of $Co_4(CO)_{11}(PPh_3)$ has been undertaken (81).

Solutions of $Co_4(CO)_{12}$ are quite unstable when exposed to air or light. However, stock solutions prepared by weighing under N_2 in very dim light, and by using "Schlenk-tube" techniques and thoroughly deoxygenated solutions, were stable for weeks when stored in a refrigerator under CO. Thoroughly deoxygenated solutions of $Co_4(CO)_{12}$ and the appropriate ligand were mixed in foil-wrapped Schlenk tubes under controlled, inert atmospheres and in dim light, and were thermostatted at appropriate temperatures. Samples for spectroscopic analysis were expelled through stainless steel tubes by a positive pressure of the inert gas.

Spectroscopic Changes in Solution. Reaction of ca. 10^{-3}M $Co_4(CO)_{12}$ with a tenfold excess of PPh_3 or $AsPh_3$ in dichloroethane (DCE) or n-heptane at room temperature led immediately to spectra characteristic of $Co_4(CO)_{11}L$ (78, 79). An attempt was made to measure the rate of reaction with PPh_3 by monitoring the spectroscopic changes at 305 nm with a stopped-flow apparatus but the reaction was too fast to measure even when $[PPh_3]$ was as low as 5×10^{-4}M. Solutions of $Co_4(CO)_{11}(AsPh_3)$ underwent no further spectroscopic changes over 75 min. at room temperature and a further 2 h at 45°C. However, $Co_4(CO)_{11}(PPh_3)$ reacted over 15-20 min. in n-heptane at room temperature to form a new species. Reaction was somewhat faster in DCE and in both cases the rates were observed to be greater at higher values of $[PPh_3]$. The product could be isolated from either n-heptane or DCE by evaporation of solvent with a stream of Ar and was recrystallized from benzene-pentane. The solid was not very stable to light and air but could be kept unchanged in the dark under an inert atmosphere. Analysis was in agreement with that expected for $Co_4(CO)_8(PPh_3)_4$ (Calc.: Co, 15.62; C, 63.68; H, 4.00; P, 8.21%. Found: Co, 15.78; C, 62.52; H, 4.40; P, 8.41%). The instability of solutions of the complex other than in the dark and under a vigorously deoxygenated atmosphere vitiated attempts to measure the molecular weight. However, the analysis was supported by measurement of the IR spectra of solutions of $Co_4(CO)_{12}$ in DCE which had been allowed to come to equilibrium with varying amounts of PPh_3. Bands due to $Co_4(CO)_{11}(PPh_3)$ grew in intensity with increasing $[PPh_3]$ and then decreased as bands due to the product complex grew. Close to 4 times as much PPh_3 was required to remove all of the $Co_4(CO)_{11}(PPh_3)$ as was required to produce it in the absence of more than traces of $Co_4(CO)_{12}$ or new product, and the spectrum of the solutions remained unchanged when $[PPh_3]$ was increased above this amount. The IR spectra of the product in various solvents are shown in Table II.

Reaction of $Co_4(CO)_{12}$ with dppm ($Ph_2PCH_2PPh_2$) led within 40 min at room temperature in DCE to a new species, the spectrum of which remained unchanged after 4 h at 60°C. After evaporation of the solvent and recrystallization from benzene-pentane the product showed a spectrum in CS_2 very similar to that of $Rh_4(CO)_8dppm_2$ (82). An equally similar spectrum was obtained after the thermal reaction of $Ir_4(CO)_{12}$ with dppm in toluene or their photochemical reaction in DCE (Table II). The product of the reaction with $Co_4(CO)_{12}$ must, therefore, be $Co_4(CO)_8dppm_2$. Its spectrum shows bands almost identical with those assigned to $Co_4(CO)_8(PPh_3)_4$, albeit with different relative intensities, and this provides additional support for the latter's formulation. The bands for $Co_4(CO)_8(PPh_3)_4$ are also in the expected positions relative to those for $Co_4(CO)_8\{P(OMe)_3\}_4$ prepared *in situ* in DCE.

The solubility of $Co_4(CO)_8(PPh_3)_4$ in pentane or heptane was low so that it usually began to precipitate soon after being prepared *in situ*. Solubility in DCE was appreciably greater and further reactions were followed in this solvent. It was stable at room temperature for some time and further reaction required temperatures above ca. 45°C for it to occur at rates convenient for kinetic study. This further reaction was accompanied by a decrease in the intensity of the bands at 2010 and 1890 cm^{-1} and an erratic growth of the band at 1955 cm^{-1}. This was because the product $Co_2(CO)_6(PPh_3)_2$ also absorbs at this wavelength but begins to precipitate from solution after the reaction is ca. 60% complete. The kinetics were therefore followed by monitoring the decreasing intensity of the band at 2010 cm^{-1}.

Kinetics of the Reaction $Co_4(CO)_8(PPh_3)_4 \xrightarrow{PPh_3} Co_2(CO)_6(PPh_3)_2$. The variation of the observed pseudo-first-order rate constant

Table II

IR Spectra of Some $M_4(CO)_8L_4$ Complexes

$Co_4(CO)_8(PPh_3)_4$ [a]		2005s	1972w	1946s	1912w	1882s	1810w		
$Co_4(CO)_8(PPh_3)_4$ [b]		2005s	1965w	1945s	1910w	1880s	1810m		1755w
$Co_4(CO)_8(PPh_3)_4$ [c]		2010s	1970w	1950s		1890s	1820w		1765w
$Co_4(CO)_8\{P(OMe)_3\}_4$ [d]		2018m	1980s	1950s				1792s	
$Co_4(CO)_8(etpb)_4$ [e]		2030m	1992vs					1805m	
$Co_4(CO)_8(dppm)_2$ [f]	2070w	2016s	1975s	1950m		1890w	1830w		1780s
$Rh_4(CO)_8(dppm)_2$ [g]	2066m	2026s	1996s				1818m		1792m
$Ir_4(CO)_8(dppm)_2$ [h]		2010s	1980s	1960sh	1892w	1812m	1790s		1765sh
$Ir_4(CO)_8(dppm)_2$ [i]		2015s	1977s	1950sh	1848w	1780m	1770s		1730w
$Ir_4(CO)_8(dppe)_2$ [j]	2064vw	2004vs	1979vs	1958vs		1830m	1776s		1752s

[a] In Nugol. [b] In heptane. [c] In $C_2H_2Cl_4$. [d] In hexadecane (ref. 79). [e] In CH_2Cl_2 (ref. 79). [f] In CS_2. [g] In CS_2 (ref. 82). [h] Prepared *in situ* by photochemical reaction in $C_2H_2Cl_4$. [i] Prepared *in situ* by thermal reaction in toluene. [j] In $CHCl_3$. Cattermole, P.E.; Orell, K.G.; Osborne, A.G. *J. Chem. Soc., Dalton Trans.*, 1974, 328.

with [PPh_3] are shown in Figure 4 for reactions under N_2 and CO. The increase in rate when [PPh_3] is decreased below ca. 0.04M suggests that a PPh_3-dissociative path is available as a rate-determining step under both N_2 and CO. The data in this region were not very reproducible and the kinetics were not followed in detail. When [PPh_3] >> 0.05M this path is fully suppressed and reaction can occur by a path inhibited by CO which presumably involves rate-determining CO dissociation. This reaction can be completely suppressed by CO, the rates with 0.05M PPh_3 under 40 and 100% CO in CO-N_2 mixtures being the same. The remaining reaction is independent of [CO] and [PPh_3] and the rate must be characteristic of spontaneous fragmentation. The activation parameters $\Delta H^{\ddagger} = 146.8 \pm 6.5$ kJ mol^{-1} and $\Delta S^{\ddagger} = 114.3 \pm 19.3$ J K^{-1} mol^{-1} are therefore a good kinetic measure of the susceptibility of this Co_4 cluster towards spontaneous fragmentation. The activation parameters for the path inhibited by CO were obtained by subtracting the rate constants for the spontaneous fragmentation from those found for reaction under N_2: $\Delta H^{\ddagger}_{CO} = 118.4 \pm 7.7$ kJ mol^{-1} and $\Delta S^{\ddagger}_{CO} = 39.8 \pm 23.3$ J K^{-1} mol^{-1}.

Although there are no directly comparable data it is clear that $Co_4(CO)_8(PPh_3)_4$ is much less stable than $Co_4(CO)_8dppm_2$ and $Co_4(CO)_8\{P(OMe)_3\}_4$ which are both virtually unaffected after several hours at 60°C whereas $Co_4(CO)_8(PPh_3)_4$ has a half-life of only 50 min at 60°C in DCE in an excess of PPh_3 and under N_2. It would appear that steric effects play an important role in determining the stability of these Co_4 clusters.

Summary

(1) Thorough mechanistic studies are capable of demonstrating when rate-determining fragmentation of metal-metal-bonded carbonyls is occurring. It is then possible to obtain activation parameters for spontaneous fragmentation processes that are excellent estimates of the reactivity of the complexes. The activation enthalpies provide precise and useful estimates of the strengths of the metal-metal bonds and enable the effects of substituents to be determined. Substituents can act either by weakening the metal-metal bonds in the reactant complexes (so increasing the reactivity) or by weakening bonds in intermediates (and so decreasing the reactivity).

(2) Spontaneous fragmentation can also occur by more than one path, i.e. to form different types of fragment, and the energetics of the two paths can differ greatly.

(3) Fragmentation is often preceded by ligand dissociation and the energetics can depend greatly on the way in which the metal-metal-bonded system can adjust to the coordinative unsaturation. Clusters have many more ways of doing this than mononuclear carbonyls and this type of behavior can also affect the energetics of substitution reactions as well as fragmentation.

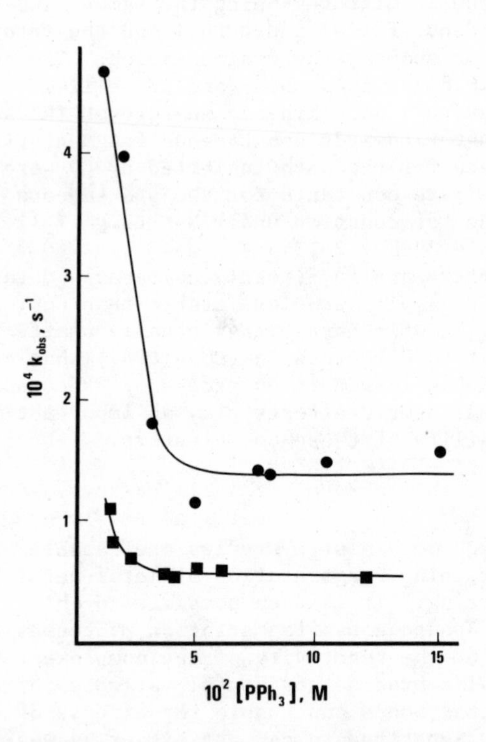

Figure 4. Values of k_{obsd} *for the reaction* $Co_4(CO)_8(PPh_3)_4 \xrightarrow{PPh_3} Co_2(CO)_6(PPh_3)_2$ *in DCE at 59.5°C: (●) reactions under* N_2*; (■) reactions under CO*

(4) Substitution reactions can occur by paths that involve fragmentation as rate-determining steps.

(5) Metal clusters are quite susceptible to nucleophilic attack, the nucleophile making use of the LUMOs on the cluster. This type of attack often simply leads to substitution but it can also lead to fragmentation of the cluster.

(6) Metal clusters are also susceptible to fragmentation through electrophilic attack (e.g. by halogens). Rates can be very rapid indeed but are very dependent on the nature of any substituents.

(7) Reactions of $Cp_2Mo_2(CO)_6$ illustrate some of the above points. Homolytic fission can be shown to occur but the overall reaction is very slow in the absence of suitable scavengers. Under these conditions it can be shown that the substitution of two CO ligands by C_2Ph_2 occurs not via radical intermediates but via $Cp_2Mo_2(CO)_4$ formed by successive dissociation of the two CO ligands before attack by the alkyne.

(8) $Co_4(CO)_{12}$ forms $Co_4(CO)_8(PPh_3)_4$ readily at room temperature. This complex is unstable to fragmentation at $\gtrsim$45°C. Spontaneous, CO-dissociative, and PPh_3-dissociative paths have been shown to lead to fragmentation and activation parameters have been obtained for the first two. They suggest that spontaneous fragmentation involves a high degree of bond weakening in the Co_4 cluster, and that this is due to steric effects. $Co_4(CO)_8(dppm)_2$ and $Co_4(CO)_8\{P(OMe)_3\}_4$ are not nearly as easily fragmented.

Literature Cited

1. Dahl, L.F.; Ishishi, E.; Rundle, R.E. J. Chem. Phys., 1957, 26, 1750.
2. Powell, H.M.; Ewens, R.V.G. J. Chem. Soc., 1939, 286.
3. Mills, O.S.; Robinson, G. Proc. Chem. Soc., 1959, 156.
4. Cotton, F.A.; Monchamp, R.R. J. Chem. Soc., 1960, 533.
5. Bidinosti, D.R.; McIntyre, N.S. J. Chem. Soc., Chem. Commun., 1966, 555.
6. Haines, L.I.B.; Hopgood, D.; Poë, A.J. J. Chem. Soc. A, 1968, 421.
7. Bidinosti, D.R.; McIntyre, N.S. Can. J. Chem., 1970, 48, 593.
8. Connor, J.A. Top. Current Chem., 1977, 71, 71.
9. Muetterties, E.L.; Sosinsky, B.A.; Zamariev, K.I. J. Amer. Chem. Soc., 1975, 97, 5299.
10. Battison, G.; Sbrignadello, G.; Bor, G.; Connor, J.A. J. Organomet. Chem., 1977, 131, 445.
11. Angelici, R.J. Organometal. Chem. Rev., 1968, 3, 173.

12. Jackson, R.A.; Poë, A.J. Inorg. Chem., 1978, 17, 997.
13. Fawcett, J.P.; Poë, A.J.; Sharma, K.R. J. Chem. Soc., Dalton Trans., 1979, 1886.
14. Fawcett, J.P.; Jackson, R.A.; Poë, A.J. J. Chem. Soc., Dalton Trans., 1978, 789.
15. Fawcett, J.P.; Poë, A.J.; Twigg, M.V. J. Organomet. Chem., 1973, 51, C17; Fawcett, J.P.; Poë, A.J.; Sharma, K.R. J. Amer. Chem. Soc., 1976, 98, 1401.
16. Fawcett, J.P.; Poë, A.J. J. Chem. Soc., Dalton Trans., 1977, 1302.
17. Fawcett, J.P.; Poë, A.J. J. Chem. Soc., Dalton Trans., 1976, 2039.
18. Wawersik, H.; Basolo, F. Inorg. Chim. Acta, 1969, 3, 113.
19. Sonnenburger, D.; Atwood, J.D. J. Amer. Chem. Soc., 1980, 102, 3484.
20. Fawcett, J.P.; Jackson, R.A.; Poë, A.J. J. Chem. Soc., Chem. Commun., 1975, 733.
21. Chowdhury, D.M.; Poë, A.J.; Sharma, K.R. J. Chem. Soc., Dalton Trans., 1977, 2352.
22. DeWit, D.G.; Fawcett, J.P.; Poë, A.J. J. Chem. Soc., Dalton Trans., 1976, 528.
23. Jackson, R.A.; Poë, A.J. Inorg. Chem., 1979, 18, 3331.
24. Cotton, F.A.; Wilkinson, G. "Advanced Inorganic Chemistry", Wiley-Interscience, New York, 3rd Edn., 1972, 547.
25. Fawcett, J.P.; Poë, A.J.; Twigg, M.V. J. Chem. Soc., Chem. Commun., 1973, 267.
26. Quicksall, C.O.; Spiro, T.G. Inorg. Chem., 1969, 8, 2363.
27. Spiro, T.G. Prog. Inorg. Chem., 1970, 11, 17.
28. Levenson, R.A.; Gray, H.B.; Ceasar, G.P. J. Amer. Chem. Soc., 1970, 92, 3653.
29. Poë, A.J.; Jackson, R.A. Inorg. Chem., 1978, 17, 2330.
30. Haines, L.I.B.; Poë, A.J. J. Chem. Soc. A, 1969, 2826.
31. Kramer, G.; Patterson, J.; Poë, A.J.; Ng, L. Inorg. Chem., 1980, 19, 1161.
32. Kramer, G.; Patterson, J.; Poë, A.J. J. Chem. Soc., Dalton Trans., 1979, 1165.
33. Hopgood, D.J. Ph.D. Thesis, London University, 1966.
34. Wegman, R.W.; Brown, T.L. J. Amer. Chem. Soc., 1980, 102, 2495.
35. Brown, T.L.; Forbus, N.P.; Wegman, R.W. Abstract 239, Division of Inorganic Chemistry, Proceedings of the Second Chemical Congress of the North American Continent, Las Vegas, 1980.
36. Ellgen, P.C. Inorg. Chem., 1972, 11, 691.
37. Basato, M.; Poë, A.J. J. Chem. Soc., Dalton Trans., 1974, 607.
38. Klingler, R.J.; Butler, W.; Curtis, M.D. J. Amer. Chem. Soc., 1975, 97, 3535.
39. Colton, R.; Commons, C.J. Aust. J. Chem., 1975, 28, 1673.
40. Cobb, M.A.; Poë, A.J. Unpublished observations.

41. Ref. 24, p. 777.
42. Tyler, D.R.; Schmidt, M.A.; Gray, H.B. J. Amer. Chem. Soc., 1979, 101, 2753.
43. Barrett, P.F.; Poë, A.J. J. Chem. Soc. A, 1968, 429.
44. Barrett, P.F. Can. J. Chem., 1974, 52, 3773, and references therein.
45. Basato, M.; Fawcett, J.P.; Fieldhouse, S.A.; Poë, A.J. J. Chem. Soc., Dalton Trans., 1974, 1856.
46. Basato, M.; Fawcett, J.P.; Poë, A.J. J. Chem. Soc., Dalton Trans., 1974, 1350.
47. Bor, G.; Dietler, U.K. Unpublished work.
48. Keeton, D.P.; Malik, S.K.; Poë, A.J. J. Chem. Soc., Dalton Trans., 1977, 233.
49. Malik, S.K.; Poë, A.J. Inorg. Chem., 1978, 17, 1484.
50. E.g., Calderon, J.L.; Fonatana, S.; Frauendorfer, E.; Day, V. W.; Iske, S.D.A. J. Organomet. Chem., 1974, 64, C16.
51. Malik, S.K.; Poë, A.J. Inorg. Chem., 1979, 18, 1241.
52. Keeton, D.P.; Malik, S.K.; Poë, A.J. J. Chem. Soc., Dalton Trans., 1977, 1392.
53. Malik, S.K.; Poë. A.J. Unpublished observations.
54. Poë, A.J.; Twigg, M.V. Inorg. Chem., 1974, 13, 2982.
55. Candlin, J.P.; Shortland, A.C. J. Organomet. Chem., 1969, 16, 289.
56. Poë, A.J.; Twigg, M.V. J. Chem. Soc., Dalton Trans., 1974, 1860.
57. Cotton, F.A.; Wilkinson, G. "Advanced Inorganic Chemistry", Wiley-Interscience, New York, 4th Edn., 1980, p. 1204.
58. Basolo, F.; Pearson, R.G. "Mechanisms of Inorganic Reactions", Wiley, New York, 2nd Edn., 1967, pp. 571-578.
59. Basato, M.; Poë, A.J. J. Chem. Soc., Dalton Trans., 1974, 456.
60. Cobb, M.A.; Hungate, B.; Poë, A.J. J. Chem. Soc., Dalton Trans., 1976, 2226.
61. Aime, S.; Gervasio, G.; Rossetti, R.; Stanghellini, P.L. Inorg. Chim. Acta, 1980, 40, 131, and references therein.
62. Karel, K.J.; Norton, J.R. J. Amer. Chem. Soc., 1974, 96, 6812.
63. Sonnenburger, D.C.; Atwood, J.D. Abstract 124, Division of Inorganic Chemistry, Proceedings of the Second Chemical Congress of the North American Continent, Las Vegas, 1980.
64. Drakesmith, A.J.; Whyman, R. J. Chem. Soc., Dalton Trans., 1973, 362.
65. Adkins, H.; Krsek, G. J. Amer. Chem. Soc., 1948, 70, 383.
66. Bor, G.; Dietler, U.K.; Pino, P.; Poë, A.J. J. Organomet. Chem., 1978, 154, 301.
67. Darensbourg, D.J.; Incorvia, J.J. J. Organomet. Chem., 1979, 171, 89.
68. Ungermann, C.; Landis, V.; Moya, S.A.; Cohen, C.; Walker, H.; Pearson, R.G.; Rinker, R.G.; Ford, P.C. J. Amer. Chem. Soc., 1979, 101, 5923.

69. Walker, H.W.; Kresge, C.T.; Ford, P.C.; Pearson, R.G. J. Amer. Chem. Soc., 1979, 101, 7428.
70. Fox, J.R.; Gladfelter, W.L.; Geoffroy, G.L. Inorg. Chem., 1980, 19, 2574.
71. Cutler, A.R.; Rosenblum, M. J. Organomet. Chem., 1976, 120, 87.
72. Krause, J.R.; Bidinosti, D.R. Can. J. Chem., 1975, 53, 628.
73. Amer, S.; Kramer, G.; Poë, A.J. Unpublished data.
74. Adams, R.D.; Collins, D.E.; Cotton, F.A. J. Amer. Chem. Soc., 1974, 96, 749.
75. Ginley, D.S.; Bock, C.R.; Wrighton, M.S. Inorg. Chim. Acta, 1977, 23, 85.
76. Curtis, M.D.; Klingler, R.J., J. Organomet. Chem., 1978, 161, 23.
77. Bailey, W.I.; Chisholm, M.H.; Cotton, F.A.; Rankel, L.A. J. Amer. Chem. Soc., 1978, 100, 5764.
78. Cetini, G.; Gambino, O.; Rossetti, R.; Stanghellini, P.L. Inorg. Chem., 1968, 7, 609.
79. Labroue, D.; Poilblanc, R. Inorg. Chim. Acta, 1972, 6, 387.
80. Sartorelli, U.; Canziani, F.; Martinengo, S.; Chini, P. "Proceedings of 12th International Conference on Coordination Chemistry", 1970, p. 144.
81. Huq, R.; Poë, A.J. Unpublished work.
82. Carré, F.H.; Cotton, F.A.; Frenz, B.A. Inorg. Chem., 1976, 15, 381.

RECEIVED December 11, 1980.

8

Metal–Metal Bond Making and Breaking in Binuclear Complexes with Phosphine Bridging Ligands

ALAN L. BALCH

Department of Chemistry, University of California—Davis, Davis, CA 95616

Polyfunctional phosphine ligands make useful bridging ligands for constructing binuclear metal complexes. This article is concerned with the use of two ligands - bis(diphenylphosphino)methane (dpm) and 2-diphenylphosphinopyridine (Ph_2Ppy)-in promoting metal-metal bond forming and breaking reactions. Three specific topics will be covered. These are the ability of dpm to modify the stability of Rh-Rh bonds in interrelated compounds of Rh(I), Rh(II) and Rh(III), the utility of dpm bridging ligands in allowing novel metal-metal bond opening and closing in low valent rhodium and palladium complexes, and the usefulness of Ph_2Ppy in the stepwise construction of binuclear complexes.

Modification of Metal-Metal Bonding in Rhodium Complexes by a Bridging Diphosphine. The yellow, planar complexes, $(RNC)_4Rh^+$, undergo novel self-association reactions in concentrated solution to form the blue or violet dimers, $(RNC)_8Rh_2^{2+}$, via reaction (1) (1,2). The equilibrium constants for this reaction are strongly

(1) $$2\,[\mathrm{RhL_4}]^{+} \longrightarrow [\mathrm{L_4Rh\text{-}\text{-}\text{-}RhL_4}]^{2+}$$

L = RNC

influenced by solvent and by the size of the substituent on the isocyanide ligand. Not unexpectedly $(t\text{-}BuNC)_4Rh^+$ with its bulky substituents is most reluctant to dimerize. Substitution of the bidentate dpm for half of the isocyanide ligands produces the blue, dimeric cation 1. (3) These dimers, which can be prepared with various isocyanide substituents including t-butyl, show no tendency to dissociate into monomers. Similar stabilization of the dimeric structure can be achieved by the use of bifunctional isocyanides like 1,3-diisocyanopropane. (4,5) These form qua-

0097-6156/81/0155-0167$05.00/0

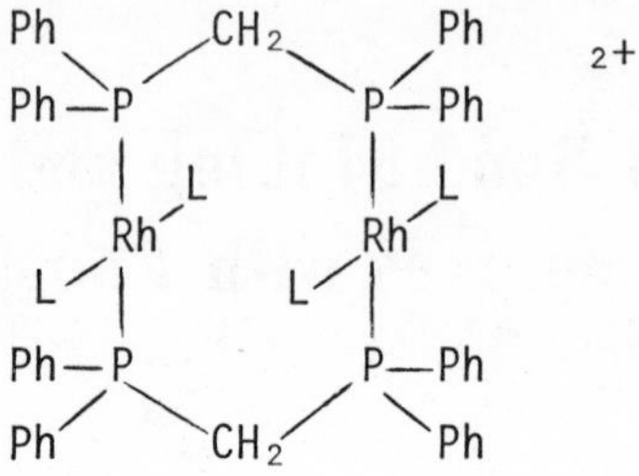

1

druply bridged cations 2. The color and stability of the Rh-Rh bonds in $(RNC)_8Rh_2^{2+}$, 1 and 2 may be understood by reference to

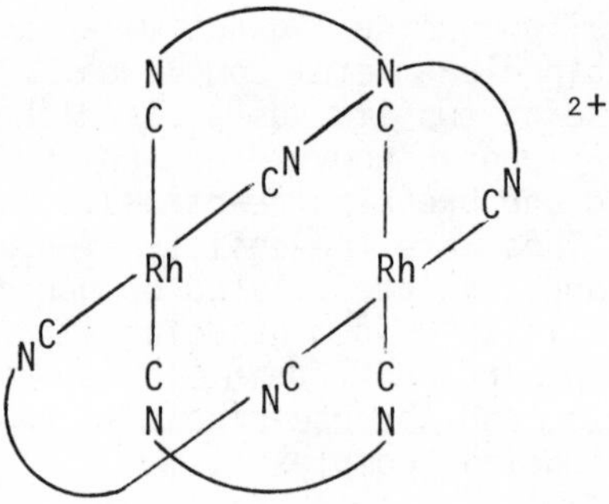

2

the molecular orbital diagram in Figure 1. This diagram shows the interaction of the out-of-plane orbitals, which are the orbitals of the complex most strongly affected by self-association. The occupied orbitals of the dimer are stabilized since the upper and lower sets of dimer orbitals, which are derived from the monomer a_{1g} and a_{2u} orbitals respectively, have the same symmetries. Consequently, the net Rh-Rh interaction is bonding; although the formally antibonding Rh-Rh σ orbital is doubly occupied. The blue color results from a lowering of the HOMO-LUMO gap upon dimerization. The spectra of a related pair of complexes, monomeric $(MeNC)_2Rh(PPh_3)_2^+$ and dimeric $(MeNC)_4Rh_2(dpm)_2^{2+}$, are compared in Figure 2. The lowest allowed transition in the monomer results from the $d_{z^2} \rightarrow L\pi^*(a_g \rightarrow 1b_{1u})$ excitation. In the dimers, the transition $1b_{1u} \rightarrow 2a_g$ is involved, and the energy gap is smaller.

Not only does the incorporation of the dpm ligand into 1 inhibit dissociation, it also prohibits further Rh-Rh bond formation. The cations $(RNC)_4Rh^+$ can further associate to form trimers via reaction (2). Additionally the isocyanide bridge complex 2 also self associates via reaction (3) to form a tetrarhodium species. However the dpm-bridged dimers, 1, show no evidence for self association even at the highest concentrations. Undoubtedly

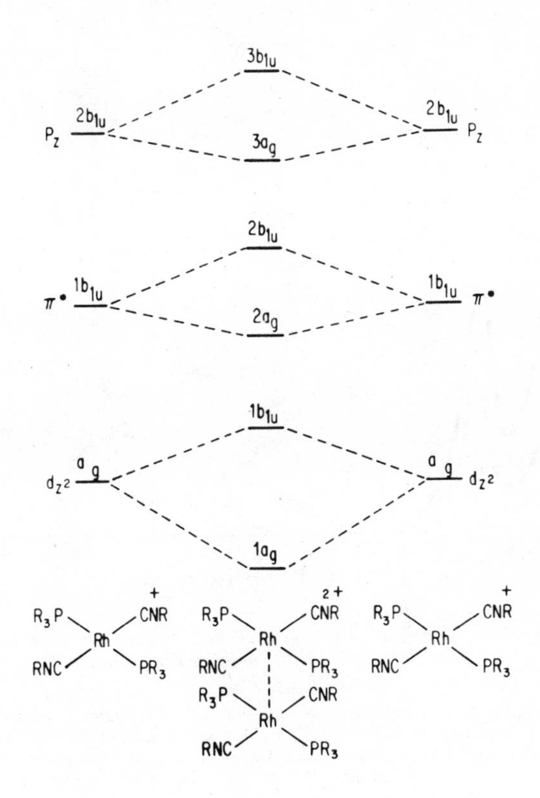

Figure 1. Partial molecular orbital diagram for the interaction of two planar Rh(I) complexes

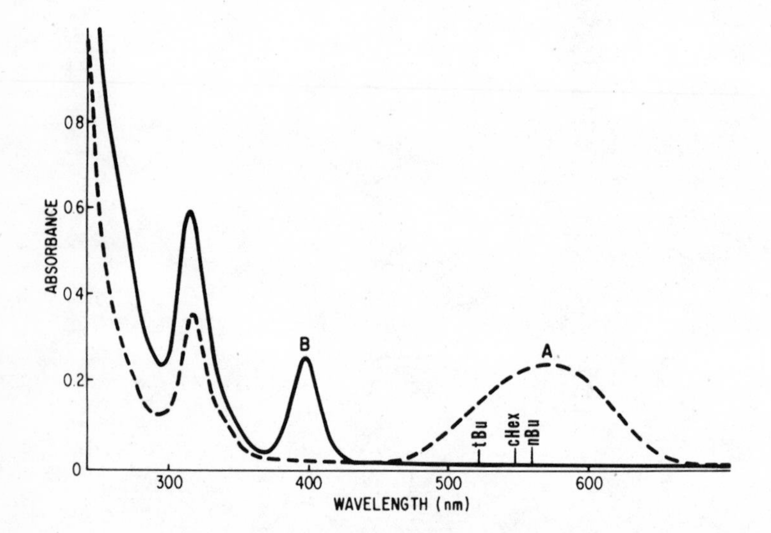

Figure 2. Electronic spectrum of (A) a 0.20-mL acetonitrile solution of $[(CH_3NC)_4Rh_2(dpm)_2][PF_6]_2$ and (B) a 0.40-mL acetonitrile solution of $[(CH_3NC)_2Rh(PPh_3)_2[PF_6]_2$.

Each solution is contained in a 1.0-mm pathlength cell. The markers show the position of the low-energy maxima for the cases where the isocyanide substituent is changed to n-*butyl, cyclohexyl, and* t-*butyl. Increasing the bulk of the isocyanide substituent increases the Rh · · · Rh separation and increases the energy of the $^1b_{1u} \rightarrow 2a_g$ transition.*

$$(2) \qquad [L_4Rh]^{+} + [L_4Rh\text{---}RhL_4]^{2+} \rightleftharpoons [L_4Rh\text{---}RhL_4\text{---}RhL_4]^{3+}$$

the bulky phenyl groups at each end of these dimers prohibits the approach of a unit along the Rh-Rh vector.

$$(3) \qquad 2\,[Rh\text{---}Rh]^{2+} \rightleftharpoons [Rh\text{---}Rh\text{---}Rh\text{---}Rh]^{4+}$$

The same pattern of dimer stabilization at the expense of monomeric or more highly polymerized forms is also evident in the behavior of these rhodium complexes toward oxidation. Treatment of the dimer 1 with iodine, bromine, or trifluoromethyl disulfide results in trans-annular, oxidative-addition to form the brown dimers, 3, via reaction (4) (3,6). As inspection of Figure 1 shows

$$(4) \qquad [L_2Rh(\mu\text{-}Ph_2PCH_2PPh_2)_2RhL_2]^{2+} + X_2 \rightleftharpoons [XL_2Rh(\mu\text{-}Ph_2PCH_2PPh_2)_2RhL_2X]^{2+}$$

$X = I_2, Br_2, (SCF_3)_2$

3

such two-electron oxidation should strengthen and shorten the Rh-Rh bond by removing electrons from the primarily antibonding $1b_{1u}$ orbital. These dimers, once formed, are stable to the presence of additional quantities of oxidant. That is, they cannot be readily converted into Rh(III) complexes. In contrast oxidation of $(RNC)_4Rh^+$ gives a variety of products whose formation can be controlled by limiting the amount of oxidant used. (7-10) The reactions involved are shown in equations 5-7. The dinuclear and trinuclear rhodium cations have been structurally characterized by X-ray crystallography. (8,10) The structure of the trinuclear cation is shown in Figure 3. These metal-metal bonded products

$$(5)\quad \mathrm{[L{-}RhL_2{-}L]^{+} + X_2 \longrightarrow [X{-}RhL_4{-}X]^{+}}$$

$$(6)\quad \mathrm{2\,[L{-}RhL_2{-}L]^{+} + X_2 \longrightarrow [X{-}RhL_4{-}RhL_4{-}X]^{2+}}$$

$$(7)\quad \mathrm{3\,[L{-}RhL_2{-}L]^{+} + X_2 \longrightarrow [X{-}RhL_4{-}RhL_4{-}RhL_4{-}X]^{3+}}$$

are subject to dissociative disproportionation via reactions (8) and (9). The complexity of equations 5-9 contrasts with the

$$(8)\quad \mathrm{[X{-}RhL_4{-}RhL_4{-}RhL_4{-}X]^{3+} \rightleftharpoons [X{-}RhL_4{-}RhL_4{-}X]^{2+} + [L{-}RhL_2{-}L]^{+}}$$

$$(9)\quad \mathrm{[X{-}RhL_4{-}RhL_4{-}X]^{2+} \rightleftharpoons [X{-}RhL_4{-}X]^{+} + [L{-}RhL_2{-}L]^{+}}$$

simplicity of oxidative chemistry shown by reaction (4) for the dpm-bridged dimer 1. The oxidation of the isocyanide-bridged dimer 2 also is more complex than that of 1 since 2 lacks substituents to block intraionic association. Consequently both dinuclear and tetranuclear complexes, 4 (11) and 5 (12), have been obtained upon oxidation of 2.

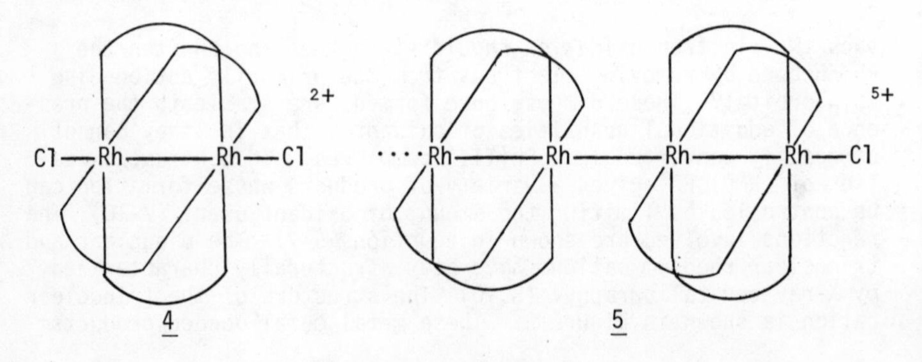

Metal-Metal Bond Formation and Rupture in Diphosphine Bridged Dinuclear Metal Complexes

The flexibility of the bridging dpm ligand allows for considerable variation in the metal-metal separation. The range of metal-metal separations runs from 2.138(1) Å in the quadruple metal-metal bonded $Mo_2(dpm)_2Cl_4$ (13) to 3.492(1) Å in the A-frame $Pd_2(dpm)_2(\mu\text{-}C_2\{CF_3\}_2)_2Cl_2$. (14,15) This flexibility allows metal-metal bond making and breaking to occur without other disruptions within a variety of binuclear complexes.

The dimeric Pd(I) complex $Pd_2(dpm)_2X_2$ involves two planar metal centers linked by a metal-metal bond. The Pd-Pd separation in these molecules is about 2.7 Å. (16,17) These dimers readily undergo insertion of small molecules to give molecular A-frames via reaction (10). (14,18-22) Carbon monoxide and sulfur dioxide

(10) $X\text{—}Pd(\mu\text{-}Ph_2PCH_2PPh_2)_2Pd\text{—}X + Y \longrightarrow XPd(\mu\text{-}Y)(\mu\text{-}Ph_2PCH_2PPh_2)_2PdX$

Pd-Pd ∿ 2.7 Å Pd···Pd ∿ 3.2 Å

insert reversibly into the Pd-Pd bond, whereas the insertion of isocyanides, diazonium ions and sulfur atoms (from cyclooctasulfur or an episulfide) are irreversible processes. In these reactions one atom of the inserted ligand is placed between the two metals. A representative structure of one such adduct, $Pd_2(dpm)_2(\mu\text{-}SO_2)Cl_2$ is shown in Figure 4. As a result of reaction (10) the metal-metal bond breaks and the Pd-Pd separation increases by about 0.5 Å. These dimers can also tolerate the insertion of a two atom fragment into the Pd-Pd bond. Acetylenes bearing electron withdrawing substituents add to $Pd_2(dpm)_2Cl_2$ via reaction (11)

(11) $Cl\text{—}Pd(\mu\text{-}Ph_2PCH_2PPh_2)_2Pd\text{—}Cl + CF_3C{\equiv}CCF_3 \longrightarrow ClPd(\mu\text{-}C(CF_3){=}C(CF_3))(\mu\text{-}Ph_2PCH_2PPh_2)_2PdCl$

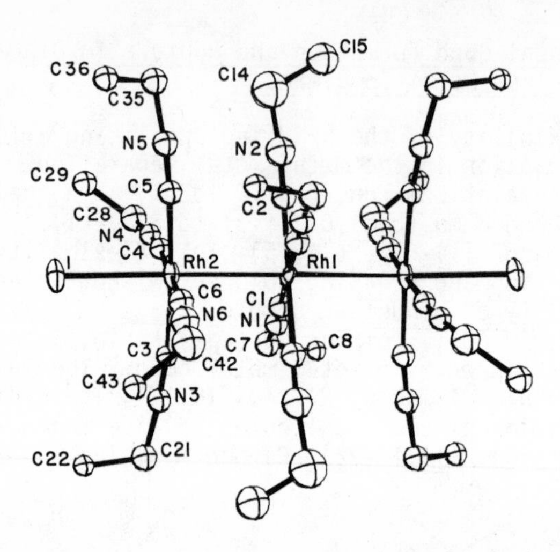

Journal of the American Chemical Society

Figure 3. An ORTEP drawing of $[(C_6H_5CH_2NC)_{12}Rh_3I_2]^{3+}$ showing 50% thermal ellipsoids (10). Bond lengths follow: Rh—Rh, 2.796(1), Rh—I, 2761(1) Å. The Rh—Rh—I angle is 175.5(1)°. To avoid cluttering of the drawing, only the first carbon atom of each phenyl group has been shown.

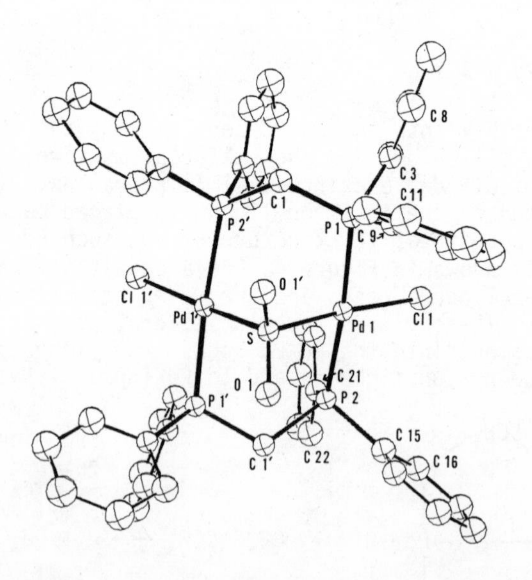

Inorganic Chemistry

Figure 4. ORTEP drawing of $Pd_2(dpm)_2(\mu\text{-}SO_2)Cl_2$, Molecule B, showing 50% thermal ellipsoids (20). The Pd—Pd separation is 3.220(4) Å.

to give a dpm-bridged product with the largest metal-metal separation yet found. Some similar reactions have been reported for the isoelectronic platinum dimer, $Pt_2(dpm)_2Cl_2$. (20,23,24)

The behavior of Rh(I) dpm-bridged complexes offers an interesting counterpoint to the Pd(I) based chemistry. For example carbon monoxide addition to the A-frame $Rh_2(dpm)_2(\mu\text{-}Cl)(CO)_2$ (15, 25,26) and to the face-to-face dimer $Rh_2(dpm)_2(CO)_2Cl_2$ (27) results in shortening the Rh-Rh separations as shown in reactions (12) and (13). Structural characterization of the common product

(12) 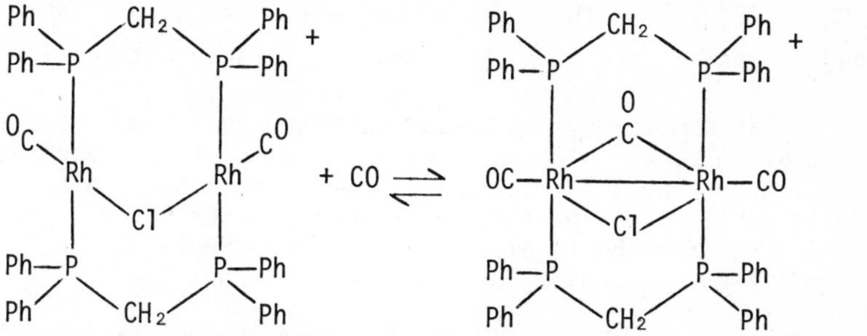 + CO ⇌

Rh···Rh, 3.153(8) Å (25) Rh—Rh, 2.838(1) Å (15)

(13) 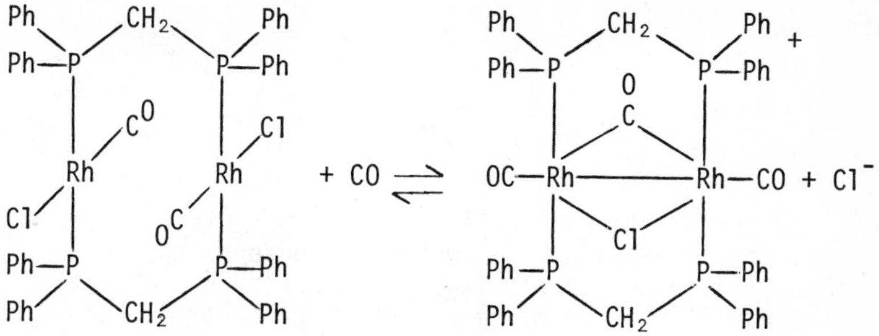 + CO ⇌ + Cl^-

Rh····Rh, 3.2386(5) Å (28) Rh—Rh, 2.838(1) Å (15)

of these reactions indicates the presence of a strong Rh-Rh interaction as a result of adding a bridging carbonyl ligand. However it is also possible to prepare a binuclear rhodium complex which, like $Pd_2(dpm)_2(\mu\text{-}CO)Cl_2$, has a long M···M separation and lacks a metal-metal bond. Addition of dimethyl acetylenedicarboxylate or hexafluoro-2-butyne to $Rh_2(dpm)_2(\mu\text{-}CO)X_2$ proceeds via reaction (14). (29) In the process the Rh-Rh separation increases by 0.60 Å.

Thus in binuclear dpm bridged complexes, bridging carbonyl groups span metal-metal separations as close as 2.8 Å and as distant as 3.35 Å. This variation is unparalleled in any other group

(14) 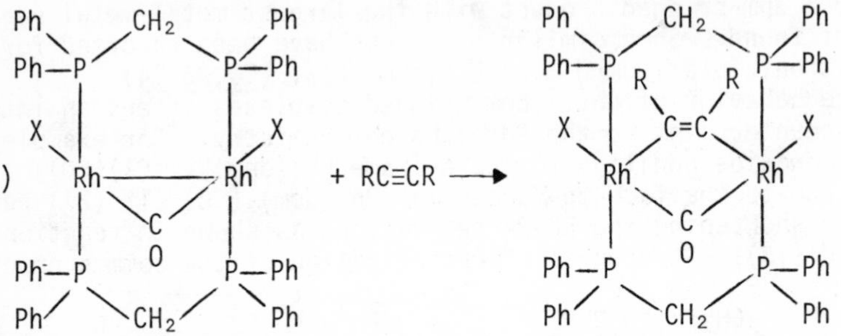 $Rh_2(\mu\text{-}CH_2(PPh_2)_2)_2(\mu\text{-}CO)X_2$ + RC≡CR ⟶ $Rh_2(\mu\text{-}CH_2(PPh_2)_2)_2(\mu\text{-}CO)(\mu\text{-}RC{=}CR)X_2$

Rh-Rh, 2.7566(8) Å for X = Br(30) Rh···Rh, 3.3542(9)Å for X = Cl (29)

of binuclear complexes. It should be noted that there is a rather poor relation between the infrared absorption due to the bridging carbonyl group and the metal-metal separation. Some data are given in Table 1. Of particular interest is the 67 cm^{-1} difference in ν(CO) for the isoelectronic and isostructural complexes $Pd_2(dam)_2(\mu\text{-}CO)Cl_2$ and $Pt_2(dpm)_2(\mu\text{-}CO)Cl_2$. (23)

2-Diphenylphosphinopyridine, a Ligand for the Stepwise Construction of Binuclear Complexes

2-Diphenylphosphinopyridine, 6, is a convenient building

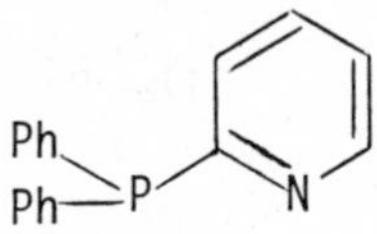

$Ph_2P\text{-}C_5H_4N$

6

block for constructing binuclear complexes. Like dpm it has only a single atom separating the two donor centers. Consequently it should prefer to act as a bridging, rather than a chelating, ligand. Since it has two different donor centers, it should be particularly useful in making heterobinuclear metal complexes. Because of its similarity to triphenylphosphine it should be capable of substituting this ligand. Thus a wide range of complexes should be available, and these have the potential for binding other metal ions. For example a trans metal complex should be capable of chelating a second metal via reaction (15). In this fashion the metal complex should resemble the trans chelating diphosphine 7. (31)

A number of complexes containing monodentate, P-bound Ph_2Ppy ligands have been prepared recently in our laboratory. (31) These include: trans-$Rh(CO)Cl(Ph_2Ppy)_2$, $Pd(Ph_2Ppy)_2Cl_2$, $Ru(CO)_2Cl_2(Ph_2Ppy)_2$, and $Ru_3(CO)_9(Ph_2Ppy)_3$. The ability of these species to bind a second metal is under active investigation.

Table I

Properties of Complexes with Bridging Carbonyl Ligands

Compound	M···M distance (Å)	ν(CO) cm^{-1}	Reference
$Pd_2(dpm)_2(\mu\text{-}CO)Cl_2$	-----	1705	19
$Pd_2(dam)_2(\mu\text{-}CO)Cl_2$	3.274(8)	1723	19, 22
$Pt_2(dpm)_2(\mu\text{-}CO)Cl_2$	3.162(4)	1638	23
$Rh_2(dpm)_2(\mu\text{-}CO)(\mu\text{-}Cl)(CO)_2$	2.838(1)	1865	15, 26
$Rh_2(dpm)_2(\mu\text{-}CO)Br_2$	2.7566(8)	1745	28
$Rh_2(dpm)_2(\mu\text{-}CO)(\mu\text{-}C_2\{CO_2CH_3\}_2)Cl_2$	3.3542(9)	~ 1700	29
$Ir_2(dpm)_2(\mu\text{-}CO)(\mu\text{-}S)(CO)_2$	2.843(2)	1760	34
$Rh_2(Ph_2Ppy)_2(\mu\text{-}CO)Cl_2$	2.612(1)	1797	33

(15) [reaction scheme: $(Ph_2Ppy)_2M$ + M' ⟶ $M(\mu\text{-}Ph_2Ppy)_2M'$]

[structure 7: phenanthrene bearing two $CH_2\text{-}PPh_2$ groups]

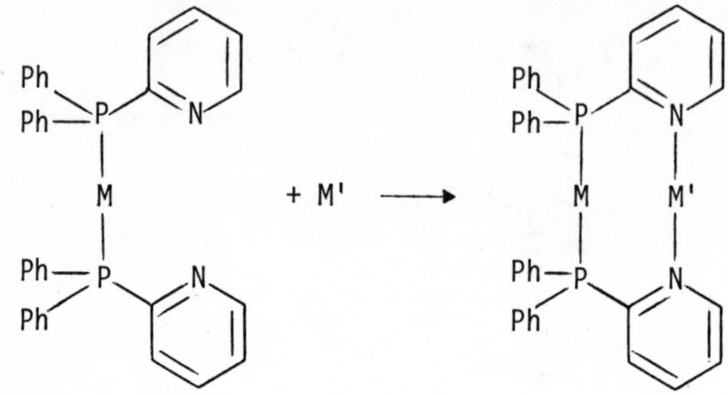

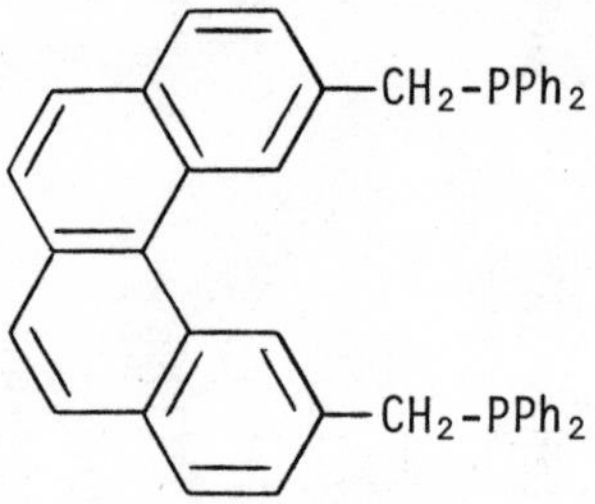

7

The reactions of $Rh(CO)Cl(Ph_2Ppy)_2$ offer an idea of what can be expected of complexes of this type. (32,33) Some pertinent reaction chemistry is summarized in Chart 1. The reaction of $Rh(CO)Cl(Ph_2Ppy)_2$ with $Rh_2(\mu\text{-}Cl)_2(CO)_4$ yields $Rh_2(Ph_2Ppy)_2(\mu\text{-}CO)\text{-}Cl_2$, with $\nu(CO)$ at 1797 cm^{-1}. The structure of this binuclear complex as determined by an X-ray crystallographic study is shown in Figure 5. The geometry is that of a distorted molecular A-frame with a nearly planar $Rh_2(\mu\text{-}CO)Cl_2$ unit. Each rhodium center possesses 16-valence electrons and consequently this complex is a rare example of an electron deficient complex containing a bridging carbonyl group and a metal-metal bond. This complex differs from the expectations of reaction 15 in that the Ph_2Ppy ligands have become reoriented so that they have a head-to-tail orientation A rather than the head-to-head orientation B.

Reaction of $Rh(CO)Cl(Ph_2Ppy)_2$ with (1,5-cyclooctadiene)$PdCl_2$ yields $RhPd(Ph_2Ppy)_2(CO)Cl_3$ with a terminal carbonyl ligand indicated by infrared absorption at 2040 cm^{-1}. Extension of the concept of reaction (15) to this system would predict structure 8 for this complex. The demonstrated tendency for the trans-Ph_2Ppy ligands to adopt the head-to-tail arrangement might lead to the expectation of structure 9. However neither of these structures would predict a carbon absorption on the Rh(I) center to be great-

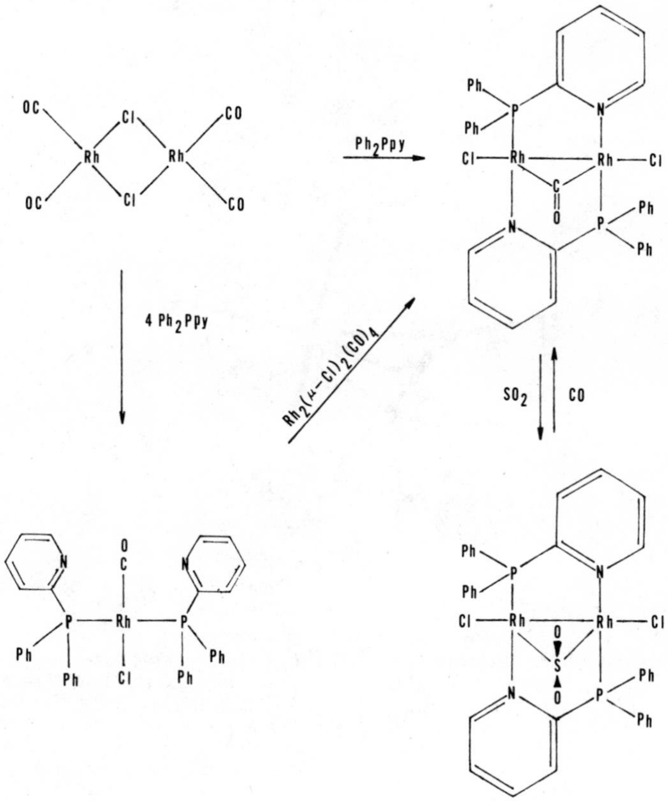
OC
Cl
CO
Rh
Rh
OC
Cl
CO
Ph2Ppy
4 Ph2Ppy
Rh2(μ-Cl)2(CO)4
SO2
CO
Ph
P
N
Cl
Rh
Rh
Cl
C
O
Ph
Ph
S
O
O

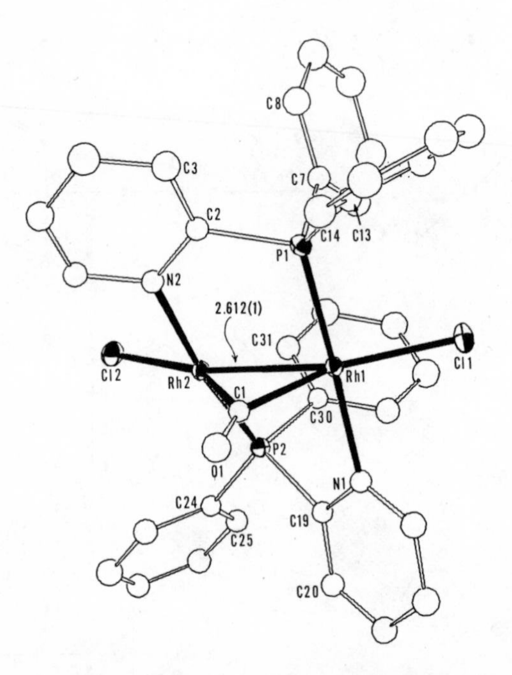

Figure 5. Perspective drawing of $Rh_2(Ph_2Ppy)_2(\mu\text{-}CO)Cl_2$. Some bond lengths are: Rh—Rh, 2.612(1); Rh(1)—Cl(1), 2.355(1); Rh(2)—Cl(2), 2.355(1); Rh(1)—P(1), 2.206(1); Rh(2)—P(2), 2.215(1); Rh(1)—N(1), 2.116(5); Rh(2)—N(2), 2.114(5) Å. The Rh—C—Rh angle is 84.3° (.2)°.

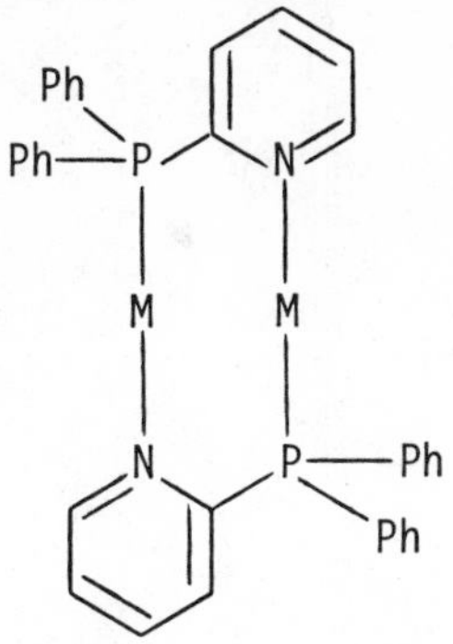

A 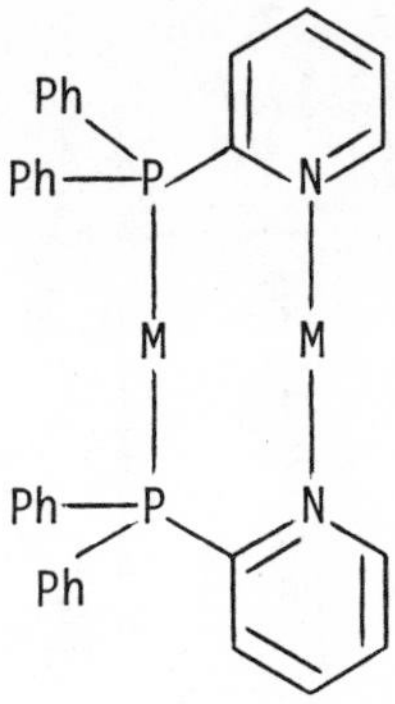 B

er than 2000 cm^{-1}. Although the ^{31}P-nmr spectrum of this complex shown in Figure 6 indicates that there are two distinct phosphorus environments and is grossly consistent with structure 9, the value of $^{1}J(Rh,P)$, 112.7 Hz, is unusually small for a Rh(I) complex. These spectroscopic inconsistencies arise because the true structure of $RhPd(Ph_2Ppy)_2(CO)Cl_2$ is 10. In forming this product,

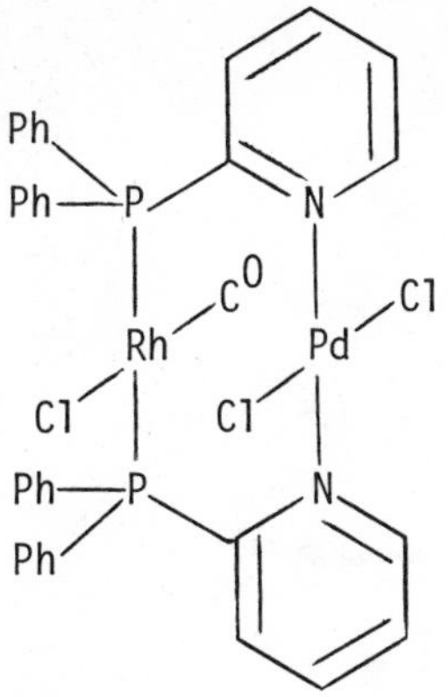

8 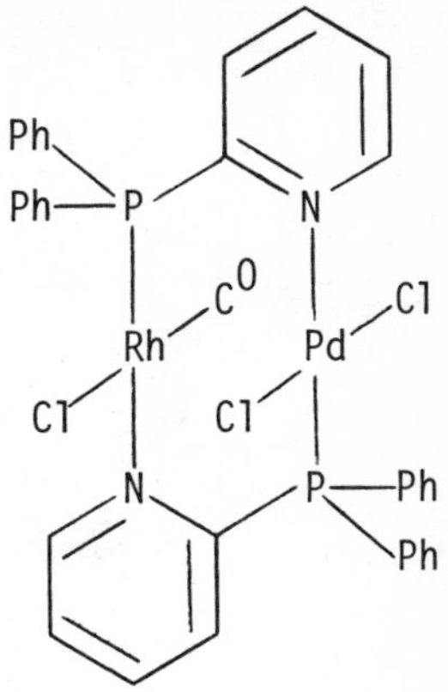 9

oxidative addition of a Pd-Cl bond to the Rh center has occurred. This is a novel case of oxidative addition of one d^8 center to another d^8 complex. A perspective drawing of the complex is shown in Figure 7. The Rh-Pd bond length (2.594(1) Å) is close to that of the Rh-Rh bond in $Rh_2(Ph_2Ppy)_2(\mu\text{-}CO)Cl_2$ (2.612(1) Å). We believe that the metal-metal separation in these binuclear complexes is determined to a considerable extent by the bridging Ph_2Ppy ligand. Because of the incorporation of the nitrogen do-

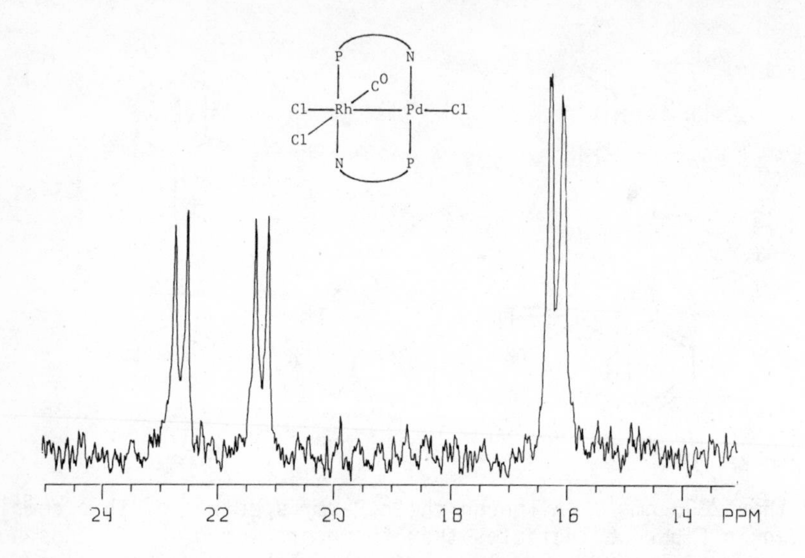

Figure 6. The $^{31}P\{^1H\}$ NMR spectrum of $RhPd(Ph_2Ppy)_2(CO)Cl_3$. The spectrum has been analyzed in terms of the following parameters: δ_1, 21.89 ppm; δ_2, 16.15 ppm; $^1J(Rh,P)$, 112.7 Hz; $^3J(P,P)$, 17.4 Hz; $^2J(Rh,P)$, 2.3 Hz.

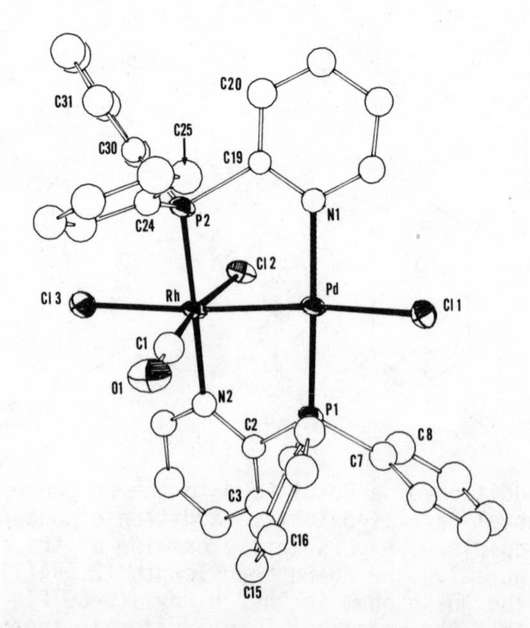

Figure 7. Perspective drawing of $RhPd(Ph_2Ppy)_2(CO)Cl_3$. Some bond lengths are: Rh—Pd, 2.594(1); Rh—Cl(2), 2.399(3); Rh—Cl(3), 2.499(4); Pd—Cl(1), 2.393(4); Rh—P(2), 2.243(3); Pd—P(1), 2.220(4); Rh—N(2), 2.16(1); Pd—N(1), 2.13(1); Rh—C(1), 1.82(1) Å.

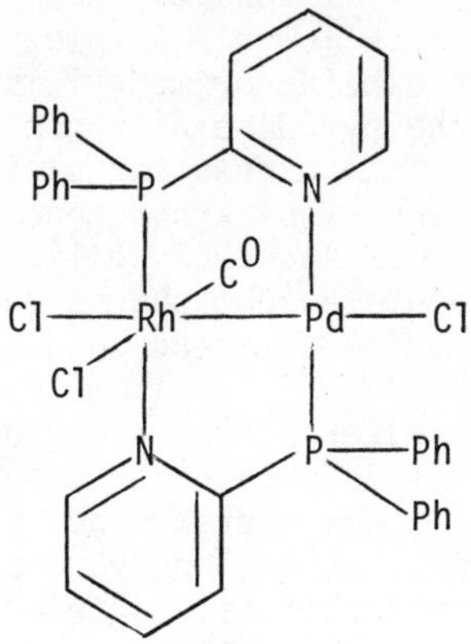

10

nor into a relatively rigid pyridine ring, this ligand lacks some of the articulation of dpm. In particular we suspect that Ph_2Ppy will have difficulty in spanning metal-metal separations in excess of 2.8 Å. Consequently face-to-face dimers such as 8 and 9 may be disfavored with respect to binuclear derivatives with direct metal-metal bonds.

This tendency of the Ph_2Ppy ligand to retain metal-metal bonding is borne out in the chemistry of the Pd(I) dimer, $Pd_2(Ph_2Ppy)_2Cl_2$. This complex is readily prepared via the conproportional reaction (16). Based on the previous results we

$$(16) \quad Pd(Ph_2Ppy)_2Cl_2 + \tfrac{1}{2}Pd_2(dba)_3 \rightarrow Pd_2(Ph_2Ppy)_2Cl_2 + \tfrac{3}{2}dba$$

(dba = dibenzylideneacetone)

suspect that the product has the head-to-tail geometry 11, with geometric features similar to that of $Pd_2(dpm)_2Cl_2$.

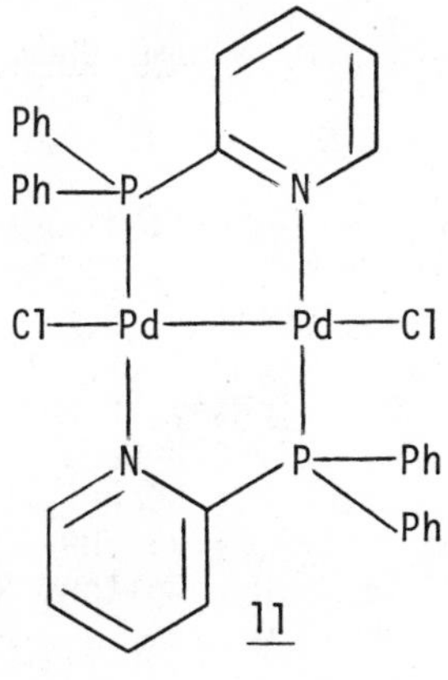

11

However, the reactivity of 11 toward insertion of small molecules into the Pd-Pd bond is much lower. Complex 11 does not insert carbon monoxide, sulfur dioxide or methyl isocyanide into the Pd-Pd bond. However, the pyridine nitrogen appears readily displaced. Thus 11 reacts with excess methyl isocyanide to form $Pd_2(CNCH_3)_4(PPh_2py)_2{}^{2+}$ and with carbon monoxide to form $Pd_2(CO)_2$-$Cl_2(PPh_2py)_2$ (ν(CO) at 2019 and 1994 cm^{-1}). These two products appear to contain monodentate, phosphorus-coordinated 2-diphenylphosphinopyridine and an intact Pd-Pd bond.

Acknowledgments. Research reported from U.C. Davis has been supported by the National Science Foundation (CHE-7681721 and CHE-7924575). I thank my coworkers, Linda Benner, Jim Farr, Hakon Hope, Chung-Li Lee, Katie Lindsay, Andre Maisonnat and Marilyn Olmstead for their critical contributions.

Literature Cited

1. Balch, A.L. Ann. N.Y. Acad. Sci., 1978, 313, 651-622.
2. Mann, K.R.; Lewis, N.S.; Williams, R.M.; Gray, H.B.; Gordon II, J.G. Inorg. Chem., 1978, 17, 828-834.
3. Balch, A.L. J. Am. Chem. Soc., 1976, 98, 8049-8054.
4. Lewis, N.S.; Mann, K.R.; Gordon II, J.G.; Gray, H.B. J. Am. Chem. Soc., 1976, 98, 7461-7463.
5. Yaneff, P.V.; Powell, J. J. Organometal. Chem., 1979, 179, 101-113.
6. Balch, A.L.; Labadie, J.W.; Delker, G. Inorg. Chem., 1979, 18, 1224-1227.
7. Balch, A.L.; Miller, J. J. Organometal. Chem., 1971, 32, 263-268.
8. Balch, A.L.; Olmstead, M.M. J. Am. Chem. Soc., 1976, 98, 2354-2356.
9. Olmstead, M.M.; Balch, A.L. J. Organometal. Chem., 1978, 148, C15-C18.
10. Balch, A.L.; Olmstead, M.M. J. Am. Chem. Soc., 1979, 101, 3128-3129.
11. Mann, K.R.; Ball, R.A.; Gray, H.B. Inorg. Chem., 1979, 18, 2671-2673.
12. Mann, K.R.; Di Pierro, M.J.; Gill, T.P. J. Am. Chem. Soc., 1980, 102, 3965-3967.
13. Abbott, E.H.; Base, K.S.; Cotton, F.A.; Hall, W.T.; Sekutowski, J.C. Inorg. Chem., 1978, 17, 3240-.
14. Balch, A.L.; Lee, C.-H.; Lindsay, C.H.; Olmstead, M.M. J. Organometal. Chem., 1979, 177, C22-C26.
15. Olmstead, M.M.; Lindsay, C.H.; Benner, L.S.; Balch, A.L. J. Organometal. Chem., 1979, 179, 289-300.
16. Holloway, R.G.; Penfold, B.R.; Colton, R.; McCormick, M.J. Chem. Commun., 1976, 485-486.

17. Olmstead, M.M.; Benner, L.S.; Hope, H.; Balch, A.L. Inorg. Chim. Acta, 1979, 32, 193-198.
18. Olmstead, M.M.; Hope, H.; Benner, L.S.; Balch, A.L. J. Am. Chem. Soc., 1977, 99, 5502-5503.
19. Benner, L.S.; Balch, A.L. J. Am. Chem. Soc., 1978, 106, 6099-6106.
20. Balch, A.L.; Benner, L.S.; Olmstead, M.M. Inorg. Chem., 1979, 18, 2886-3003.
21. Rattray, A.D.; Sutton, D. Inorg. Chim. Acta, 1978, 27, L85-L86.
22. Colton, R.; McCormick, M.J.; Pannan, C.D. Chem. Commun., 1977, 823-824.
23. Brown, M.P.: Keith, A.N.; Manojlovic-Muir, Lj.; Muir, K.W.; Puddephatt, R.J.; Seddon, K.R. Inorg. Chim. Acta, 1979, 34, L223-L224.
24. Brown, M.P.; Fisher, J.R.; Puddephatt, R.J.; Seddon, K.R. Inorg. Chem., 1979, 18, 2808-2813.
25. Cowie, M. Inorg. Chem., 1979, 18, 286-292.
26. Cowie, M.; Dwight, S.K. Inorg. Chem., 1979, 18, 2700-2706.
27. Mague, J.T.; Sanger, A.R. Inorg. Chem., 1979, 18, 2060-2066.
28. Dwight, S.K.; Cowie, M. Inorg. Chem., in press.
29. Cowie, M.; Southern, T.G. J. Organometal. Chem., 1980, 193, C46-C50.
30. Dwight, S.K.; Cowie, M. Inorg. Chem., in press.
31. Farr, J.P.; Maisonnat, A.; Balch, A.L., unpublished results.
32. Farr, J.P.; Olmstead, M.M.; Balch, A.L. J. Am. Chem. Soc., in press.
33. Farr, J.P.; Olmstead, M.M.; Lindsay, C.H.; Balch, A.L. Inorg. Chem., in press.
34. Kubiak, C.P.; Woodcock, C.; Eisenberg, R. Inorg. Chem., 1980, 19, 2733-2739.

RECEIVED December 1, 1980.

9

Binuclear Hydridoplatinum Complexes with Platinum–Platinum Bonds

R. J. PUDDEPHATT

Department of Chemistry, University of Western Ontario, London, Canada N6A 5B7

There has been considerable interest in binuclear and polynuclear metal complexes as models for intermediates proposed to be formed during reactions which are heterogeneously catalysed by transition metals (1). Since platinum is one of the most versatile catalysts, we have begun an investigation into the synthesis, and chemical and catalytic properties of some binuclear organoplatinum complexes. In this article some hydrido and methyl complexes will be described, and a preliminary account of catalysis with binuclear complexes given. In addition, structural studies indicate that Pt-Pt bonding interactions may take several different forms in these complexes and so the nature of the Pt-Pt bond will also be discussed.

Synthesis and Characterisation of Electron-Deficient Hydrido and Methyl Diplatinum Complexes

The ligand bis(diphenylphosphino)methane, dppm, has been used to stabilise binuclear complexes. Owing to the small bite angle, this ligand is known to bridge between two transition metal centers in many cases although it can also act as a simple chelate ligand (2).

Reduction of monomeric $[PtCl_2(dppm)]$ with sodium borohydride first gave the binuclear hydride, $[Pt_2H_3(dppm)_2]Cl$, which was crystallised more conveniently as the hexafluorophosphate salt, $[Pt_2H_3(dppm)_2][PF_6]$. This complex has structure (I) as deduced from detailed studies of the ^{1}H and ^{31}P NMR spectra (3). The hydride region of the NMR spectrum is particularly informative, since a hydride ligand bridging between two equivalent platinum atoms gives a 1:8:18:8:1 quintet pattern with separation of lines equal to 1/2 $^{1}J(PtH)$ due to coupling with ^{195}Pt ($I = 1/2$, natural abundance 33.8%) whereas a terminal hydride H-Pt-Pt-H gives a 1:1:4:1:1 pattern with satellites about the central peak showing different couplings $^{1}J(PtH)$ and $^{2}J(PtH)$. Complex(I) gives two hydride resonances at low temperatures with relative intensities of 1:2 due to bridging and terminal hydrides respectively, assigned

0097-6156/81/0155-0187$05.00/0

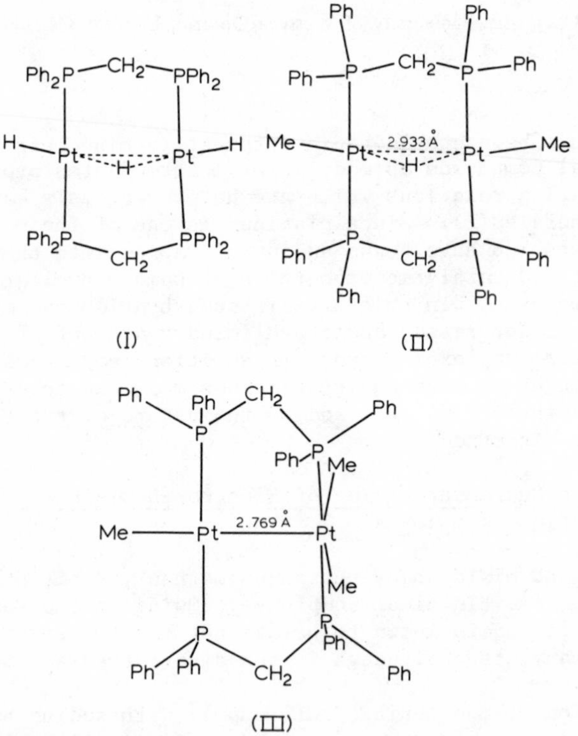

Figure 1. Structures of the complex cations I, II, and III

using the criteria given above. Exchange of bridging and terminal hydride ligands on the NMR time scale occurs at temperatures above 65°C. Other complexes of similar molecular formula but with different structures are now known. For example, the complex ions $[Pt_2H_2(\mu\text{-}H)(diphos)_2]^+$, with chelating ligands disphos = $Ph_2PCH_2CH_2PPh_2$ or $^tBu_2PCH_2CH_2CH_2P^tBu_2$, have the bridging and terminal hydride ligands in mutually cis co-ordination sites and exchange is rapid on the NMR time scale even at low temperatures (4). Clearly the presence of μ-dppm ligands in (I) causes the bridging and terminal hydrides to occupy mutually trans co-ordination sites and exchange is then considerably more difficult. Complex(I) is electron deficient since each platinum(II) centre can attain the 16-electron configuration only by sharing the electrons in the 3-centre-2-electron $Pt_2(\mu\text{-}H)$ bond (Figure 1).

Reduction of either $[Pt_2Me_2(\mu\text{-}Cl)(\mu\text{-}dppm)_2][PF_6]$ or $[Pt_2Me_3(\mu\text{-}dppm)_2][PF_6]$ with sodium borohydride gives the binuclear methyl(hydrido) derivative $[Pt_2Me_2(\mu\text{-}H)(\mu\text{-}dppm)_2][PF_6]$, (II). This complex has been characterised crystallographically as shown in (II), although the position of the hydride ligand was not found in the structure determination (5). That the hydride is symmetrically bridging between the two platinum atoms is clearly indicated by the hydride resonance in the ^1H NMR spectrum (Fig. 2) which clearly shows the inner lines of the expected 1:8:18:8:1 quintet due to coupling with ^{195}Pt. Methyl(hydrido) derivatives of transition metals are rare because reductive elimination of methane usually occurs very readily. For example, cis-$[PtHMe(PPh_3)_2]$ decomposes at -20°C by reductive elimination of methane (6). However, complex (II) has remarkable thermal stability and is not decomposed on heating to 70°C in the solid or solution, although slow elimination of methane is induced by added PMe_2Ph at 70°C. It is suggested that the rigid A-frame structure of (II) with methyl and hydrido groups held in mutually trans positions makes reductive elimination difficult in this case.

The Pt-Pt distance in (II) is 2.933 Å, which may be compared with a value of 2.652 Å in $[Pt_2Cl_2(\mu\text{-}dppm)_2]$, a complex having a covalent $Pt^I\text{-}Pt^I$ bond (7). The longer Pt-Pt distance for (II) is expected since protonation of metal-metal bonds always leads to bond lengthening. However, the bond length is indicative of weak residual Pt-Pt bonding in (II) and suggests a closed $Pt_2(\mu\text{-}H)$ system, in which the metal atom orbitals involved in bonding with the hydrido ligand overlap significantly with one another (8).

The third member of this series of electron deficient hydride and methyl complexes is the per(methyl) derivative $[Pt_2Me_3(\mu\text{-}dppm)_2][PF_6]$, (III). This can be prepared in a number of ways of which the reaction of equation (1) is perhaps simplest (9).

$$2\,[PtMe_2(dppm)] + H[PF_6] \longrightarrow CH_4 + [Pt_2Me_3(\mu\text{-}dppm)_2][PF_6] \quad ...(1)$$

Of particular interest is the unique structure of (III) which has been determined crystallographically (10). Presumably the A-frame structure adopted by (I) and (II) is unfavorable because the methyl group is unable to act as a bridging ligand and so the

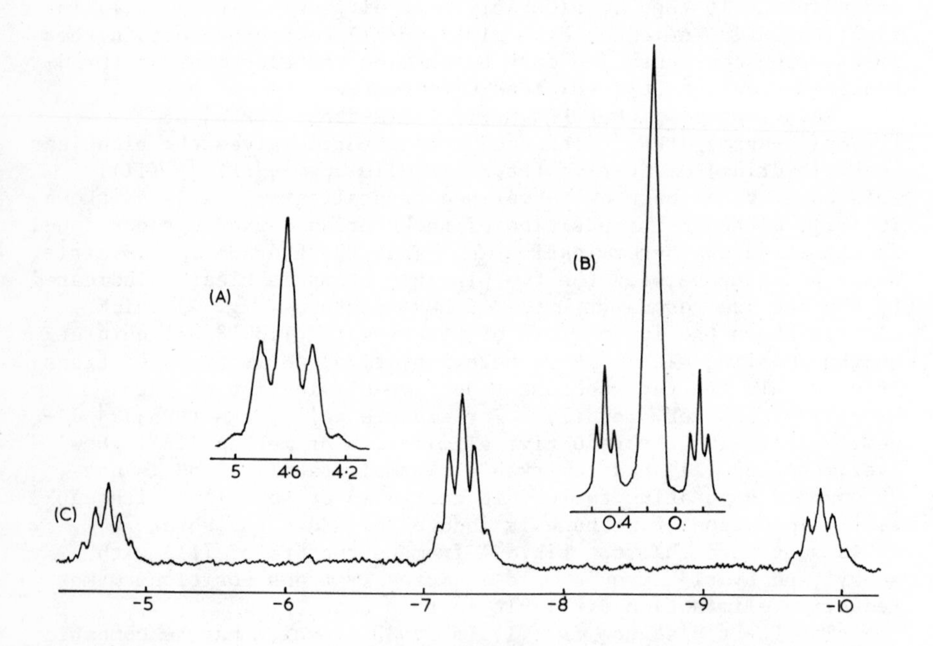

Figure 2. *^{1}H NMR spectrum (100 MHz) of complex (II) in CD_2Cl_2.*

(A) $C\underline{H}_2P_2$ resonance with 1:8:18:8:1 quintet indicating μ-dppm groups. (B) $\underline{Me}Pt$ resonance with 1:1:4:1:1 splittings due to coupling with ^{195}Pt indicating terminal MePtPtMe groups. The inner satellites overlap with the center peak giving the broad central resonance shown, and the fine structure of the outer satellites is due to coupling with ^{31}P. (C) The $Pt_2(\mu\text{-}\underline{H})$ resonance showing the inner part of the 1:8:18:8:1 quintet due to coupling with ^{195}Pt. The fine structure is due to coupling with 4 equivalent ^{31}P atoms.

unique structure shown in (III) is adopted. We suggest that there is a co-ordinate Pt → Pt bond, with r(Pt-Pt) 2.769 Å, to explain the stereochemistry of this product. Thus, the monomethylplatinum(II) centre achieves the 16-electron count by accepting a lone-pair of electrons from the electron-rich dimethylplatinum(II) centre. This electron pair is presumed to be donated from the filled $5\underline{d}z^2$ orbital of the dimethylplatinum centre. However, coupling constants $^2J(PtMe)$ and $^1J(PtP)$ about the dimethylplatinum centre from the NMR spectra of (III) are smaller than expected on this basis (Table). Thus, it seems clear that some rehybridisation about the dimethylplatinum centre occurs so that the orbital used in forming the Pt → Pt bond has some 6s-character mixed with the $5\underline{d}z^2$, thus leading to reduced s-character in the bonds to methyl and phosphine ligands. The co-ordinate Pt-Pt bond is not broken by reaction of (III) with ligands such as ammonia and pyridine, and unchanged (III) can be recovered from such reactions.

Table

Some coupling constants from the ^1H and ^{31}P NMR spectra of (III)

Complex	$^2J(PtMe)$ Hz	$^1J(PtP)$ Hz
cis-[$Me_2Pt(\mu\text{-dppm})_2PtMe_2$]	70	1830
(III)	60^a	1459^a
	80.5^b	3009^b

[a] values for the Me_2Pt centre [b] values for the MePt centre

Binuclear Reductive Elimination Reactions

The complex (I) undergoes binuclear reductive elimination reactions induced by tertiary phosphine and other ligands (equation 2) (11).

$$(I) + L \longrightarrow H_2 + \left[H-\overset{\overset{P\frown}{|}}{\underset{\underset{P\smile}{|}}{Pt}}-\overset{\overset{P}{|}}{\underset{\underset{P}{|}}{Pt}}-L \right][PF_6] \quad (2)$$

(IV), L = PMe_2Ph
(V), L = PPh_3
(VI), L = η^1-dppm
(VII), L = $4\text{-}MeC_6H_4NC$

When $L = \eta^1_o$-dppm, an X-ray structure determination shows r(PtPt) 2.770 Å, which is close to the value found for the proposed co-ordinate Pt—Pt bond in (III). In this regard, it may be noted that these complexes could be formulated as $Pt^{II}-Pt^{0}$ complexes containing a co-ordinate Pt → Pt bond:

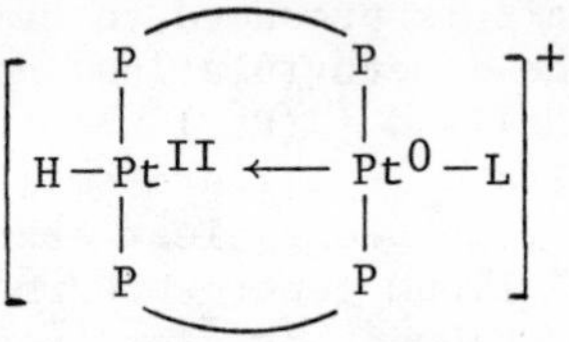

$$\left[H-Pt^{II} \longleftarrow Pt^{0}-L \right]^{+}$$

However, in this case the formulation as a diplatinum(I) complex is more reasonable and the long Pt—Pt bond probably results from the high <u>trans</u>-influence of the hydride ligand.

The reactions (2) are irreversible but the corresponding reaction when L = CO is easily and quantitatively reversible (equation 3) (<u>12</u>).

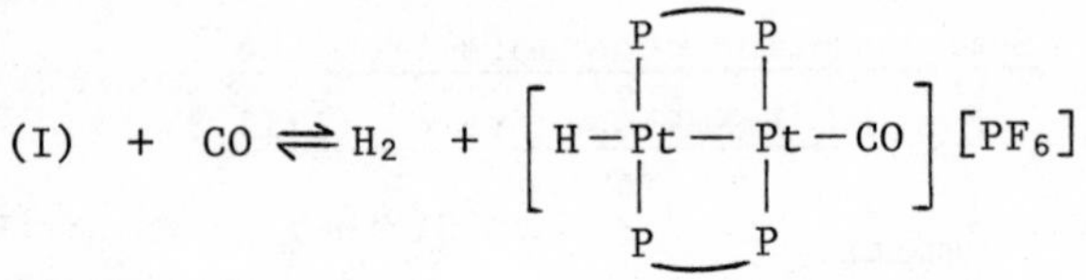

$$(I) + CO \rightleftharpoons H_2 + \left[H-Pt-Pt-CO \right][PF_6] \qquad (3)$$

(VIII)

Both the forward and backward reactions are complete in about one day at room temperature and one atmosphere pressure in methylene chloride solution.

We suggest that binuclear reductive elimination also occurs as the initial step in reactions of (I) with methanethiol or diphenylphosphine (equation 4) (<u>13</u>).

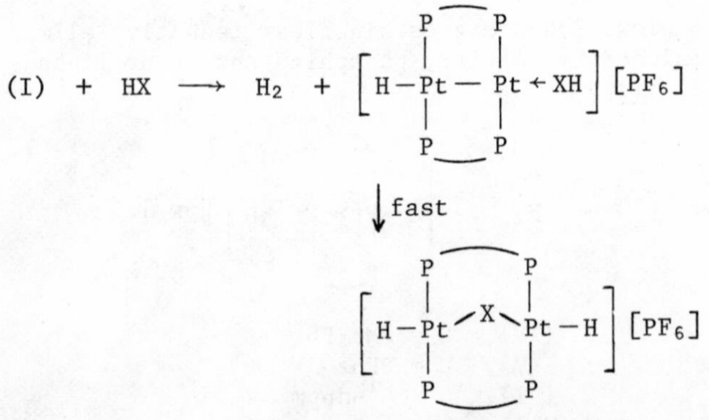

$$(I) + HX \longrightarrow H_2 + \left[H-Pt-Pt \leftarrow XH \right][PF_6] \qquad (4)$$

$$\downarrow \text{fast}$$

$$\left[H-Pt \diagup^{X} \diagdown Pt-H \right][PF_6]$$

(IX), X = SMe
(X), S = PPh_2

In this case the proposed hydridodiplatinum(I) intermediates could not be isolated or detected, and it is likely that the subsequent intramolecular binuclear oxidative addition is very fast.

A final example of the apparent occurrence of binuclear reductive elimination is found in reactions of (I) with alkynes bearing electronegative substituents. These reactions occur as shown in equation (5)

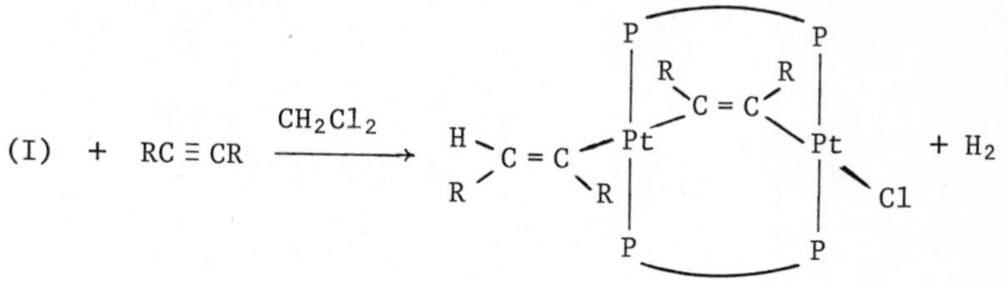

$$\text{(I)} + RC\equiv CR \xrightarrow{CH_2Cl_2} \text{H(R)C=C(R)–Pt}(\mu\text{-P–P})_2(\mu\text{-RC=CR})\text{Pt–Cl} + H_2 \quad (5)$$

(XI), $R = CF_3$

(XII), $R = CO_2Me$

It is tentatively suggested that the first step involves reductive elimination of H_2 generating a hydridodiplatinum(I) species. This is followed by *cis*-insertion of alkyne into the remaining Pt-H bond, *cis*-insertion of a second molecule of alkyne into the Pt-Pt bond and abstraction of chloride from the solvent. The hydrogen evolution is observed immediately on mixing the reagents and yellow intermediates are formed which have not yet been characterised. Hence it seems that the reductive elimination is the first step, and recent evidence indicates that the $Pt_2(\mu\text{-}C_4F_6)$ group is formed next. More work is required to determine the nature of the final steps.

Since reductive elimination of H_2 appears to be a key reaction in the chemistry of complex(I), a brief study of the kinetics of the reaction of equation (2) when $L = PPh_3$ has been made. The reaction followed overall second-order kinetics in 1,2-dichloroethane at 21°C, the rate expression being:

$$\frac{-d}{dt}[I] = (1.85 \pm 0.4\ \ell\ mol^{-1}\ sec^{-1})\,[I][L]$$

In this sense the reaction is qualitatively different from the usual reductive elimination of H_2 from mononuclear complexes, such as equation (6) (14).

$$[IrH_2(CO)_2L_2]^+ + L' \longrightarrow [Ir(CO)_2L_2L']^+ + H \quad (6)$$

In this case the starting complex is co-ordinatively saturated and the reductive elimination is the initial slow step and is followed by rapid addition of L'. Hence the rate is independent of the concentration or nature of the ligand L'. Complex (I) is electron deficient and it seems that the rate-determining step involves attack of L on (I), to give an intermediate or transition state which will not be electron-deficient and which should therefore contain no bridging hydrido ligand. Structures (XIII) and (XIV), $L = PPh_3$, are likely for this proposed species, but there are a number of ways in which the subsequent rapid reductive elimination of H_2 and formation of complex (V) could then occur.

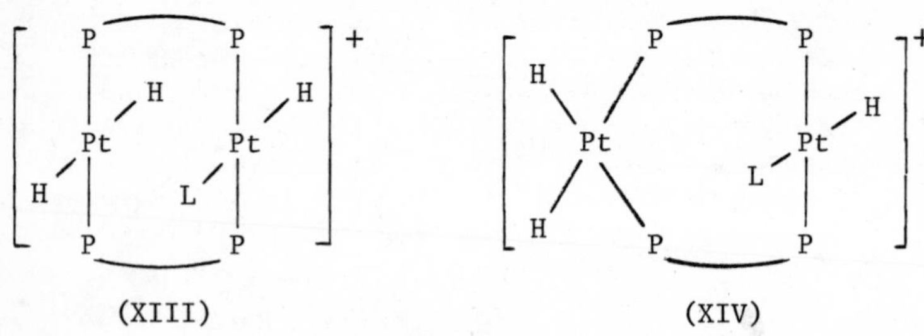

Catalysis of the Water Gas Shift Reaction

There have been several recent studies of homogeneous catalysis of the water gas shift reaction (equation 7) by mononuclear and cluster catalysts, including mononuclear platinum complexes (15).

$$CO + H_2O \rightleftharpoons CO_2 + H_2 \quad (7)$$

Often, mechanisms are suggested in which a cationic carbonyl reacts with water according to equation (8) to generate CO_2 (16).

$$[L_nM-CO]^+ + OH^- \rightleftharpoons [L_nM-CO_2H] \downarrow [L_nMH] + CO_2 \quad (8)$$

Thus, a catalytic cycle can easily be envisaged involving formation of H_2 from (I) and CO according to equation (3), followed by formation of CO_2 with regeneration of (I) by the reaction of equation (9).

$$(VIII) + H_2O \rightleftharpoons (I) + CO_2 \quad (9)$$

Indeed, catalysis of the water gas shift reaction is observed using (I) dissolved in methanol (50 ml) and water (25 ml) as catalyst in a 300 mL bomb at 100°C and with pressures of CO from 80 - 160 psi. Under these conditions the rate of reaction is independent of CO pressure and first order with respect to concentration of catalyst (I), giving a turnover rate of 3.6 ± 0.6 (moles CO_2 or H_2)/(mole catalyst)(hour). Good linear kinetics are observed for at least one day and solutions remain homogeneous during this period. The rate is not affected by added lithium chloride (0.02 M) nor by sodium hydroxide (0.27 M). From the solutions at the end of the catalytic runs, complex (VIII) together with an uncharacterised complex having a strong band in the infrared spectrum at 1760 cm^{-1}, most likely due to a bridging carbonyl group, can be isolated.

The observation that the rate of reaction is linear with concentration of (I) strongly suggests that the catalytic intermediates are binuclear but more work is needed before a detailed catalytic cycle can be deduced with confidence. It is likely that the mechanism defined by equations (3) and (9) is an oversimplification. Attempts to prove directly that reaction (11) can occur have been unsuccessful since, in the absence of CO, (VIII) decomposes faster than it reacts with water. However, the dicationic $[Pt_2(CO)_2(\mu\text{-dppm})_2]^{2+}$ reacts readily with water according to equation (10).

$$[OC{-}Pt(\mu\text{-}P\frown P)_2Pt{-}CO]^{2+} + H_2O \longrightarrow [H{-}Pt(\mu\text{-}P\frown P)_2Pt{-}CO]^{+} + H^{+} + CO_2 \quad (10)$$

The lower reactivity of (VIII) towards water is no doubt due to the single positive charge being delocalised over both platinum centres, thus making the carbonyl group less susceptible to nucleophilic attack by water.

Acknowledgments

This work was carried out in collaboration with Dr. M.P. Brown, University of Liverpool, and Dr. Lj. Manojlović-Muir, University of Glasgow. Much of the credit belongs to them and to their graduate students whose names are given in the references. Dr. M.A. Thomson and Mr. R.H. Hill made major contributions to this work. We thank NSERC (Canada) for financial assistance and NATO for a travel grant.

Literature Cited

1. Muetterties, E.L. and Stein, J., Chem. Rev., 1979, 79, 479.

2. Kubiak, C.P. and Eisenberg, R., J. Am. Chem. Soc., 1977, 99, 6129.

3. Brown, M.P., Puddephatt, R.J., and Rashidi, M., J.C.S. Dalton, 1978, 516.

4. Minghetti, G., Banditelli, G. and Bandini, A.L., J. Organomet. Chem., 1977, 139, C80. Tulip, T.H., Yamagata, T., Yoshida, T., Wilson, R.D., Ibers, J.A., and Otsuka, S., Inorg. Chem., 1979, 18, 2239.

5. Brown, M.P., Cooper, S.J., Frew, A.A., Manojlović-Muir, Lj., Muir, K.W., Puddephatt, R.J. and Thomson, M.A., J. Organomet. Chem., 1980, 198, C33.

6. Abis, L., Sen, A., and Halpern, J., J. Am. Chem. Soc., 1978, 100, 2915.

7. Manojlović-Muir, Lj., Muir, K.W. and Solomun, T., Acta Cryst., 1979, B35, 1237.

8. Bau, R., Teller, R.G., Kirtley, S.W. and Koetzle, T.F., Acc. Chem. Res., 1979, 12, 176.

9. Brown, M.P., Cooper, S.J., Puddephatt, R.J., Thomson, M.A., and Seddon, K.R., J.C.S. Chem. Comm., 1979, 1117.

10. Frew, A.A., Manojlović-Muir, Lj., and Muir, K.W., J.C.S. Chem. Comm., in press.

11. Brown, M.P., Fisher, J.R., Manojlović-Muir, Lj., Muir, K.W., Puddephatt, R.J., Thomson, M.A., and Seddon, K.R., J.C.S. Chem. Comm., 1979, 931.

12. Brown, M.P., Fisher, J.R., Mills, A.J., Puddephatt, R.J. and Thomson, M.A., Inorg. Chim. Acta, 1980, 44, L271.

13. Brown, M.P., Fisher, J.R., Puddephatt, R.J. and Seddon, K.R., Inorg. Chem., 1979, 18, 2808.

14. Mays, M.J., Simpson, R.N.F. and Stefanini, F.P., J.C.S.A., 1970, 3000.

15. Yoshida, T., Ueda, Y., and Otsuka, S., J. Am. Chem. Soc., 1978, 100, 3941.

16. Catellani, M., and Halpern, J., Inorg. Chem., 1980, 19, 566.

RECEIVED December 1, 1980.

Thermochemistry of Metal–Metal Bonds

J. A. CONNOR and H. A. SKINNER

Department of Chemistry, University of Manchester, Manchester M13 9PL, England

The relationships between bond enthalpy, bond length and bond order which appear relatively simple in the case of a main group element such as carbon and its compounds, are more difficult to establish when the *d*-transition metal elements and their compounds are considered. Progress in establishing these relationships for metals is severely hindered by a lack of relevant thermochemical data. This paper reviews some of the more useful information that is available for diatomic molecules, for polynuclear binary carbonyls and for binuclear complexes of the *d*-transition elements.

Diatomic Molecules of Transition Metal Elements.

a) Homonuclear. This subject was treated in detail recently (*1*). The dissociation energies of the *d*-transition metal diatomic molecules (*2*) are given in Table I. The pattern of variation of $D^o_o(M_2,g)$ for 3*d*-metals is similar to that of the enthalpy of sublimation of the metal $\Delta H^o_{vap}(M,c)$

Table I

Dissociation Energies, D^o_o/kJ mol^{-1} of Homonuclear Diatomic *d*-Transition Metal Molecules

Sc	159±21	Y	156±21	La	243±21
Ti	126±17				
V	238±21	Nb	503±10		
Cr	151±21	Mo	404±20		
Mn	42±29				
Fe	100±21				
Co	167±25	Rh	281.6±20.9		
Ni	230±21	Pd	105±21	Pt	358±7
Cu	190.2±5.4	Ag	159.0±6.3	Au	221.3±2.1

The low value of $D^o_o(Mn_2)$ may indicate that this is a van der Waals molecule. Calculations indicate that the diatomic molecules of

0097-6156/81/0155-0197$05.00/0

the $3d$-metals are bound by s–s single bonds between the atoms in the $3d^n4s^1$ valence state, although configurations of higher formal bond order may contribute to the description of the interatomic bond. For example, a sextuple bond configuration may contribute to the description of bonding in dichromium (*3*). An increasing d-orbital contribution to the metal-metal bonding results in bond orders greater than one for elements of the $4d$ and $5d$ series. The precise interpretation of the bonding in these molecules is sensitive to the type of calculation which is carried out. This is exemplified by dimolybdenum (Table II), of which there exist several calculations (*4*) in addition to thermochemical (*5*) and spectroscopic (*6*) data.

Table II

Experimental and Computational Results for Dimolybdenum, Mo_2

<u>Experiment</u>

$D^o_0(Mo_2,g)$ (404±20) kJ mol^{-1}					(*5*)
r (Mo-Mo) 192.9 pm; ω_e 477.1 cm^{-1} D^o_0(397±63) kJ mol^{-1}					(*6*)

<u>Computation</u>

Method	r_e/pm	D_o/kJ mol^{-1}	ω_e/cm^{-1}	bond order	
Hückel	210	326		6	(a)
Ab initio	197		475	5	(c)
	201	83	388	4.8	(d)

The value of the maximum dissociation energy in a homonuclear diatomic transition metal molecule has been predicted (*2*) to be (600±40) kJ mol^{-1} for ditantalum. The dissociation energies of homonuclear transition metal molecules have been predicted by various methods of which the most recent is the cell model of Miedema (*7*). This proposes a relation between the enthalpy of vaporisation of the solid metal, ΔH^o_{vap}, the dissociation energy, D^o, of the diatomic molecule and the surface energy of the metal, γ^o as follows,

$$\frac{(2\Delta H^o_{vap} - D^o)}{\gamma^o\ V_m^{2/3}} = \text{constant}$$

where, V_m is the molar volume, a parameter for atomic size. The success of this and other methods can be judged by their predictions (*2*) of $D^o(Nb_2)$ as 371, 335 and 448-360 kJ mol^{-1} compared with the experimental (*8*) value, (503±10) kJ mol^{-1}.

b) <u>Heteronuclear</u>. Experimental data are available for 115 molecules, among which those involving silver or gold or an f-block element as one component form a majority (*2*). The values for heteronuclear diatomic d-transition metal molecules are given in Table III.

Table III

Dissociation Energies, D_0^o/kJ mol^{-1} of Heteronuclear Diatomic *d*-Transition Metal Molecules.

Molecule	D_0^o	Molecule	D_0^o
AgAu	200.8±10.5	CoCu	163±21
AgCu	169.5±10.5	CrCu	155±25
AgMn	96±21	CuNi	201±21
AuCo	218±17	IrLa	573±12
AuCr	209±17	IrY	452.8±16
AuCu	224.3±5.1	LaPt	500.0±8.0
AuFe	188±21	LaRh	524.7±16.7
AuMn	188±13	LaY	197±17
AuNi	251±21	MoNb	488±25
AuPd	151±21	PtTi	394±11
AuRh	228.9±29	PtY	470±8
AuSc	276.6±17	RhSc	440.3±10.5
AuV	238±12	RhTi	387.0±14.6
AuY	304.1±8.2	RhV	360±29
RhY	441.8±10.5	RuV	410±29

Two models have been used to predict dissociation energies for heteronuclear diatomic transition metal molecules, the valence bond model (*9*), which proposes a polar single bond, and the atomic cell model (*7*). Their success when compared with experiment is indicated by the following examples:

Molecule	D_0^o(expt)	valence bond	atomic cell
PtTi	394±11	360	494
RhV	360±29	460	362
RuV	410±29	569	382

The maximum bond energy of a heteronuclear diatomic transition metal molecule has been suggested (*2*) as (670±84) kJ mol^{-1}. The dissociation energies of heteronuclear diatomic molecules can be used to estimate values for related homonuclear systems. For example, the experimental values (*5,10*) of D_0^o(RhY) (441.8±10.5) kJ mol^{-1} and the isoelectronic molecule Mo_2 ($D_0^o(Mo_2)$(404±20) kJ mol^{-1}) suggests that $D_0^o(W_2)$ is not larger than that (*11*) of the isoelectronic molecule IrLa (D_0^o(IrLa)(573±12) kJ mol^{-1}). Similarly, from experimental values (*5*) of $D_0^o(Mo_2)$ and D_0^o(MoNb) (449±25) kJ mol^{-1}, a value of $D_0^o(Nb_2)$(502 kJ mol^{-1}) was predicted (*5*). This was subsequently shown (*8*) to be in agreement with experiment (503±10) kJ mol^{-1}. The observation that $D_0^o(V_2)$ > $D_0^o(Cr_2)$ and $D^o(Nb_2)$ > $D_0^o(Mo_2)$ suggests that $D_0^o(Ta_2)$ will be found to be greater than $D_0^o(W_2)$, notwithstanding the fact that ΔH^o_{vap}(Ta,*c*) < H^o_{vap}(W,*c*). In summary, there is a steadily increasing amount of data on the dissociation energies of diatomic transition metal molecules, but the description of the bonding and in particular of the bond order in these molecules is a subject of contention.

Binary metal carbonyl clusters

The thermochemical data obtained, for the most part, by high temperature microcalorimetry of thermal decomposition or of halogenation of polynuclear metal carbonyls have been summarized recently (*12*). The disruption enthalpy, ΔH_D referring to the process

$$[M_m(CO)_n, g] \rightarrow mM, g + nCO, g$$

can be interpreted in two ways. The simpler relies on a two-centre electron pair bond description of the structure and leads to empirical relations between the enthalpy contributions of the metal-metal bonds, $\overline{M}$, the terminal ($\overline{T}$) and bridging ($\overline{B}$) metal-ligand bonds and the enthalpy of atomisation of the particular metal as follows

$$2\overline{T} \sim 3\overline{M} \sim 4\overline{B} \sim 6\Delta H_f^o(M,g)/Z$$

in which Z is the coordination number of the bulk metal. The second method (*13*), relates the enthalpy contribution of a metal-metal bond, $\overline{M}$, to the length d, of that bond taking the bulk metal as standard, as follows

$$\overline{M} = A\, d^{-4.6}$$

where A is a constant. An indication of the different results of these two methods in relation to the enthalpy contributions of metal-metal bonds in polynuclear metal carbonyls is given in Table IV.

Table IV

Metal-Metal bond enthalpy contributions, $\overline{M}$/kJ mol^{-1} in polynuclear metal carbonyls derived from a) two-centre electron pair bond description, b) bond enthalpy/bond length relation.

Compound	$\overline{M}$(a)	$\overline{M}$(b)
$Mn_2(CO)_{10}$	67	35
$Fe_2(CO)_9$	80	70
$Fe_3(CO)_{12}$	80	52,65
$Co_2(CO)_8$	88	70
$Co_4(CO)_{12}$	88	74
$Ru_3(CO)_{12}$	115	78
$Rh_4(CO)_{12}$	110	86
$Rh_6(CO)_{16}$	110	80
$Re_2(CO)_{10}$	127	80
$Os_3(CO)_{12}$	128	94
$Ir_4(CO)_{12}$	127	117

In a diatomic transition metal molecule there is no ambiguity concerning the existence of a bond between the metal atoms. Recent experimental work (*14*) has found practically no features of electron density significantly different from zero in the region of the Fe-Fe bond of $[(\eta\text{-}C_5H_5)Fe(CO)_2]_2$, and calculations (*15*) of $Fe_2(CO)_9$ and $Co_2(CO)_8$ have concluded that in these molecules, where the metal-metal bond is bridged by carbonyl ligands, there is no evidence for a direct metal-metal interaction. Instead, there is evidence that the bonding arises from a strong interaction between the metal *d*-orbitals and the π^*-levels of the bridging CO ligand. A similar conclusion regarding $Fe_3(CO)_{12}$, that there is no evidence for a direct Fe-Fe bond along the short, bridged Fe-Fe direction, has been reached (*16*) on the basis of molecular orbital calculations.

When this information is accounted for in the description of bond type, the calculated bond enthalpy contributions for the iron carbonyls are $\overline{T}$ = 117, $\overline{B}$ = 79 and $\overline{M}$ = 94 kJ mol^{-1}. The increase in the value of $\overline{B}$ (from 64 kJ mol^{-1}) relative to $\overline{T}$ is particularly marked, It remains to be established whether there are direct cobalt-cobalt bonds along those directions containing bridging carbonyl ligands in $Co_4(CO)_{12}$, that is to say, whether the bond description is $9\overline{T} + 6\overline{B} + 6\overline{M}$ as at present understood, or whether it is $9\overline{T} + 6\overline{B} + 3\overline{M}$, which would lead to $\overline{T}$ = 136, $\overline{B}$ = 86 and $\overline{M}$ = 127 kJ mol^{-1}.

Compounds containing multiple metal-to-metal bonds

Presuming that calorimetric data are available, the central problem is that of interpretation. This will be clear from the brief description of some systems which have been studied recently.

a) Triple bonds. The heat of formation of the compounds $Mo(NMe_2)_4$, $W(NMe_2)_4$, $Mo_2(NMe_2)_6$ and $W_2(NMe_2)_6$ was determined (*17*) by various calorimetric methods (combustion; oxidative hydrolysis with $K_2Cr_2O_7/H_2SO_4$; acid hydrolysis). These are shown in Table V, together with the derived enthalpy of disruption, ΔH_D for the process

$$M_x(NMe_2)_y,g \rightarrow x M,g + y\ Me_2N,g$$

Table V

Standard heat of formation and derived values of enthalpy of disruption (kJ mol^{-1}) for dimethylamide compounds of Mo and W.

compound	$\Delta H_f^o(g)$	ΔH_D
$Mo(NMe_2)_4$	131.4±8	1021.6±19
$W(NMe_2)_6$	268.0±14	1332.5±29
$Mo_2(NMe_2)_6$	128.2±13	1928.6±28
$W_2(NMe_2)_6$	132.5±11	2327.9±29

For the mononuclear compounds, the average metal-ligand bond enthalpy $\bar{D}(M\text{-}NMe_2) = \Delta H_D/n$ so that $\bar{D}(Mo\text{-}NMe_2) = (255.4\pm5)$ kJ mol^{-1} in $Mo(NMe_2)_4$ and $\bar{D}(W\text{-}NMe_2) = (222.1\pm5)$ kJ mol^{-1} in $W(NMe_2)_6$. It is possible to estimate $\bar{D}(W\text{-}NMe_2) \sim (295\pm5)$ kJ mol^{-1} in the unknown $W(NMe_2)_4$ molecule from a comparison with the variation in the values of $\bar{D}(M\text{-}X)$ in MX_4(M=Ti,Zr,Mo;X=Cl,Br,NMe_2). Similarly, a value of $\bar{D}(Mo\text{-}NMe_2) \sim (190\pm5)$ kJ mol^{-1} in the unknown $Mo(NMe_2)_6$ molecule can be estimated from a comparison with $\bar{D}(M\text{-}X)$ in TaX_5 and WX_6(X=F,OMe,Cl,Br,NMe_2,Me). The average bond enthalpy $\bar{D}(M\text{-}NMe_2)$ can be estimated (*18*) for other formal oxidation numbers (3,5) of M, on the basis of comparison with data for halides of other metals.

For molecules $M_2(NMe_2)_6$, $\Delta H_D = 6D(M\text{-}NMe_2) + D(M\overset{3}{-}M)$; the nature of the problem is immediately apparent. What value of $\bar{D}(M\text{-}NMe_2)$ should be transferred into the equation to derive $D(M\overset{3}{-}M)$? Any error in $\bar{D}(M\text{-}NMe_2)$ enters sixfold into the derived value. For example, the formal oxidation number of the metal, M(III) in $M_2(NMe_2)_6$ yields values of $\bar{D}(Mo\overset{3}{-}Mo) = 200$, and $D(W\overset{3}{-}W) = 340$ kJ mol^{-1} which are close to the respective metal-ligand bond enthalpy contributions. On the other hand, each metal atom has a valency of six, which yields values of $D(M\overset{3}{-}M)$ (Mo,(788±24); W,(995±18) kJ mol^{-1}) which seem unacceptably high relative to the enthalpy contributions of the metal-metal single bond (*19*) in $[(\eta\text{-}C_5H_5)_2W_2(CO)_6]$(234 kJ mol^{-1}), and in relation to the dissociation energy of the diatomic molecules Mo_2(404 kJ mol^{-1}) and W_2(<570 kJ mol^{-1}).

If ancillary information from the crystal structures (*20*) of the compounds is taken into consideration, it would seem that the value of $D(W\text{-}NMe_2)$ in $W(NMe_2)_6$ might be too low because the W-N bond is longer and the CNC angle is compressed compared to $W_2(NMe_2)_6$; on the other hand the Mo-N bond is shorter and there is no evidence of steric strain in $Mo(NMe_2)_4$ compared to $Mo_2(NMe_2)_6$. Until more information on other systems containing triple bonds becomes available it is preferable to recognize the metal atom as four-coordinate in both $M_2(NMe_2)_6$ and $M(NMe_2)_4$ and to transfer the value of $\bar{D}(M\text{-}NMe_2)$ in the latter to obtain $D(Mo\overset{3}{-}Mo) = (396\pm18)$ and $D(W\overset{3}{-}W) = (558\pm20)$ kJ mol^{-1}.

A generalized SCF-MO with configuration interaction calculation (*21*) on the hypothetical molecule Mo_2H_6 has estimated $D(Mo\overset{3}{-}Mo)$ = (526±63) kJ mol^{-1} which is close to the mean value (592±196) kJ mol^{-1} tentatively suggested (*18*) for $D(Mo\overset{3}{-}Mo)$ in $Mo_2(NMe_2)_6$.

b) Quadruple bonds. The heat of oxidation of $Cs_2[Re_2Br_8]$ in acidic aqueous bromate to form perrhenate has been measured (*22*), and leads to the enthalpy of formation, $\Delta H_f^o(Cs_2Re_2Br_8,c)$ = - (1171±35) kJ mol^{-1}. The lattice enthalpy of the solid

$$Cs_2Re_2Br_8,c \rightarrow 2Cs^+,g + Re_2Br_8^{2-},g$$

is calculated to be (1029±20) kJ mol^{-1}, so that a value $\Delta H_f^o(Re_2Br_8^{2-},g)$ = - (1046±40) kJ mol^{-1} is obtained. Reference to other tetrahalometallate anions and neutral osmium (IV) halides permits an estimate of $\Delta H_f^o(ReBr_4^-,g)$ = - 303 or - 314 kJ mol^{-1}. In this way, the enthalpy contribution of the rhenium-rhenium quadruple bond in the octabromodirhenate (III) ion is estimated as (408±50) kJ mol^{-1}. This can be compared with an earlier estimate (*23*) of $D(Re\overset{4}{-}Re)$ ~ 500 kJ mol^{-1} which is based on a questionable (*22*) Birge-Sponer extrapolation applied to the frequencies of the vibrational progression in ν(Re-Re) observed in the resonance Raman spectrum of the $Re_2Br_8^{2-}$ ion in the solid state.

c) Multiple bonds in binuclear molybdenum compounds. The enthalpy of formation of several binuclear compounds of chromium (II) and molybdenum (II) containing multiple metal-metal bonds have been measured by solution reaction calorimetry (*24*). For the present discussion attention will be focussed on compounds of molybdenum. The relevant thermochemical data together with the metal-metal bond length in each compound are as follows:

Table VI

Standard enthalpy of formation in the gas phase, $\Delta H_f^o(g)$/kJ mol^{-1} and metal-metal bond length, r(Mo-Mo)/pm for binuclear compounds of molybdenum.

Compound	$\Delta H_f^o(g)$/kJ mol^{-1}	r(Mo-Mo)/pm
$Mo_2(OAc)_4$	- (1806±10)	209.3
$Mo_2(OAc)_2(acac)_2$	- (1650±10)	212.9
$Mo_2(OPr^i)_6$	- (1549±14)	(222.2)
$Mo_2(OPr^i)_8$	- (2156±18)	252.3

To obtain the enthalpy contribution of the metal-metal bond in each case, it is suggested (*25*) that there is a logarithmic relation between bond length and bond enthalpy contribution both for the molybdenum-molybdenum bonds and for the molybdenum-oxygen bonds. The data base includes molybdenum metal and $[Mo(acac)_3]$ and MoO_3. The enthalpy contribution E(Mo-Mo) of a

(Mo-Mo) bond is found to be related to the bond length r by the equation:

$$E(\text{Mo-Mo}) = 4.42 \times 10^{12} \; [r(\text{Mo-Mo})]^{-4.29}$$

with the results shown in Table VII.

Table VII

Metal-metal bond enthalpy contributions in binuclear molybdenum compounds.

Compound	$E(\text{Mo-Mo})/\text{kJ mol}^{-1}$
$Mo_2(OAc)_4$	489
$Mo_2(OAc)_2(acac)_2$	454
$Mo_2(OPr^i)_6$	378
$Mo_2(OPr^i)_8$	220

Using the equation, we obtain E(Mo-Mo) = 384 kJ mol^{-1} in $Mo_2(NMe_2)_6$ which may be compared with the value, (396±18) kJ mol^{-1}, derived earlier. The enthalpy contribution of the metal-metal bond in $[(\eta\text{-}C_5H_5)_2Mo_2(CO)_6]$($r$(Mo-Mo) 323.5 pm (*26*) is calculated to be ~75 kJ mol^{-1} which is only slightly greater than that (~65 kJ mol^{-1}) calculated for $[Mo_2(OPr^i)_6(NO)_2]$(r(Mo-Mo) 333.5 pm (*27*)) in which there is no metal-metal bond. Decarbonylation of $[(\eta\text{-}C_5H_5)_2Mo_2(CO)_6]$ produces (*28*) $[(\eta\text{-}C_5H_5)_2Mo_2(CO)_4]$($r$(Mo-Mo) 244.8 pm) for which E(Mo-Mo) is calculated to be ~250 kJ mol^{-1}. The shortest molybdenum-molybdenum bond (203.7 pm) known at the present time is found (*29*) in $[Mo_2\{N(2\text{-pyridyl})acetamide\}_4]$ for which E(Mo-Mo) is calculated to be ~550 kJ mol^{-1}.

Literature cited

1. Diatomic Metals and Metallic Clusters, Symposium. Faraday Div. Chem. Soc. *14*, (1980) in press.
2. K.A. Gingerich, in ref. 1 and refs. therein.
3. C. Wood, M. Doran, I.H. Hillier, in ref. 1 and refs. therein.
4. a) W. Klotzbucher and G.A. Ozin, Inorg. Chem. *16*, 984 (1977).
   b) J.G. Norman, H.J. Kolari, H.B. Gray and W.C. Trogler, Inorg Chem., *16*, 987 (1977).
   c) B.E. Bursten, F.A. Cotton and M.B. Hall, J. Amer. Chem. Soc. *102*, 6348 (1980).
   d) P.M.Atha, I.H. Hillier, and M.F. Guest, Chem. Phys. Letters, *75*, 84 (1980).
5. S.K. Gupta, R.M. Atkins and K.A. Gingerich, Inorg. Chem. *17*, 3211 (1978).
6. Y.M. Efremov, A.N. Samoilova, V.B. Kozhukhovsky and L.V. Gurvich, J. Mol. Spectr. *73*, 430 (1980).

7. A.R. Miedema, in ref. 1 and refs. therein.
8. S.K. Gupta and K.A. Gingerich, J. Chem. Phys.*70*, 5350 (1979).
9. L. Pauling, The Nature of the Chemical Bond, Cornell University Press, Ithaca, N.Y., 3rd edn., 1960.
10. R. Haque and K.A. Gingerich, J. Chem. Thermodyn. *12*, 439 (1980).
11. R. Haque, M. Pelino and K.A. Gingerich, J. Phys. Chem. *71*, 2929 (1979).
12. J.A. Connor in 'Transition Metal Clusters' (editor, B.F.G. Johnson) Wiley, London. 1980. p. 345.
13. C.E. Housecroft, K. Wade and B.C. Smith, J.C.S. Chem. Comm., 1978, 765; J. Organometallic Chem., *170*, C1 (1979).
14. A. Mitschler, B. Rees and M.S. Lehmann, J. Amer. Chem. Soc., *100*, 3390 (1978).
15. W. Heijser, E.J. Baerends and P. Ros, in ref. 1.
16. M. Benard, private communication.
17. F.A. Adedeji, K.J. Cavell, S. Cavell, J.A. Connor, G. Pilcher, H.A. Skinner, and M.T. Zafarani-Moattar, J.C.S. Faraday 1, *75*, 603 (1979).
18. J.A. Connor, G. Pilcher, H.A. Skinner, M.H. Chisholm and F.A. Cotton, J. Amer. Chem. Soc. *100*, 7738 (1978).
19. J.P. Krause and D.R. Bidinosti, Canad. J. Chem., *53*, 628 (1975).
20. M.H. Chisholm and F.A. Cotton, Ace. Chem. Res.*11*, 356 (1978).
21. M.B. Hall, J. Amer. Chem. Soc., *102*, 2104 (1980).
22. L.R. Morss, R.J. Porcja, J.W. Nicoletti, J. San Fillippo and H.D.B. Jenkins, J. Amer. Chem. Soc., *102*, 1923 (1980).
23. R.J.H. Clark and N.R. D'Urso, J. Amer. Chem. Soc., *100*, 3088 (1978).
24. K.J. Cavell, C.D. Garner, G. Pilcher and S. Parkes, J.C.S. Dalton, 1979, 1714; M.T. Zafarani-Moattar, Ph.D. thesis Manchester 1979.
25. K.J. Cavell, J.A. Connor, G. Pilcher, M.A.V. Ribeiro da Silva, M.D.M.C. Ribeiro da Silva, H.A. Skinner, Y. Virmani, and M.T. Zafarani-Moattar, to be published.
26. R.D. Adams, D.M. Collins and F.A. Cotton, Inorg. Chem., *13*, 1086 (1974).
27. M.H. Chisholm, F.A. Cotton, M.W. Extine and R.L. Kelly, J. Amer. Chem. Soc., *100*, 3354 (1978).
28. R.J. Klingler, W. Butler and M.D. Curtis, J. Amer. Chem. Soc., *97*, 3535 (1975).
29. F.A. Cotton, W.H. Ilsley and W. Kaim, Inorg. Chem., *18*, 2171 (1979).

RECEIVED December 11, 1980.

11

Breaking Metal–Metal Multiple Bonds

Their Use as Synthetic Starting Materials

RICHARD A. WALTON

Department of Chemistry, Purdue University, West Lafayette, IN 47907

With the explicit recognition of the existence of the first metal-metal quadruple bond in 1964(1), efforts over the intervening years have been directed towards establishing which classes or organometallic and non-organometallic complexes of the transition metals possess metal-metal multiple bonds. Presently, such bonds have been recognized in compounds of a majority of the transition metals(2) ranging from species as diverse as the vanadium (II) dimer $V_2(DMP)_4$ (DMP is the 2,6-dimethoxyphenyl ligand)(3,4) to the hydrido-bridged iridium complex $[Ir_2H_5(PPh_3)_4](PF_6)$(5). Through theoretical and spectroscopic studies(6,7) on selected examples of key classes of dimers containing triple or quadruple bonds, a fairly clear picture of the bonding in dimers of the types M_2L_8 and M_2L_6 has now emerged.

As synthetic routes to multiply bonded dimers have been devised (some serendipitously, others by design)(2) so parallel developments in the reaction chemistry of many of these molecules have taken place. To date, the molecules of most interest in this latter regard have been those which contain metal-metal quadruple or triple bonds and which fall into categories (A)-(D).

(A) M_2L_8 skeleton; bond order = 4(i.e. $(\sigma)^2(\pi)^4(\delta)^2$ ground state configuration), M = Cr(II), Mo(II), W(II) or Re(III).

(B) M_2L_8 skeleton; bond order = 3(i.e. $(\sigma)^2(\pi)^4(\delta)^2(\delta^*)^2$ configuration), M = Re(II).

(C) M_2L_6 skeleton; bond order = 3(i.e. $(\sigma)^2(\pi)^4$ ground state configuration), M = Mo(III) or W(III).

(D) $(\eta^5\text{-}C_5H_5)_2M_2(CO)_4$, bond order = 3 (on the basis of the EAN rule), M = Mo or W.

The above classifications will serve as a convenient means of identifying these molecules in our subsequent discussions.

Efforts to explore and categorize the reaction patterns of triply-bonded dimers of molybdenum and tungsten of the types M_2L_6 (L = R, NR_2 or OR) and $(\eta^5\text{-}C_5H_5)_2M_2(CO)_4$, i.e., those of classes (C) and (D), have been pursued(8-13). Reactions which may be viewed as examples of ligand substitution, Lewis base association,

0097-6156/81/0155-0207$05.00/0

dimer to cluster transformation, oxidative-addition, reductive-elimination and metal-metal bond cleavage have been well documented. In the case of the Lewis base associations, these reactions may be reversible or irreversible and may (class (D) dimers) or may not (class (C) dimers) be accompanied by a decrease in M-M bond order (8,9,10). In addition, there is an example of an interesting "carbene-like" addition of X: to M≡M, namely, the reversible formation of the carbonyl-bridged dimer $Mo_2(OBu^t)_6CO$ (Mo-Mo double bond) upon reacting $Mo_2(OBu^t)_6$ with carbon monoxide in hydrocarbon solvents(14).

While certain reactions of complexes which contain the M_2L_8 skeleton, and M-M quadruple or triple bonds (*i.e.* molecules of classes (A) and (B), *vide supra*), resemble those of dimers of classes (C) and (D), other aspects of their chemical reactivity are noticeably different. They resemble dimers (C) and (D) in undergoing ligand substitution reactions(2), Lewis base associations (particularly in the case of carboxylate-bridged dimers of the type $M_2(O_2CR)_4$)(2) and dimer to cluster transformations, recent examples of the latter being $M_2 \longrightarrow M_3$ (triangular)(15,16), $M_2 \longrightarrow M_4$ (planar)(17,18) and $M_2 \longrightarrow M_4$ (distorted tetrahedron)(19,20). Also, oxidative-addition and reductive-elimination reactions have been documented. The best example of the former is the conversion of $Mo_2(O_2CCH_3)_4$ to $Mo_2Cl_8H^{3-}$, by warm concentrated hydrochloric acid(21), a reaction which involves formally the substitution of acetate by Cl^- plus the oxidative-addition of the elements of HCl. The reaction of $Mo_2Cl_8H^{3-}$ with pyridine to produce $Mo_2Cl_4(py)_4$ (22), may be viewed (2) as the base induced reductive-elimination of HCl accompanied by the partial substitution of Cl^- by pyridine.

Dimers of types (A) and (B) differ from those of (C) and (D) in exhibiting a very rich redox chemistry, often reversible, in which the products retain a dimeric multiply-bonded structure. Several examples of this are given in Figure 1 where closely interrelated electrochemical and chemical oxidations and reductions of dimers containing the Re_2^{n+} core (n = 4,5,6 or 8) are represented in the "grid". Most of these reactions have been discovered in our laboratory(23, 24, 25, 26) while others stem from earlier studies by Cotton and co-workers(27,28). Those reactions which remain to be accomplished are represented by a dashed line. These studies are important because they demonstrate the ability of the Re_2^{n+} core to withstand at least a 4-electron oxidation or reduction. While powerful oxidants can promote the complete disruption of the dimer unit, it is clear that many of these dimeric species are not at all fragile under many oxidizing and reducing conditions. An even more extensive redox chemistry may yet be discovered.

While the structural integrity of dimers of classes (A)-(D) under a variety of reaction conditions has been illustrated in our preceding discussion, nonetheless there are many instances where facile M-M bond cleavage is encountered. It is this type of

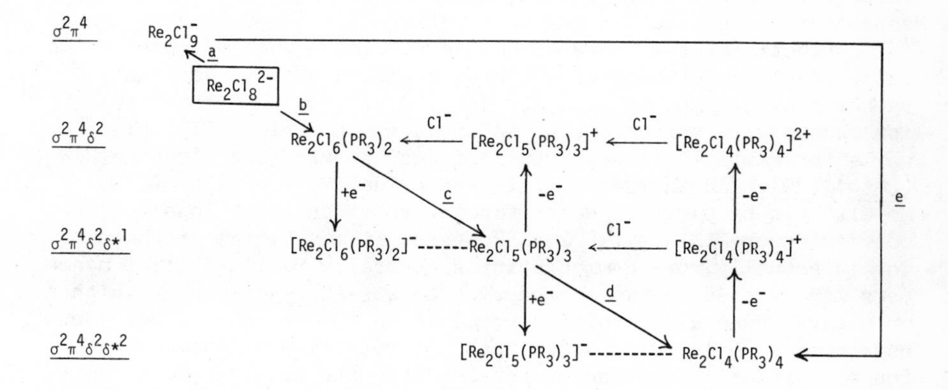

Figure 1. Electrochemical and chemical oxidations and reductions of Re_2^{n+} complexes.

Electrochemical oxidations and reductions are represented as $-e^-$ and $+e^-$, respectively, and reaction with chloride ion by Cl^-. Other reactions are: (a) Cl_2 oxidation; (b) reaction with PR_3 at room temperature; (c) reaction with $PRPh_2$(R = Me or Et) under reflux; (d) reaction with PR_2Ph (R = Me or Et) or PR_3(R = Me, Et, Pr^n, or Bu^n) under reflux; (e) reaction with PEt_3.

reaction which has been of recent interest to us, and which we have found may be exploited to synthetic advantage. Specific examples of reactions of this type as we have encountered between class (A) and (B) dimers and σ-donor and π-acceptor ligands will be considered in the following discussion. This account constitutes a status report as of mid-1980.

Cleavage Reactions by σ-Donors.

Generally speaking it is the reactions of σ-donors with quadruply-bonded dimers of class (A), particularly those containing the Re_2^{6+} core, where examples of M-M bond cleavage are encountered. To date such behavior with dimers of types (B), (C) and (D) is rare. However, even with Re_2^{6+} and other isoelectronic species this reaction course is the exception rather than the rule, is unpredictable and as yet generally offers little promise as a useful synthetic route to new monomeric species. Such examples as do exist have more often than not been stumbled upon by accident rather than design, as with our early discovery that the $Re_2Cl_8^{2-}$ ion reacts with thiourea (tu) to produce monomeric $ReCl_3(tu)_3$, while tetramethylthiourea(tmtu) affords the expected dimer $Re_2Cl_6(tmtu)_2$ (29). Another example is the oxidative cleavage of $Mo_2Cl_4(py)_4$ by pyridine under forcing reaction conditions to give monomeric *mer*-$MoCl_3(py)_3$(22). However, it may be that such behavior is actually more common than is generally realized since monomers may be unsuspected secondary products in reactions in which a multiply-bonded dimer unit is retained in the major product. We have, for example, discovered that the reduction of the $Re_2Cl_8^{2-}$ ion by tertiary phosphines which leads to the triple-bonded $Re_2Cl_4(PR_3)_4$(23), is accompanied by the formation of soluble monomers. These are isolable as *trans*-$ReCl_4(PR_3)_2$ upon the addition of 6N hydrochloric acid to the reaction filtrates(30).

There are, however, two systems where metal-metal bond cleavage by σ-donors is clearly the favored reaction pathway. The addition of tertiary phosphines to the salt $(Bu_4N)_2Re_2(NCS)_8$, itself prepared by the reaction of $(Bu_4N)_2Re_2Cl_8$ with NaSCN(31), gives the insoluble green thiocyanate-bridged dimers $(Bu_4N)_2Re_2(\mu\text{-}NCS)_2(NCS)_6(PR_3)_2$ (I) which contain magnetically dilute rhenium(III)(32).

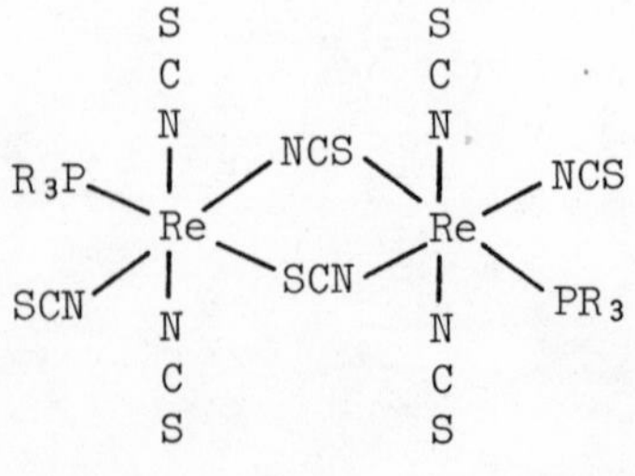

```
            S           S
            C           C
            N    NCS    N
   R3P      |  /     \  |     NCS
        \   | /       \ |   /
           Re           Re                  I
        /   | \       / |   \
   SCN      |  \ SCN /  |     PR3
            N           N
            C           C
            S           S
```

Why this product is formed rather than the metal-metal bonded dimer $Re_2(NCS)_6(PR_3)_2$, which would be analogous to the $Re_2Cl_6(PR_3)_2$ dimers that are formed initially upon reacting $Re_2Cl_8^{2-}$ with monodentate tertiary phosphines(23,27), remains a puzzle. Dimers of type I may of course be the thermodynamically favored products (whereas $Re_2X_6(PR_3)_2$ dimers may not) and it is certainly true that spectroscopic evidence(32) favors a weaker Re-Re bond in $Re_2(NCS)_8^{2-}$ compared to $Re_2Cl_8^{2-}$, perhaps a hint that it is more susceptible to fission. Unfortunately, the mechanistic features of the reactions which lead to I are unknown.

The second (and final) set of reactions which will concern us in this section are those in which acetonitrile solutions of the salts $(Bu_4N)_2Re_2X_8$ (X = Cl or Br) are converted to the centrosymmetric halogen-bridged dimers $Re_2X_6(LL)_2$(II) by 1,2-bis(diphenylphosphino)ethane and 1-diphenylphosphino-2-diphenylarsinoethane (33,34,35). This behavior contrasts with the substitution using monodentate tertiary phosphines and arsines which gives rise to $Re_2X_6(PR_3)_2$.

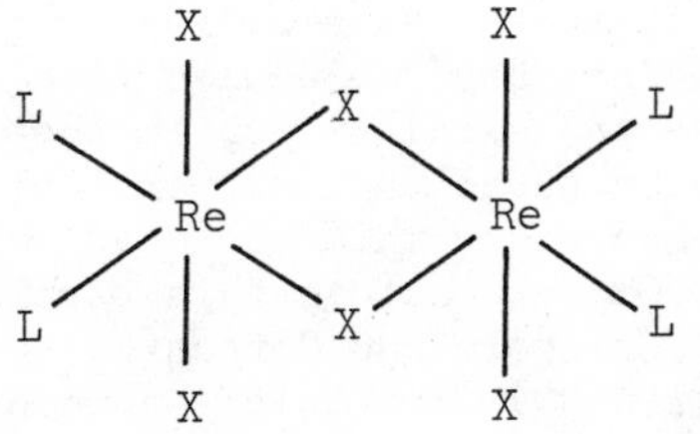

X X
L X L
Re Re II
L X L
X X

The question as to why II is formed rather than a dimer which retains a multiple bond has been addressed by Shaik and Hoffmann (36) through a comparison with other d^4-d^4 dimers, specifically the "Vahrenkamp compound" $(CO)_4V(\mu\text{-}PMe_2)_2V(CO)_4$, possessing a V-V double bond(37), and the "Cotton structure" $Re_2Cl_8L_2^{2-}$, an unbridged quadruply-bonded dimer(6). Shaik and Hoffmann(36) have contended that the possibility of the "Walton complexes" possessing the alternative Re-Re double bonded di-μ-Cl structure is unlikely when LL = $Ph_2PCH_2CH_2PPh_2$ or $Ph_2PCH_2CH_2AsPh_2$, because with these particular ligands the small Re-Cl-Re angle needed to ensure formation of such a double-bond is inaccessible. Of particular significance in this latter regard is our isolation(34) of diamagnetic $Re_2Cl_6(dppm)_2$ (dppm = bis(diphenylphosphino)methane), which we have proposed possesses structure III. Unfortunately, crystals

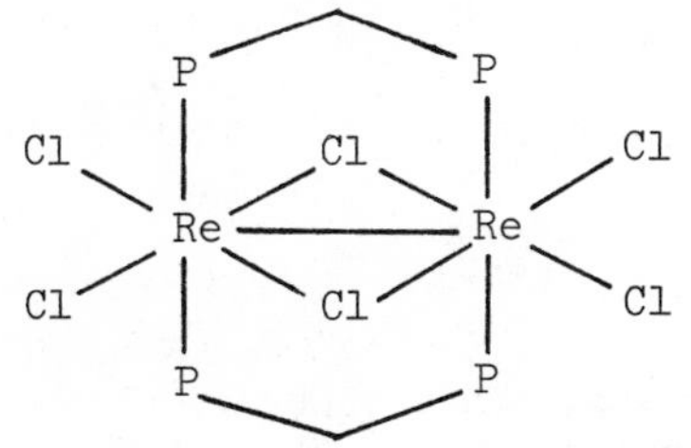

P P
Cl Cl Cl
Re Re III
Cl Cl Cl
P P

suitable for an x-ray structure analysis have not yet been obtained, so our structure conclusions have been based principally upon spectroscopic measurements, including x-ray photoelectron spectroscopy(34). The possibility that this molecule is an example of a doubly-bonded Vahrenkamp type structure(36) is an attractive one. The bridging mode of dppm as depicted in III might relieve some of the steric problems which would develop in II upon forcing a closer approach of the metal atoms.

Cleavage Reactions by π-Acceptor Ligands.

Since the majority of this work has been carried out in our laboratory it is perhaps appropriate to consider the results in a historical perspective. In the light of the rather confusing and largely ill-understood nature of those reactions of quadruply-bonded dimers with σ-donors which result in M-M bond breaking, an obvious question concerned the related behavior of sterically undemanding π-acceptors such as CO, NO and the alkyl and aryl isocyanides. Our interest in these systems arose in 1975 because of the possibility of binding these molecules to class (A) and (B) dimers either in the vacant axial sites, in the form of M–L–M bridges or through ligand substitution (particularly of PR_3). Our objective had been to learn how this would affect the nature of the M-M multiple bonding. Of interest in this particular context is the later discovery by Chisholm and Cotton(14) of the reversible formation of $Mo_2(OBu^t)_6(\mu\text{-}CO)$ from triply-bonded $Mo_2(OBu^t)_6$.

Carbon Monoxide, Nitric Oxide and the Nitrosonium Cation.

The first system we looked at was that involving the reaction of $Re_2X_4(PR_3)_4$ (X = Cl or Br; R = Et or Pr^n) with CO(38,39). No special significance was attached to this choice other than the fact that, at the time, $Re_2X_4(PR_3)_4$ constituted a new class of triply-bonded dimers(23) whose reactivity we were particularly interested in. As it turned out, the reactions between CO and $Re_2X_4(PR_3)_4$ in refluxing ethanol, toluene or acetonitrile proved to be extremely complicated. The paramagnetic 17-electron monomers $ReX_2(CO)_2(PR_3)_2$ (assigned the all *trans*-geometry IV on the basis of their spectroscopic properties) were the primary reaction

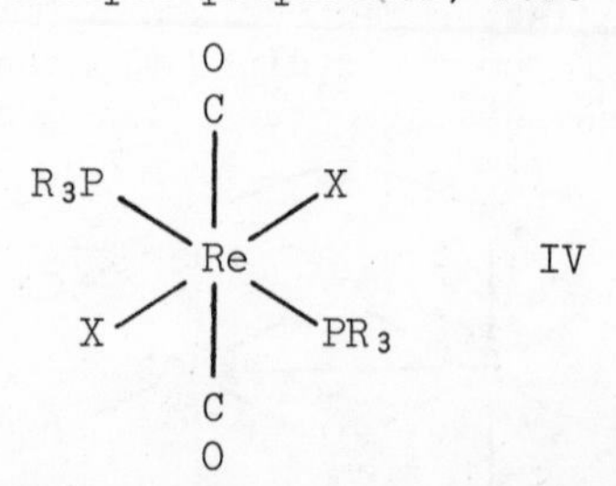

products(38,39). However, since these complexes were themselves converted to other species (*e.g.* $ReX(CO)_3(PR_3)_2$) under our reaction conditions and other competing reactions (not involving CO) were also taking place, such as the reaction of $Re_2Cl_4(PPr^n_3)_4$ with ethanol to produce a mixture of $Re_2Cl_5(PPr^n_3)_3$ and $ReCl(CO)_3(PPr^n_3)_2$ (39), it was some time before we were able to fully understand these complex systems. However, the important point was that trans-$ReX_2(CO)_2(PR_3)_2$ is formed rapidly by cleavage of the Re-Re triple bond. Unfortunately, we were unable to isolate any dimeric carbonyl-containing intermediates.

The complicated course of the reactions involving $Re_2X_4(PR_3)_4$ coupled with our failure to induce any significant reaction between CO and the quadruply-bonded dimers $Re_2Cl_6(PR_3)_2$(39) and $Mo_2X_4(PR_3)_4$ (X = Cl or NCS)(40) led us to ponder the question of whether any significance could be attached to this apparent difference in reactivity between systems containing the $(\sigma)^2(\pi)^4(\delta)^2$ and $(\sigma)^2(\pi)^4(\delta)^2(\delta^*)^2$ electronic configurations. Actually, a plausible explanation for the absence of reaction between CO and $Re_2Cl_6(PR_3)_2$ and $Mo_2X_4(PR_3)_4$ may be the low solubilities of the reagents in the reaction solvents and the more sluggish nature of these reactions compared to $Re_2X_4(PR_3)_4$. An increase in the pressure of CO beyond atmosphere may enhance its reactivity towards these dimers since it is apparent that other π-acceptors (vide infra) react rapidly with dimers containing metal-metal quadruple bonds.

The addition of NO to dichloromethane solutions of $Mo_2X_4L_4$, where X = Cl or Br and L = $PEtPh_2$, PEt_3 or PBu^n_3, and $Mo_2X_4(LL)_2$, where X = Cl or NCS and LL = $Ph_2P(CH_2)_nPPh_2$ (n = 1 or 2), gives the 18-electron dinitrosyl monomers $Mo(NO)_2X_2L_2$ and $Mo(NO)_2X_2(LL)$(40). These reactions constitute the first general synthetic route to a wide range of phosphine derivatives of this type. Actually, our study (40) had followed an earlier one by Kleinberg and co-workers (41) who found that the treatment of $K_4Mo_2Cl_8$ and $Mo_2(O_2CCH_3)_4$ with NOCl gave only monomeric products (based upon the $Mo(NO)Cl_3$ moiety).

There are now three well-defined systems where NO cleaves M-M triple or quadruple bonds, the other two examples being the conversion of $(\eta^5\text{-}C_5Me_5)_2M_2(CO)_4$(M = Cr or Mo) to $(\eta^5\text{-}C_5Me_5)M(CO)_2(NO)$(11) and of $Mo_2(OR)_6$ to the alkoxy-bridged dimers $Mo_2(OR)_6(NO)_2$ (R = CMe_3, $CHMe_2$ or CH_2CMe_3)(42), none of which contain a Mo-Mo bond. Each of these three groups of reactions is quite different, the production of $Mo(NO)_2X_2L_2$ and $Mo(NO)_2X_2(LL)$ involves an increase in the electron count of the metal from 16 to 18, whereas no change is encountered in the conversion of $(\eta^5\text{-}C_5Me_5)_2M_2(CO)_4$ to $(\eta^5\text{-}C_5Me_5)M(CO)_2(NO)$. In all instances the fission of the M-M multiple bonds is accompanied formally by the formation of Mo-N bonds which exhibit some degree of multiple character.

An interesting difference exists when NO^+ (as its hexafluorophosphate salt) is used in place of CO and NO. With $Re_2X_4(PEt_3)_4$ (X = Cl or Br) and $Re_2Cl_4(LL)_2$ (LL = $Ph_2PCH_2CH_2PPh_2$ or $Ph_2PCH_2CH_2AsPh_2$), which possess a very accessible oxidation(25,43), electron-transfer proceeds extremely rapidly to afford stable $[Re_2X_4(PEt_3)_4]PF_6$ and $[Re_2Cl_4(LL)_2]PF_6$ with the concomitant release of NO gas. With $Mo_2X_4(PR_3)_4$, oxidation also occurs but we have been unable to isolate well-defined products(40), the problem being the rather fragile nature of the $[Mo_2X_4(PR_3)_4]^+$ cations(44) under the reaction conditions we originally used(40).

The question which now arises concerns the mechanism whereby the M-M bond of class (A) and (B) dimers is cleaved. With species of the type which possess the M_2L_8 skeleton, the formation of a "weak" adduct $M_2L_8L'_2$ (L' = CO, NO or NO^+) is feasible both sterically and electronically. However, no such complexes have been isolated nor detected spectroscopically since it seems that reaction proceeds rapidly beyond this stage. If L' is a strongly π-accepting ligand (as is CO and NO) then it will compete for metal-based d_π orbital density. This will weaken the M-M bond by diminishing (perhaps drastically) the magnitude of the π-contribution to both the quadruply and triply-bonded dimers of classes (A) and (B). It would seem that the consequence of this is to activate the dimers to attack by two additional molecules of L' thereby producing monomeric $ML_4L'_2$. With the NO^+ ligand, initial coordination is apparently followed by electron-transfer to give $M_2L_8^+$ and NO, the cation crystallizing as its PF_6^- salt before the more sluggish reaction with further NO^+ (or the released NO) can occur.

Alkyl and Aryl Isocyanides. Perhaps the most interesting and useful results concerning M-M bond cleavage are those which have emerged from the reactions between the alkyl and aryl isocyanides and dimers containing the Cr_2^{4+}, Mo_2^{4+}, W_2^{4+} and Re_2^{6+} cores, *i.e.*, those of class (A). Long before the existence of multiple metal-metal bonds was proposed(1), Malatesta(45) had found that chromium(II) acetate $Cr_2(O_2CCH_3)_4 \cdot 2H_2O$ was converted to $Cr(CNPh)_6$ by phenyl isocyanide. Much more recently Gray and co-workers(46) used $Mo_2(O_2CCH_3)_4$ as a means of preparing $Mo(CNPh)_6$ by stirring a suspension of the acetate with an excess of phenyl isocyanide in methanol. In neither study(45,46) was the objective to investigate the general phenomenon of cleavage reactions involving M-M multiple bonds *per se*.

Our initial foray into the isocyanide reactions was the discovery(47) that the alkyl isocyanides RNC (R = CH_3, CMe_3 or C_6H_{11}) cause fission of the Mo-Mo quadruple bond of $Mo_2(O_2CCH_3)_4$ and $K_4Mo_2Cl_8$ to generate the 18-electron seven coordinate cations $[Mo(CNR)_7]^{2+}$ which were isolated as their BF_4^- and PF_6^- salts. These "one-pot" reactions, using methanol as the solvent, are an especially convenient entry into this class of homoleptic isocyanide complexes, species which had first been discovered by Lippard (48). They have an interesting structural and reaction chemistry which has proved worth pursuing in its own right(47,48,49,50), but these particular features will not concern us here.

The generality of the reaction Mo_2^{4+} + RNC $\longrightarrow$ $2[Mo(CNR)_7]^{2+}$, has been further explored by extending the systems studied to include the phosphine dimers $Mo_2Cl_4(PR_3)_4$, $Mo_2Cl_4(dppm)_2$ and $Mo_2Cl_4(dppe)_2$(49). Again cleavage of the Mo-Mo bond occurs but in these instances mixed phosphine-alkyl isocyanide complexes are obtained. These are of the types $[Mo(CNR)_6(PR_3)]^{2+}$, $[Mo(CNR)_5(PR_3)_2]^{2+}$, $[Mo(CNR)_5(dppm)]^{2+}$ and $[Mo(CNR)_5(dppe)]^{2+}$ all of which have been isolated as their PF_6^- salts. These same complexes can also be obtained by phosphine substitution of the $[Mo(CNR)_7]^{2+}$ cations(49).

In a closely parallel study to that of ourselves(47,49), Girolami and Andersen(51) claimed that the addition of t-butyl isocyanide to solid $Mo_2(O_2CR)_4$ (R = CH_3 or CF_3) gave $Mo(CNCMe_3)_4(O_2CCH_3)_2$ and $Mo(CNCMe_3)_5(O_2CCF_3)_2$, whereas solid $K_4Mo_2Cl_8$ and $Mo_2Cl_4(PBu^n_3)_4$ are both converted to $Mo(CNCMe_3)_5Cl_2$. Note the apparent dependence of the latter reactions upon the presence or absence of solvent, since in our hands(47,49) we isolate $[Mo(CNR)_7]^{2+}$ and $[Mo(CNR)_6(PR_3)]^{2+}$, respectively, when methanol is used as the reaction solvent. A summary of the important reactions between alkyl isocyanides and Mo_2^{4+} species is given in Figure 2. The data we draw upon is that from our own work(47,49).

At the end of our earlier discussion of the reactions between class (A) and (B) dimers and CO, NO and NO^+, a mechanism for the M-M bond cleavage was outlined. We proposed that the most likely first step is the conversion of M_2L_8 to $M_2L_8L'_2$ through axial coordination of L'. In the case of molybdenum(II) carboxylates and other ligand-bridged dimers, an alternative possibility is non-axial coordination of L' through the opening of the carboxylate bridge, some of the carboxylato ligands now functioning as monodentate donors(52).

The development of this chemistry to include the related tungsten(II) systems is at present restricted by the non-existence of such dimers as $W_2(O_2CR)_4$ and $W_2Cl_8^{4-}$(2). We have circumvented this problem by resorting to the quadruply-bonded dimers $W_2(mhp)_4$ (mhp is the anion of 2-hydroxy-6-methylpyridine)(53) and $W_2(dmhp)_4$ (dmhp is the anion of 2,4-dimethyl-6-hydroxypyrimidine)(54) as starting materials. The homoleptic isocyanide complexes $[W(CNR)_7](PF_6)_2$ are formed upon heating $W_2(mhp)_4$ with a mixture of RNC (R = CMe_3 or C_6H_{11}) and KPF_6 in acetone(50). The red phenyl isocyanide complex $W(CNPh)_6$ is formed upon stirring a mixture of $W_2(dmhp)_4$ and PhNC at room temperature or below(55). This synthesis of $W(CNPh)_6$ is certainly preferable to the existing low yield literature procedure(46) and we are currently working on methods to improve the yields from $W_2(dmhp)_4$(55). Since dimers such as $W_2(mhp)_4$ and $W_2(dmhp)_4$ are very susceptible to decomposition via oxidation, particularly through reaction with adventitious oxygen, we are currently interested in ways of increasing the oxidation potential for the process $W_2L_8 \longrightarrow [W_2L_8]^+$ (probably the first step in their decomposition) by changing the bridging ligand. These changes can be monitored by cyclic voltammetry and will hopefully lead to more stable starting materials.

One final point which will concern us before we leave the

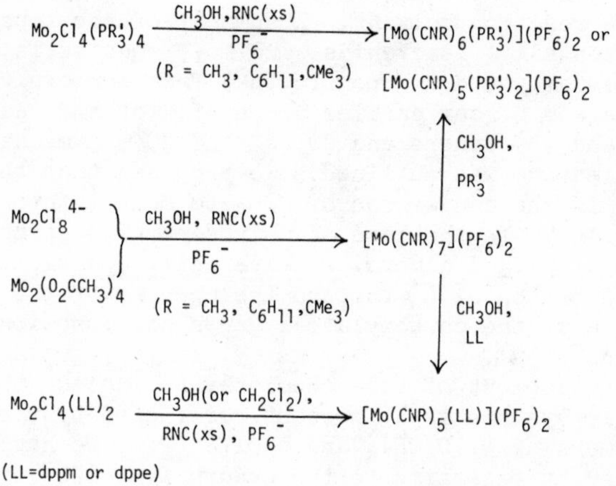

$$Mo_2Cl_4(PR'_3)_4 \xrightarrow[\substack{PF_6^- \\ (R = CH_3,\ C_6H_{11}, CMe_3)}]{CH_3OH, RNC(xs)} [Mo(CNR)_6(PR'_3)](PF_6)_2 \text{ or } [Mo(CNR)_5(PR'_3)_2](PF_6)_2$$

$$\left.\begin{matrix} Mo_2Cl_8^{4-} \\ Mo_2(O_2CCH_3)_4 \end{matrix}\right\} \xrightarrow[\substack{PF_6^- \\ (R = CH_3,\ C_6H_{11}, CMe_3)}]{CH_3OH,\ RNC(xs)} [Mo(CNR)_7](PF_6)_2$$

$$[Mo(CNR)_7](PF_6)_2 \xrightarrow{CH_3OH,\ PR'_3} [Mo(CNR)_5(PR'_3)_2](PF_6)_2$$

$$[Mo(CNR)_7](PF_6)_2 \xrightarrow{CH_3OH,\ LL} [Mo(CNR)_5(LL)](PF_6)_2$$

$$Mo_2Cl_4(LL)_2 \xrightarrow[RNC(xs),\ PF_6^-]{CH_3OH(\text{or } CH_2Cl_2),} [Mo(CNR)_5(LL)](PF_6)_2$$

(LL=dppm or dppe)

Figure 2. Formation of isocyanide complexes of molybdenum(II) starting from Mo_2^{4+} derivatives

topic of multiply-bonded dimers of the Group IV elements, is the reaction of Cr_2^{4+} dimers. We have already mentioned how $Cr(CNPh)_6$ is formed in good yield from $Cr_2(O_2CCH_3)_4 \cdot 2H_2O$(45). However, our preliminary attempts to prepare the previously unknown alkyl isocyanide complexes $[Cr(CNR)_6]^{2+}$ by cleaving the Cr-Cr bonds of $Cr_2(O_2CCH_3)_4 \cdot 2H_2O$ and $Cr_2(mhp)_4$(53) were unsuccessful. It appears that these systems are quite susceptible to decomposition. Since we have now discovered a very simple alternative route to $[Cr(CNR)_6]^{2+}$ and their seven coordinate analogs $[Cr(CNR)_7]^{2+}$ (R = C_6H_{11} or CMe_3)(56), we are not at present pursuing studies on reactions between isocyanides and Cr_2^{4+} dimers.

The quadruple bond in rhenium(III) dimers is likewise cleaved by alkyl isocyanides. The treatment of the acetate complex $Re_2(O_2CCH_3)_4Cl_2$ with neat t-butyl isocyanide is said to give red $ReCl(CNCMe_3)_5$ in 80% yield(51), while such a reaction when carried out in refluxing methanol produces a brown solution from which $[Re(CNCMe_3)_6]PF_6$ may be precipitated in 60% yield upon the addition of KPF_6(57). The benzoate dimer $Re_2(O_2CPh)_4Cl_2$ reacts in a similar fashion to give $[Re(CNCMe_3)_6]PF_6$(57). The formation of the homoleptic 18-electron $[Re(CNCMe_3)_6]^+$ is not unexpected in view of the conversion of $Mo_2(O_2CCH_3)_4$ to $[Mo(CNR)_7]^{2+}$ since the alternative seven-coordinate 18-electron system $[Re(CNCMe_3)_7]^{3+}$ is unknown and $[Re(CNR)_6]^+$ species like their manganese analogs $[Mn(CNR)_6]^+$ appear to be quite stable(58). The reactions of the rhenium(III) carboxylates $Re_2(O_2CR)_4Cl_2$ with isocyanides must proceed by dissociation of a terminally bound chlorine atom (*i.e.* $Re_2(O_2CR)_4Cl_2 + RNC \longrightarrow [Re_2(O_2CR)_4Cl(CNR)]^+ + Cl^-$) or through attack of RNC at an 'equatorial' position with concomitant opening of a carboxylate bridge.

The reactions between $(Bu_4N)_2Re_2X_8$ (X = Cl or Br) and alkyl isocyanides proceeds, so it seems, in a different fashion to that described above for $Re_2(O_2CR)_4Cl_2$. The one product isolated from the reaction between $(Bu_4N)_2Re_2Cl_8$ and methyl isocyanide in methanol is the green rhenium(IV) complex $(Bu_4N)[ReCl_5(CNCH_3)]$(59), but the nature of the 'reduction product' is unknown, as also is the fate of the major proportion of the rhenium. We have found (57) that a mixture of t-butyl isocyanide and $(Bu_4N)_2Re_2X_8$(X = Cl or Br) when stirred in methanol at room temperature produces a red-brown solution which gives $[Re(CNCMe_3)_5X_2]PF_6$ in 70% yield upon the addition of an acetone solution of KPF_6. These seven-coordinate cations appear to bear a close relationship to the isoelectronic neutral monomers $Re(CNR)_4Br_3$ which were prepared and structurally characterized by Treichel *et al*(60) by a different method.

Conclusions.

While metal-metal multiple bonds are occasionally cleaved by σ-donors, these reactions are sufficiently unpredictable that little advantage can yet be taken of them to design synthetic routes to desired monomeric complexes. In contrast, the facile cleavage of the metal-metal multiple bonds of dimers containing the M_2L_8 skeleton and the Cr_2^{4+}, Mo_2^{4+}, W_2^{4+}, Re_2^{6+} and Re_2^{4+} cores by the π-acceptor ligands CO, NO, CNR (R = alkyl) and CNAr (Ar = aryl) provides an excellent synthetic route to monomeric species, many of which cannot yet be made readily by other methods.

Acknowledgements

I thank the National Science Foundation for support of our research and wish to acknowledge the research efforts of my many capable co-workers. The contributions of Dr. Patrick Brant, Dr. Tayseer Nimry, Ms. Cathy Hertzer, Mr. William S. Mialki, and Mr. Thomas E. Wood have been particularly invaluable.

Literature Cited

1. Cotton, F. A.; Curtis, N. F.; Harris, C. B.; Johnson, B. F.; Lippard, S. J.; Mague, J. T.; Robinson, W. R.; Wood, J. S. Science, 1964, 145, 1305.
2. Cotton, F. A.; Walton, R. A. "Compounds with Multiple Bonds Between Metal Atoms", J. Wiley & Sons in preparation (1980).
3. Cotton, F. A.; Millar, M. J. Am. Chem. Soc., 1977, 99, 7886.
4. Seidel, W.; Kreisel, G.; Mennenga, H. Z. Chem., 1976, 16, 492.
5. Crabtree, R. H.; Felkin, H.; Morris, G. E.; King, T. J.; Richards, J. A. J. Organomet. Chem., 1976, 113, C7.
6. Cotton, F. A. Acc. Chem. Res., 1978, 11, 225.
7. Gray, H. B.; Trogler, W. C. Acc. Chem. Res., 1978, 11, 232.
8. Chisholm, M. H.; Cotton, F. A. Acc. Chem. Res., 1978, 11, 356.
9. Chisholm, M. H. Trans. Met. Chem., 1978, 3, 321.
10. Chisholm, M. H. Adv. Chem. Ser., 1979, No. 173, 396.
11. King, R. B.; Efraty, A.; Douglas, W. M. J. Organomet. Chem., 1973, 60, 125.
12. Curtis, M. D.; Klingler, R. J. J. Organomet. Chem., 1978, 161, 23.
13. Ginley, D. S.; Bock, C. R.; Wrighton, M. S. Inorg. Chim. Acta, 1977, 23, 85.
14. Chisholm, M. H.; Cotton, F. A.; Extine, M. W.; Kelly, R. L. J. Am. Chem. Soc., 1979, 101, 7645.
15. Glicksman, H. D.; Hamer, A. D.; Smith, T. J.; Walton, R. A. Inorg. Chem., 1976, 15, 2205.
16. Glicksman, H. D.; Walton, R. A. Inorg. Chem., 1978, 17, 200.
17. McGinnis, R. N.; Ryan, T. R.; McCarley, R. E. J. Am. Chem. Soc., 1978, 100, 7900.
18. Ryan, R. R.; McCarley, R. E. Abstracts of the 179th ACS National Meeting, Houston, 1980; Abstract No. INOR. 162.
19. Glicksman, H. D.; Walton, R. A. Inorg. Chem., 1978, 17, 3197.
20. DeMarco, D.; Nimry, T.; Walton, R. A. Inorg. Chem., 1980, 19, 575.

21. Bino, A.; Cotton, F. A. Angew. Chem. Int. Ed. Engl., 1979, 18, 332.
22. San Filippo, Jr., J.; Schaefer King, M. A. Inorg. Chem., 1976, 15, 1228.
23. Ebner, J. R.; Walton, R. A. Inorg. Chem., 1975, 14, 1987.
24. Salmon, D. J.; Walton, R. A. J. Am. Chem. Soc., 1978, 100, 991.
25. Brant, P.; Salmon, D. J.; Walton, R. A. J. Am. Chem. Soc., 1978, 100, 4424.
26. Hertzer, C. A.; Walton, R. A. Inorg. Chim. Acta, 1977, 22, L10.
27. Cotton, F. A.; Curtis, N. F.; Robinson, W. R. Inorg. Chem., 1965, 4, 1696.
28. Bonati, F.; Cotton, F. A. Inorg. Chem., 1967, 6, 1353.
29. Cotton, F. A.; Oldham, C.; Walton, R. A. Inorg. Chem., 1967, 6, 214.
30. Myers, R. E. Ph.D. Thesis, Purdue University, 1977.
31. Cotton, F. A.; Robinson, W. R.; Walton, R. A.; Whyman, R. Inorg. Chem., 1967, 6, 929.
32. Nimry, T.; Walton, R. A. Inorg. Chem., 1977, 16, 2829.
33. Jaecker, J. A.; Murtha, D. P.; Walton, R. A. Inorg. Chim. Acta, 1975, 13, 21.
34. Ebner, J. R.; Tyler, D. R.; Walton, R. A. Inorg. Chem., 1976, 15, 833.
35. Jaecker, J. A.; Robinson, W. R.; Walton, R. A. J. Chem. Soc., Dalton Trans., 1975, 698.
36. Shaik, S.; Hoffmann, R. J. Am. Chem. Soc., 1980, 102, 1194.
37. Vahrenkamp, H. Chem. Ber., 1978, 111, 3472.
38. Hertzer, C. A.; Walton, R. A. J. Organomet. Chem., 1977, 124, C15.
39. Hertzer, C. A.; Myers, R. E.; Brant, P.; Walton, R. A. Inorg. Chem., 1978, 17, 2383.
40. Nimry, T.; Urbancic, M. A.; Walton, R. A. Inorg. Chem., 1979, 18, 691.
41. Voss, K. E.; Hudman, J. D.; Kleinberg, J. Inorg. Chim. Acta, 1976, 20, 79.
42. Chisholm, M. H.; Cotton, F. A.; Extine, M. W.; Kelly, R. L. J. Amer. Chem. Soc., 1978, 100, 3354.
43. Brant, P.; Glicksman, H. D.; Salmon, D. J.; Walton, R. A. Inorg. Chem., 1978, 17, 3203.
44. Zietlow, T. C.; Klendworth, D. D.; Nimry, T.; Salmon, D. J.; Walton, R. A. Inorg. Chem., accepted for publication.
45. Malatesta, L.; Sacco, A.; Ghielmi, S. Gazz. Chim. Ital., 1952, 82, 516.
46. Mann, K. R.; Cimolino, M.; Geoffroy, G. L.; Hammond, G. S.; Orio, A. A.; Albertin, G.; Gray, H. B. Inorg. Chim. Acta, 1976, 16, 97.
47. Brant, P.; Cotton, F. A.; Sekutowski, J. C.; Wood, T. E.; Walton, R. A. J. Am. Chem. Soc., 1979, 101, 6588.

48. Lippard, S. J. Prog. Inorg. Chem., 1976, 21, 91.
49. Wood, T. E.; Deaton, J. C.; Corning, J.; Wild, R. E.; Walton, R. A. Inorg. Chem., 1980, 19, 2614.
50. Mialki, W. S.; Wild, R. E.; Walton, R. A. Inorg. Chem., submitted for publication.
51. Girolami, G. S.; Andersen, R. A. J. Organomet. Chem., 1979, 182, C43.
52. Girolami, G. S.; Mainz, V. V.; Andersen, R. A. Inorg. Chem., 1980, 19, 805.
53. Cotton, F. A.; Fanwick, P. E.; Niswander, R. H.; Sekutowski, J. C. J. Am. Chem. Soc., 1978, 100, 4725.
54. Cotton, F. A.; Niswander, R. H.; Sekutowski, J. C. Inorg. Chem., 1979, 18, 1152.
55. Welters, W. W.; Walton, R. A. unpublished results.
56. Mialki, W. S.; Wood, T. E.; Walton, R. A. J. Amer. Chem. Soc., accepted for publication.
57. Wood, T. E.; Walton, R. A. unpublished results.
58. Treichel, P. M.; Williams, J. P. J. Organomet. Chem., 1977, 135, 39.
59. Cotton, F. A.; Fanwick, P. E.; McArdle, P. A. Inorg. Chim. Acta, 1979, 35, 289.
60. Treichel, P. M.; Williams, J. P.; Freeman, W. A.; Gelder, J. I. J. Organomet. Chem., 1979, 170, 247.

RECEIVED November 21, 1980.

12

The Novel Reactivity of the Molybdenum–Molybdenum Triple Bond in $Cp_2Mo_2(CO)_4$

M. DAVID CURTIS, LOUIS MESSERLE, NICEPHOROS A. FOTINOS, and ROBERT F. GERLACH

Department of Chemistry, The University of Michigan, Ann Arbor, MI 48109

The existence of multiple bonds between metal atoms in discrete complexes was first demonstrated by Cotton and coworkers in the early 1960's (1). Since that time, several hundred papers have appeared dealing with the synthesis and structural and spectroscopic (2) characterization of complexes with multiple bonds between metal atoms. However, there have been relatively few systematic investigations of the reactivity of the metal-metal multiple bonds.

Chisholm has published extensively on the chemistry of complexes of the type $L_3M{\equiv}ML_3$ (L = univalent ligand, e.g., Me_2N, R, OR, etc.; M = Mo, W), and this chemistry has been reviewed(3,4,5). By and large, the reactivity of metal-metal quadruple bonds has not been well documented in spite of the fact that complexes of the type $L_4M{\equiv}ML_4$ constitute the largest class of multiply-bonded metal dimers. To some extent, the reactivity of the metal-metal multiple bonds in $L_4M{\equiv}ML_4$ complexes is obscured by their tendency to simply add further ligands in the axial positions (eq. 1) or to undergo simple ligand exchange (eq. 2) (see 1 and 6 for leading

$$L_4MML_4 \ + \ 2L' \longrightarrow L'L_4MML_4L' \qquad (1)$$

$$L_4MML_4 \ + \ XL' \longrightarrow M_2L_{8-X}L'_X \ + \ XL \qquad (2)$$

references). In these reactions, the M-M bond order is unaltered and the dimeric units behave chemically almost as eight-coordinate mononuclear complexes, although there are exceptions which will be noted as appropriate. Reactions and structural features of a variety of complexes, including the M_2L_8 type, have been reviewed (7).

It is the intent of this article to summarize the chemistry of the M≡M triple bonds in complexes of the type $Cp(CO)_2M{\equiv}M(CO)_2Cp$ (M = Cr, Mo, W) with emphasis on the Mo-dimer. Unlike the $L_3M{\equiv}ML_3$ or $L_4M{\equiv}ML_4$ complexes, the metals in $Cp_2M_2(CO)_4$ achieve the saturated, 18-electron configuration upon forming the M≡M triple bond. Hence, the addition of further electrons (in the form of added

0097-6156/81/0155-0221$09.25/0

ligands) must cause a change in the metal-metal bond order, provided the carbonyl ligands are not displaced. It is this fact which renders the M≡M triple bond in the $Cp_2M_2(CO)_4$ complexes particularly prone to reactions under very mild conditions. Before describing the actual reactions which have been observed, some preliminary remarks on general reactivity patterns will be made as they relate to the electronic structure of M≡M triple bonds.

General Reactivity Considerations

Multiple bonds are important functional groups in organic chemistry for several reasons. The π-bonds are rich in electron density and are polarizable; hence, multiple bonds are a point of reactivity towards electrophiles. The π-bonds may also be displaced by nucleophiles (e.g., in a Michael's addition). These reactivity modes are illustrated in eqs. 3 and 4.

$$C{\equiv}C \; + \; E^+ \longrightarrow E\text{-}C{=}C^+ \quad (3)$$

$$C{\equiv}C \; + \; N{:}^- \longrightarrow N\text{-}C{=}C{:}^- \quad (4)$$

With metals in place of carbons, we might expect some similarities as well as differences, the latter due to the closer packing of energy levels and the availability of nonbonding (or nearly so) orbitals centered on the metals. One major difference between M≡M bonds and C≡C bonds is especially evident in nucleophilic addition. In eq. 4, displacement of the π-bond by a nucleophile renders the remote carbon nucleophilic as a result of its nonbonding lone pair. However, displacement of the π-bonds of M≡M apparently renders the remote metal electrophilic as seen by electron counting arguments (eq. 5). Assume that both M and M' have

$$M{\equiv}M' \; + \; N{:} \longrightarrow N\text{-}M\text{-}M' \quad (5)$$

achieved the 18-electron count by formation of the M≡M triple bond. Addition of two electrons to M means that both π-bonds must be displaced (since M counts one electron from each metal-metal bond) in order for M to remain at the 18-electron configuration. The remote metal, M', is now two electrons short of the 18-electron configuration and is coordinatively unsaturated, hence, electrophilic. Stated in another way, the transformation M≡M → M-M opens up a vacant, two-electron coordination site on each metal. Thus, we anticipate that M≡M bonds in 18-electron complexes will add two 2-electron donors or one 4-electron donor (eqs. 6 and 7). The addition of two 3-electron donors, e.g. NO, would be expected to completely displace all three metal-metal bonds to give the monomeric complexes (eq. 8). With NO, this fact occurs with $Cp_2Mo_2(CO)_4$ (1), giving good yields of $CpMo(CO)_2NO$.

Two-electron donors capable of bridging the M-M bond can react in an alternate way which leaves each metal with the eight-

$$M{\equiv}M \; + \; 2L: \longrightarrow L\text{-}M\text{-}M\text{-}L \qquad (6)$$

$$M{\equiv}M \; + \; :L: \longrightarrow M\text{—}M \text{ (bridged by L)} \qquad (7)$$

$$M{\equiv}M \; + \; 2L \longrightarrow 2\, M{\equiv}L \qquad (8)$$

een-electron configuration. In this mode, the two electrons displace only one π-bond, giving a bridged M=M double bond (eq. 9).

$$M{\equiv}M \; + \; :E \longrightarrow M{=}M \text{ (bridged by E)} \qquad (9)$$

In order to form two M-E bonds to complete the bridge, a two-electron donor also must have an accessible, vacant orbital. Carbenes and carbenoids meet this requirement so that these reactive groups might be anticipated to react as in eq. 9, or, by a continuation, as in eq. 10.

$$M{=}M \text{ (bridged by E)} \; + \; :E \longrightarrow M\text{—}M \text{ (bridged by two E)} \qquad (10)$$

Oxidative addition to mononuclear centers is well known, and, in view of the high electron density in the M≡M bond, one expects oxidative addition reactions across the metal-metal multiple bonds. These reactions are the analogues of consecutive additions of an electrophile and a nucleophile across the C=C bond, as in halogenation (eq. 11). As was the case with nucleophilic addition

$$X^+X^- \; + \; C{\equiv}C \longrightarrow XC{=}C^+ \xrightarrow{X^-} XC{=}CX \qquad (11)$$

to M≡M bonds, we also expect M≡M bonds to behave differently than C≡C bonds in electrophilic additions, and these differences will be detailed in later discussions.

Finally, in analogy with organic chemistry, "polymerization" of metal-metal multiple bonds may lead to clusters as illustrated in eqs. 12 and 13. To date, the only definitely characterized oligomerization reaction of this type has been reported by McCarley et al. (8)(eq. 14), although Chisholm et al. (9) have observed that $Mo_2(OEt)_6$ dimerizes to a tetranuclear complex of unknown structure and that the O-H bond of isopropanol oxidatively adds to the W≡W bond of $W_2(i\text{-}PrO)_6$ to give a tetranuclear complex with an "open" as opposed to a closed, or cluster, structure (10). Also, some evidence has been presented that $Cp_2Mo_2(CO)_4$ may form unstable tetrahedrane intermediates (eq. 12) (6).

Synthesis and Structure of $Cp_2M_2(CO)_4$

The complexes, $Cp_2M_2(CO)_4$, are easily prepared from the readily available dimers, $Cp_2M_2(CO)_6$, by heating their respective solutions in a slow stream of nitrogen. It has been shown that $Cp_2Mo_2(CO)_4$ is formed by the sequence involving homolysis of the Mo-Mo bond (eq. 15, 16) (6). Although it has not been established,

$$Cp_2Mo_2(CO)_6 \rightleftharpoons 2\ CpMo(CO)_3 \underset{+CO}{\overset{-CO}{\rightleftharpoons}} 2\ CpMo(CO)_2 \quad (15)$$

$$2\ CpMo(CO)_2 \longrightarrow Cp(OC)_2Mo{\equiv}Mo(CO)_2Cp \quad (16)$$

the formation of the Cr and W complexes presumably follow the same path. At a given temperature, the rate of formation of $Cp_2M_2(CO)_4$ is qualitatively Cr > Mo >> W, paralleling the strength of the M-M bonds.

The solvent of choice for the preparation of the Mo and W compounds is refluxing diglyme, whereas refluxing toluene gives a convenient rate with the Cr compound. Refluxing a 1:1 mixture of $Cp_2Mo_2(CO)_6$ and $Cp_2W_2(CO)_6$ in diglyme gives an approximately statistical mixture of $Cp_2Mo_2(CO)_4$, $Cp_2MoW(CO)_4$, and $Cp_2W_2(CO)_4$ (10). Preliminary x-ray data suggest that the W≡W and mixed Mo≡W dimers are isostructural with the Mo≡Mo compound (see below).

The structures of $Cp_2M_2(CO)_4$ (M = Cr (11), Mo (12)) and $Cp^*_2Cr_2(CO)_4$ ($Cp^* = \eta^5\text{-}C_5Me_5$) (13) have been reported, and the results are rather surprising. Both chromium species have a structure with non-linear Cp-Cr-Cr-Cp axes which is usually observed for Cp_2M_2 compounds (Fig. 1), whereas the molybdenum complex has an unprecedented, linear Cp-Mo-Mo-Cp axis (Fig. 2). Some relevant structural parameters are collected in Table I. The M≡M distances in these carbonyl complexes are considerably longer than those in the Chisholm complexes, of the type $L_3M{\equiv}ML_3$, in which the M≡M distances average about 2.21 Å (3,4,5).

One factor which may contribute to the longer M≡M bond length in the carbonyl complexes is the presence of semi-bridging carbonyls which interact with orbitals involved in M≡M multiple bonding. An "asymmetry parameter", α, may be defined as $\alpha = (d_2-d_1)/d_1$ (see Table I) where d_2 = long M····CO distance and d_1 = short M-C distance. For symmetrically bridged carbonyls, $d_2 = d_1$ and $\alpha = 0$. For terminal carbonyls, $\alpha >$ *ca*. 0.6, and the semi-bridging appellation is appropriate if $0.1 < \alpha < 0.5$.

Cotton has proposed that semi-bridging carbonyls serve to remove excess charge on M' if M' is electron rich relative to M (see I) (14). In this type of semi-bridging interaction, the carbonyl group *accepts* electron density from M' into the π^*-orbitals. We have shown that a plot of the asymmetry parameter (α) vs. the MCO angle (θ) clearly reveals two types of behavior for semi-bridging carbonyl groups (see Fig. 3). The acceptor type of behavior is

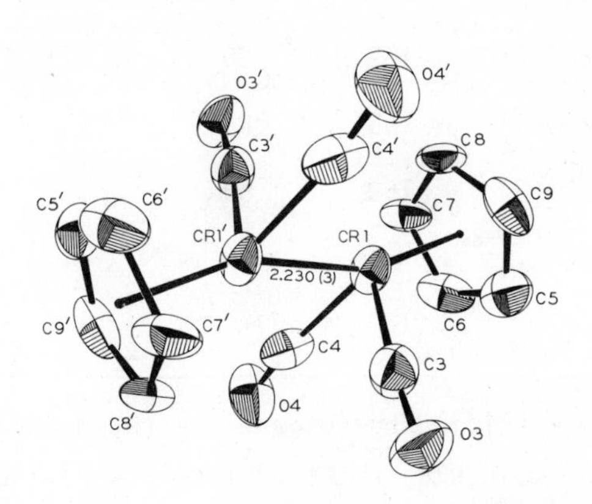

Journal of Organometallic Chemistry

Figure 1. ORTEP plot of $Cp_2Cr_2(CO)_4$ (11)

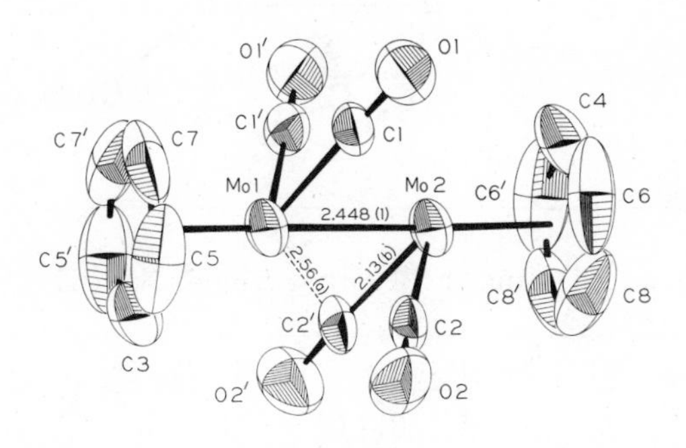

Journal of the American Chemical Society

Figure 2. ORTEP plot of $Cp_2Mo_2(CO)_4$ (1) (12)

Table I. Some Structural Parameters for $Cp_2MM'(CO)_4$.[a]

MM'	d(MM')Å	ω(deg.)	Φ(deg.)	θ(deg.)	α	ref.
MoMo	2.448(1)	180	86	176	0.20	12
CrCr	2.230(3)	159	86	171	0.33	11
CrCr[b]	2.200(3)	165	84	173	0.25	11
CrCr[c]	2.280(2)	159	89	173	0.39	13
MoW	2.44	—	—	—	—	10

a) Angles are specified below, and $\alpha = (d_2-d_1)/d_1$. b) Second, independent molecule in the unit cell. c) The Cp-ligand is $(\pi-C_5Me_5)$.

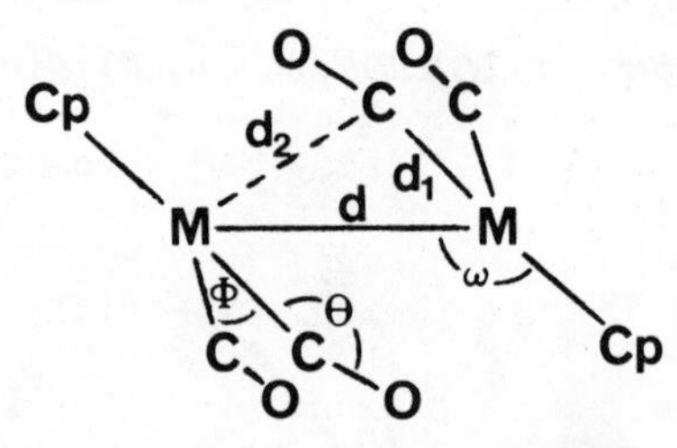

Table II. Bond Distances (Å) in $(C_5H_4R)_2Mo_2(CO)_4(R'C{\equiv}CR'')$

R	R'	R"	Mo-Mo	Mo-C[a]	C-C	ref.
H	H	H	2.980(1)	2.17	1.337(5)	25
H	H	Ph	2.972(2)	2.16	1.34(1)	26
H	Ph	Ph	2.956(1)	2.18	1.329(6)	25
H	Et	Et	2.977(1)	2.20	1.335(8)	25
Me	H	H	3.001(1)	2.20	1.35	26

a) Average of four values with $\sigma \simeq 0.03 - 0.04$.

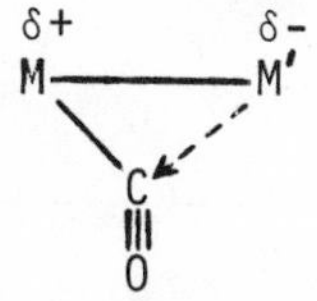

(I) 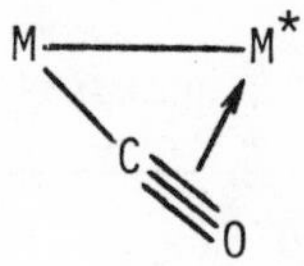 (II)

represented by the bottom line of Fig. 3 and the new type of semi-bridging is represented by the top line.

The bottom curve of Fig. 3 shows that acceptor semi-bridging carbonyls bend back as the carbon approaches the second metal (i.e., as the asymmetry decreases) until $\theta \simeq 135°$ for symmetrically bridged carbonyls. In the second type of interaction, which we denote as "donor semi-bridging", the M-C-O angle is nearly invariant with respect to the asymmetry parameter. That the carbonyls remain essentially linear when they act in a donor capacity is reasonable when one considers that the oxygen must also approach the metal, M^*, in structure II.

We have noted donor semi-bridging (or bridging) under three circumstances (15): (1), when M^* is a Lewis acid, e.g., Ga or Zn, (2), when M^* is short of the 18-electron count in the absence of the donor interaction, e.g., in $(dppm)_2Mn_2(CO)_5$ (16) or in $Cp_2Nb(\mu-\eta^2-CO)_2Mo(CO)Cp$ (17), and (3), when M^* is potentially unsaturated due to the presence of $M \equiv M^*$ multiple bonding (6). In the last case, one can consider that the CO π-electrons are donated into $M \equiv M$ π^*-orbitals and that the $M \equiv M$ π-bonds are weakened to the extent that the carbonyls donate electrons. Hence, the relatively long $M \equiv M$ bonds in $Cp_2M_2(CO)_4$ (M = Mo, Cr) or $Cp_2V_2(CO)_5$ (18) may be ascribed in part to the semi-bridging carbonyl interactions.

Jemmis et al. have recently reported the results of extended Huckel calculations on $Cp_2M_2(CO)_4$ compounds (19). Their results indicate that the carbonyls in $Cp_2Mo_2(CO)_4$ are net electron-withdrawing. However, these authors did not address the question of bent vs. linear semi-bridging carbonyls. Clearly, this topic warrants further theoretical scrutiny.

Nucleophilic Addition to M≡M Bonds

Reaction with P-Donors. In accord with the expectations discussed above, $Cp_2Mo_2(CO)_4$ reacts readily with two equivalents of soft nucleophiles, e.g., phosphines, phosphites, CO, etc., to give exclusively the trans-products indicated in eq. 17. With one equivalent of ligand, only disubstituted product (1/2 equiv.) and unreacted 1 (1/2 equiv.) are isolated. Hence, the addition of the first ligand is the slow step (eq. 18). Complex 1 does not react with hard bases, e.g., aliphatic amines, pyridine, ethers, alcohols, or ketones. Bulky phosphines, e.g., $(cyclohexyl)_3P$, and Ph_3As or Ph_3Sb also fail to react at room temperature. Rather

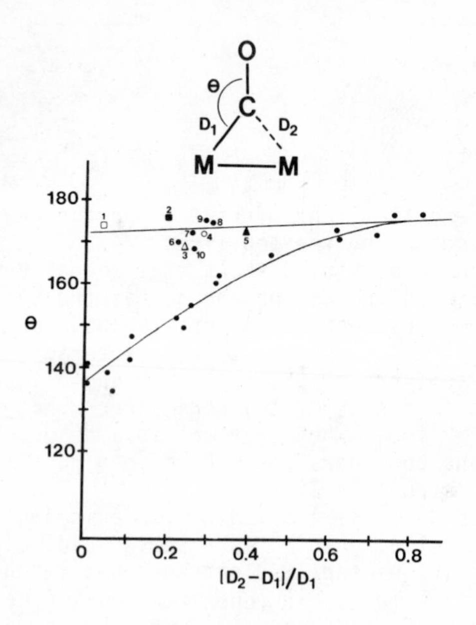

Figure 3. Plot of the asymmetry parameter, $\alpha = (d_2 - d_1)/d_1$ vs. θ, the MCO angle ((1) $(dppm)_2Mn_2(CO)_5$; (2) $Cp_2Mo_2(CO)_4$; (3) $Cp_2V_2(CO)_5$; (4) $Cp_2Cr_2(CO)_4$; (5) $Cp_2{}^*Cr_2(CO)_4$; (6) $Cp_2Mo_2(CO)_4CN^-$; (7) $CpW(CO)_3GaMe_2$; (8) $CpMo(CO)_3$-$ZnBr \cdot 2THF$; (9) $[CpMo(CO)_3]_2[ZnCl \cdot OEt_2]_2$; (10) $CpW(CO)_3AuPPh_3$)

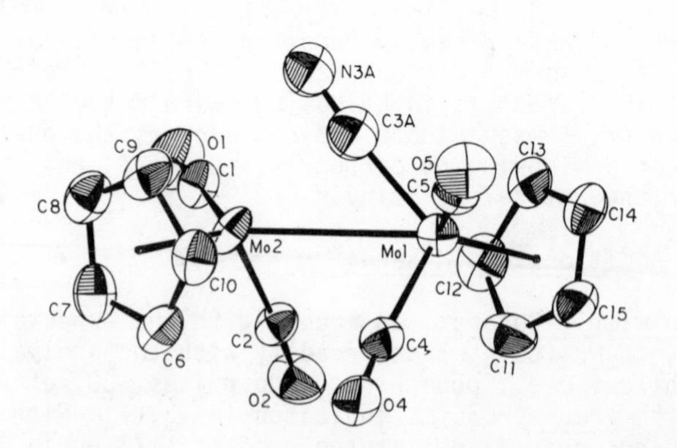

Figure 4. ORTEP plot of $Cp_2Mo_2(CO)_4CN^-$ ion (2) (15)

$$Cp_2Mo_2(CO)_4 + 2L \longrightarrow \text{(Cp)(L)(OC)}_2\text{Mo—Mo(CO)}_2\text{(L)(Cp)} \quad (17)$$

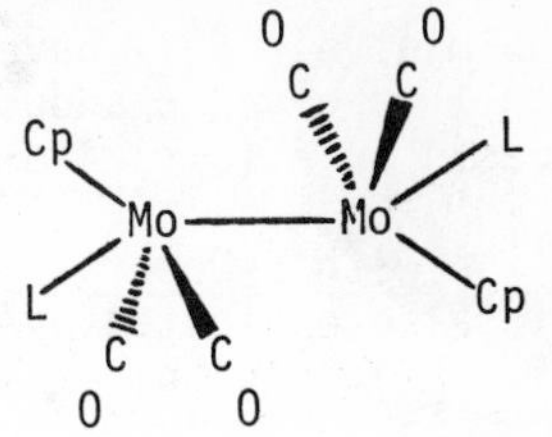

$$M{\equiv}M + L \xrightarrow[\text{slow}]{} M\text{-}M\text{-}L \xrightarrow[\text{fast}]{L} L\text{-}M\text{-}M\text{-}L \quad (18)$$

surprisingly, $Cp_2Cr_2(CO)_4$ reacts with phosphines but no products could be isolated (20).

Reaction with Cyanide Ion. Potassium cyanide in methanol also reacts rapidly with 1 to give a monocyano adduct, 2 (eq. 19).

$$Cp_2Mo_2(CO)_4 + CN^- \longrightarrow [Cp(OC)_2Mo\text{—}Mo(CO)_2Cp]^- \ (2) \quad (19)$$

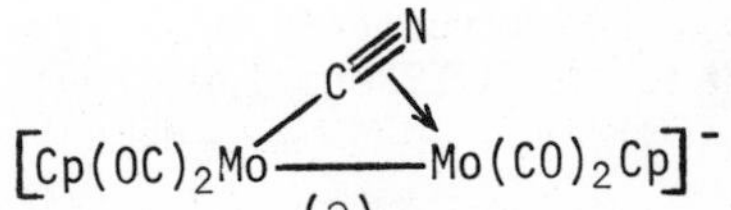

The structure of 2 has been determined by x-ray diffraction (Fig. 4) and shows that the cyanide is coordinated in an unprecedented $\sigma+\pi$ manner (15). Thus, the cyano ligand donates a total of 4 electrons ($2\sigma+2\pi$), totally displacing the M-M π-bonds. The Mo-Mo distance, 3.139(2)Å, is in the range expected for a Mo-Mo single bond. Other pertinent structural data are Mo-CN (1.95Å), Mo···· CN(2.37Å), Mo···N(2.53Å), MoMo(CN)(49°), and MoCN(170°). The asymmetry parameter for the bridging cyanide is 0.22 which, when plotted vs. θ, gives point 6 in Fig. 3 demonstrating the similarity in the bonding to the ($\sigma+\pi$)-donor carbonyls.

One further feature of the structure of the cyanide adduct is of interest. The molecule is fluxional, showing a singlet Cp resonance down to -52° which splits into two signals at -60° and below. This fluxionality is ascribed to a "wind-shield wiper" motion of the bridging cyanide (eq. 20) which averages the Cp environments. The solid state structure of 2 shows disorder for the cyanide position which was interpreted as the overlapping of the two orientations, 2a and 2b. The solid state structure is thus a "stop-action" photograph of the fluxional process (15).

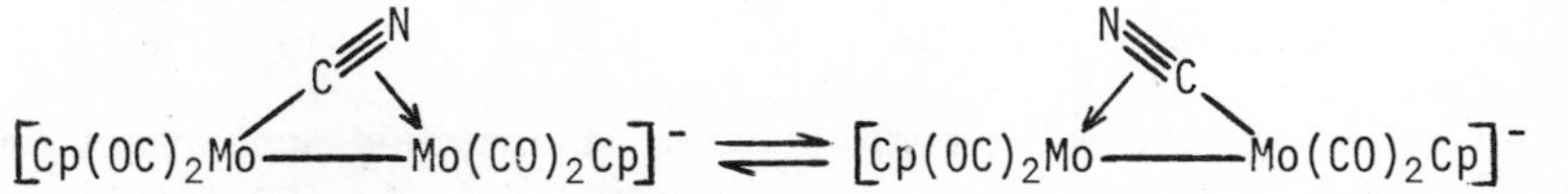

$$[Cp(OC)_2Mo\text{—}Mo(CO)_2Cp]^- \ (2a) \rightleftharpoons [Cp(OC)_2Mo\text{—}Mo(CO)_2Cp]^- \ (2b) \quad (20)$$

Chisholm et al. (21) have shown that dimethylaminocyanimide reacts with 1 to give an adduct in which the aminocyanimide ligand also acts as a (σ+π) 4-electron donor (eq. 21).

Reaction with S-Donors. A somewhat similar (σ+π) 4-electron donation is also exhibited by thioketones in their adducts with $Cp_2Mo_2(CO)_4$ (22). In this instance, one of the carbonyls trans to the R_2CS ligand adopts a semi-bridging bonding mode (eq. 22).

An unusual 1,1-bis(adduct) of 1 is formed upon reaction with Ph_3PS (eq. 23). Adduct 3 represents the only known example in which 1,1-substitution occurs instead of the usual 1,2-addition to M≡M. It is not known if 3 is a rearrangement product of a 1,2-addition or if 3 is formed by some more complex mechanism. The permethylated complex, $Cp_2{}^*Mo_2(CO)_4$ ($Cp^* = \pi\text{-}C_5Me_5$), which is generally less reactive than 1 (24) failed to react with Ph_3PS (23).

Reaction with Acetylenes. Acetylenes react with 1 at room temperature to give the dimetallatetrahedrane complexes, 4 (eq. 24) (6), the structures of several of which have been determined (25, 26). An ORTEP of the structure of the acetylene adduct of $Cp_2'Mo_2(CO)_4$ (4, Cp' = $\pi\text{-}C_5H_4Me$) is shown in Fig. 5. Common

$$Cp_2Mo_2(CO)_4 \; + \; RC{\equiv}CR \longrightarrow \underset{(4)}{Cp(OC)_2Mo(\mu\text{-}RCCR)Mo(CO)_2Cp} \qquad (24)$$

features to all the structures of type 4 are Mo-Mo bond lengths of ca. 2.98Å, Mo-C (acetylene), 2.18Å, and C-C, 1.33Å. The latter distance is consistent with a C-C bond order of approximately 1.4 (27). All the acetylene adducts also have a semi-bridging carbonyl trans to the coordinated acetylene, and adducts of type 4 are fluxional (25).

Slater and Muetterties (28) have shown that the acetylene adducts, 4, undergo acetylene exchange above about 110°, and that complexes 4 react with H_2 at 150-170° to produce 1 and cis-olefin (eq. 25). In the presence of excess H_2 and acetylene, the process

$$4 \; + \; H_2 \longrightarrow (R)(H)C{=}C(R)(H) \; + \; 1 \qquad (25)$$

becomes catalytic in 1 with turnover numbers of ca. 10^{-2} moles olefin/mole 4/min. Inhibition and labelling studies demonstrated that 4 loses CO in the rate determining step to give an intermediate postulated to be 5. Photolysis of 4 in the presence of excess acetylene gave a new complex, 6, while photolysis in a flow of H_2 gave 7a (28).

An analogous chromium complex, 7b, is obtained upon reaction of $Cp_2Cr_2(CO)_4$ with acetylenes in refluxing toluene (no reaction occurs at 25°) (29,30). The Cr≡Cr bond in 7b has a length of 2.337(4)Å, somewhat longer than in the parent $Cp_2Cr_2(CO)_4$. Knox

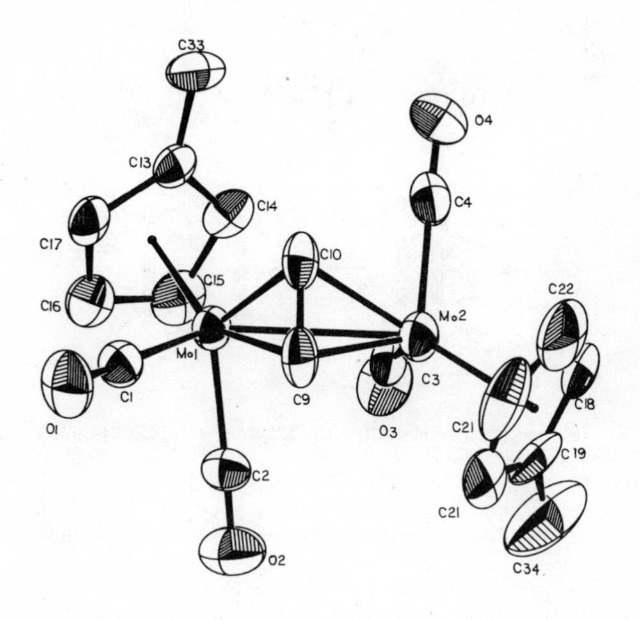

Figure 5. ORTEP plot of $Cp_2'Mo_2(CO)_4(C_2H_2)$ ($Cp' = \pi$-C_5H_4Me) (5)

$$2\ M{\equiv}M \longrightarrow M_4\ \text{(tetrahedron)}\ \text{or}\ \text{M–M / M–M (rectangle, with M=M)} \qquad [12]$$

$$3\ M{\equiv}M \longrightarrow M_6\ \text{(trigonal prism)}\ \text{or}\ M_6\ \text{(octahedron)} \qquad [13]$$

$$2\ [L(Cl)(MeOH)Mo{\equiv}Mo(Cl)(MeOH)L] \xrightarrow{-MeOH} [L(Cl)Mo(\mu\text{-}Cl)_2Mo(Cl)L]\ /\ [Cl(L)Mo(\mu\text{-}Cl)_2Mo(Cl)L] \qquad [14]$$

$$\underset{\underset{\sim}{1}}{Cp_2Mo_2(CO)_4} + Me_2NCN \rightarrow Cp(OC)_2Mo(\mu\text{-}N{\equiv}C\text{-}NMe_2)Mo(CO)_2Cp \qquad [21]$$

$$\underset{\sim}{1} + R_2CS \longrightarrow Cp(CO)_2Mo(\mu\text{-}R_2C{=}S)Mo(CO)_2Cp \qquad [22]$$

$$\underset{\sim}{1} + 2\ Ph_3PS \longrightarrow \underset{\underset{\sim}{3}}{Cp(Ph_3PS)_2(CO)Mo\text{—}Mo(CO)_3Cp} \qquad [23]$$

$$\underset{\underset{\sim}{5}}{(RC{\equiv}CR)CpMo(CO){\equiv}Mo(CO)_2Cp}$$

et al. (30) have also shown that the acetylene adducts, 4, are just the first in a series of complexes containing linked acetylene units obtained by heating 4 with acetylenes to 110-130°. Thus, 4 reacts to give 7, 7 reacts further to give 8 which in turn affords 9 (open circles denote the original acetylenic carbons of 4). The Mo-Mo distance in 9 is 2.618(1)Å, consistent with a Mo=Mo double bond. Some gyrations are necessary to assign an 18-electron configuration to each Mo-atom in 9. The structure of 9 is shown dissected in III. If each terminal carbon in 9 is considered to be a neutral, bridging carbene, then each Mo is assigned the oxidation state, $+1(d^5)$. The left Mo gains 6(Cp) + 4(2xC=C)+ $2(2x\frac{1}{2}R_2C:)$ = 12 electrons from the ligands for a total of 17 electrons (in the absence of a M-M bond). The atom Mo', however, ends up with a total of $15e^-$. A "normal" M-M bond gives $1e^-$ to each Mo so that Mo has 18, Mo' only 16 electrons. A dative Mo→Mo' bond (dashed arrow in III) then completes the valence shell of Mo' and also introduces a charge asymmetry between Mo and Mo'. In the actual structure, the "carbenic" carbons are displaced toward Mo' in response to the charge asymmetry.

Knox et al. (31) have also reported that cyclic 1,3-dienes react in a Diels-Alder fashion with the acetylene adduct, 4 (eq. 26). The Mo-Mo distance in 10 is 2.504(1)Å, commensurate with a Mo≡Mo triple bond as required by the 18-electron rule. Acyclic dienes do not give 10, but with hexa-2,4-diene, 4 yields ethyne and the hexyne complex from the isomerized diene (eq. 27).

$$4 + \text{(hexa-2,4-diene)} \longrightarrow Cp(OC)_2Mo(\mu\text{-EtCCEt})Mo(CO)_2Cp + C_2H_2 \quad (27)$$

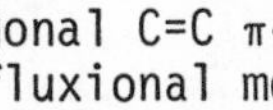

In contrast to its ready reaction with acetylenes, $Cp_2Mo_2(CO)_4$ does not react with simple olefins or dienes, e.g., C_2H_4, butadiene, C_2H_3CN, norbornadiene, etc. With TCNE ($C_2(CN)_4$), 1 is oxidized and $[CpMo(CO)_4]^+$ $[TCNE]^{\bar{\cdot}}$ may be isolated (6). Allene does react, however, to form a complex, 11, in which each of the orthogonal C=C π-bonds donates two electrons to each molybdenum. This fluxional molecule has C_2 symmetry in the solid state and has a Mo-Mo bond length of 3.117(1)Å (32).

Reaction of Hydrides. The donor semi-bridging intereaction of the carbonyls in 1 might render them more susceptible to nucleophilic attack by such reagents as hydrides. We reasoned that the formation of η^2-formyl species, e.g., IV, or alternatively, anionic metal hydrides, e.g., V, might result. The actual reaction of 1 with Et_3BH^- is consistently more complex than anticipated (33).

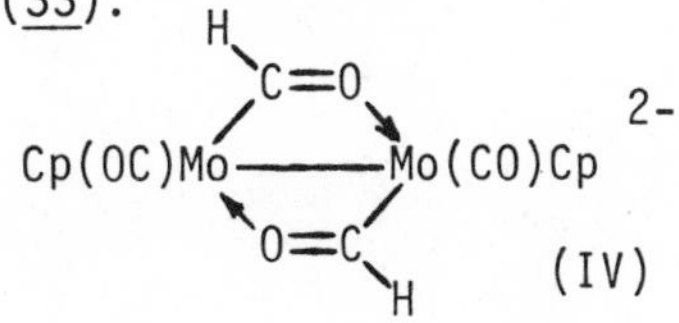

$$\left[Cp(OC)Mo(\mu\text{-}\eta^2\text{-HC=O})_2Mo(CO)Cp\right]^{2-} \quad (IV)$$

$$\left[Cp(OC)_2Mo(\mu\text{-H})Mo(CO)_2Cp\right]^{-} \quad (V)$$

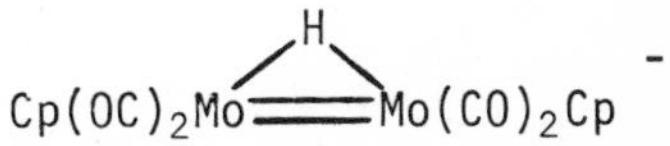

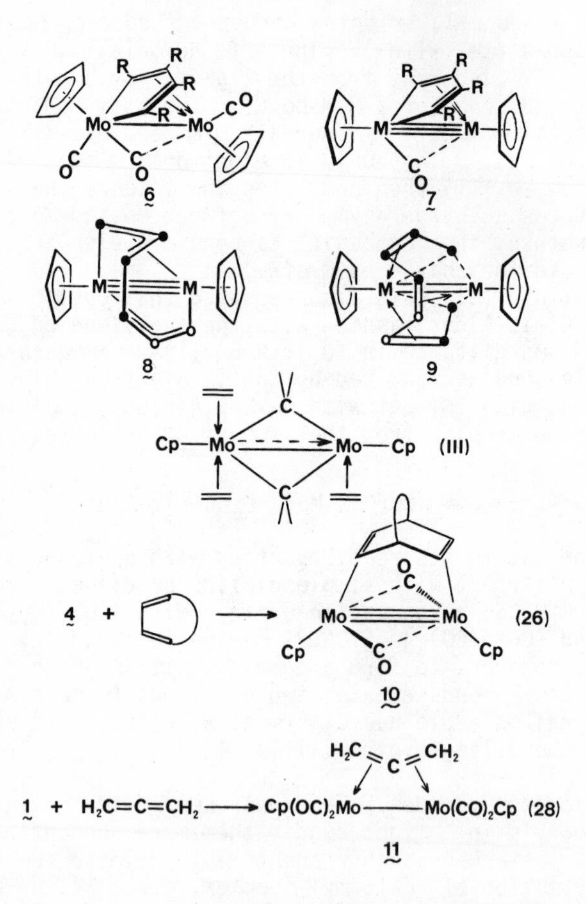
R
R
R
R
Mo
Mo
CO
C
O
C
O
6
R
R
R
R
M
M
C
O
7
M
M
8
M
M
9
C
Cp
Mo
Mo
Cp
(III)
C
4
+
Mo
Mo
(26)
O
C
Cp
C
O
Cp
10
H_2C
C
CH_2
1
+
$H_2C=C=CH_2$
$Cp(OC)_2Mo$
$Mo(CO)_2Cp$
(28)
11

Treatment of THF solutions of 1 with $LiBEt_3H$ at -15° to 0° results in the formation of $CpMo(CO)_3^-$ and various Mo-clusters, the nuclearity of which depends upon the BH^-/Mo_2 ratio (eq. 29).

$$1 + Et_3BH^- \longrightarrow \underset{(green)}{Cp_3Mo_3(CO)_xH} + \underset{(blue)}{Cp_6Mo_6(CO)_xH_2} + \underset{(brown)}{Cp_4Mo_4(CO)_xH_2} + CpMo(CO)_3^- \quad (29)$$

Ratios of 2/1, 3/1, and 4/1 (BH^-/Mo_2) give predominately the Mo_3, Mo_6, and Mo_4 clusters, respectively. As the number of reducing equivalents increases, the average CO stretching frequency in the clusters also decreases, signalling a greater reduction of the CO bond order in the more reduced clusters.

The green Mo_3 cluster shows three 'H resonances in the ratio 1:10:5. The peak of area one (δ8.9, $CDCl_3$) is in the range associated with metal formyls (34) and there are two equivalent and one different Cp groups. The carbonyl stretches of the Mo_3 cluster occur at 1763 and 1713 cm^{-1}. The mass spectrum shows a series of Mo_3 patterns starting at m/e = 580 with consecutive loss of 5 carbonyls, and a series of strong peaks due to dipositive ions corresponding to the monopositive series 524, 496, 468, and 440. It is not known if this mass spectrum is characteristic of the Mo_3 cluster or its decomposition products on the heated inlet probe.

The blue Mo_6 cluster exhibits five 'H resonances in the ratio 5:20:5:1:1. The latter two peaks are absent when 1 is reduced with Et_3BD^- and are thus attributed to metal hydride peaks. The blue complex shows peaks in the IR spectrum at 1955(w), 1900(br,w), 1690(vs), 1640(vs), and 1430(vs) (Nujol mull). We have not yet established if the higher frequency peaks are due to M-H stretches or are overtone/combination bands. Especially noteworthy is the very low 1430 cm^{-1} band which is almost in the region expected for a C-O single bond stretch.

Our attempts to fully characterize these clusters have been hampered by their relatively low solubility, poor crystallinity (we have been unable to grow crystallographic quality crystals) and high reactivity. In order to prepare derivatives with better solubility characteristics, we have allowed several small molecules to react with the blue Mo_6-cluster. Ethylene reacts to give a violet adduct, H_2 and CO black ones, and phosphites and phosphines give several products, depending on stoichiometry. None of these derivatives are as yet fully characterized.

The brown Mo_4 cluster is pyrophoric, sparingly soluble in common solvents, and has four 'H resonances in the ratio 5:15:1:1. The second resonance (δ3.3, $(CD_3)_2CO$) is highly unusual for a Mo-Cp group.

Various other hydridic reducing agents, e.g., $(RO)_3BH^-$, Vitride, $(t\text{-}BuO)_3AlH^-$, etc. also react with 1 to give different products. These are currently in various stages of characterization. Clusters of early transition metal carbonyls, possibly involving unusual coordination modes for H/CO, are of interest in connection with Fischer-Tropsch chemistry (35). At present, synthetic routes to such clusters are virtually unknown, but the preliminary results reported above show that the reactivity of the Mo≡Mo triple bond may provide access to highly unusual cluster chemistry.

Reaction with Diazoalkanes; μ-Alkylidene Formation. The reaction of diazoalkanes with transition metal complexes, an area of growing interest, has involved primarily mononuclear reactants (36). Previous to our work, there had been no reports of studies of diazoalkane reactions with M≡M multiply-bonded compounds. We have found that diazoalkanes do in fact react with $Cp_2Mo_2(CO)_4$ to give novel diazoalkane adducts which lose dinitrogen under mild conditions to produce complexes featuring bridging alkylidene groups.

Diphenyl- or di-p-tolyldiazomethane reacts at room temperature with 1 or with the π-C_5H_4Me (Cp') analogue to give the adducts, 11 (eq. 30) (37). In the solid state, these deep green complexes are stable indefinitely under N_2. On the other hand,

$$(C_5H_4R)_2Mo_2(CO)_4 \ + \ Ar_2CN_2 \ \rightarrow \ (C_5H_4R)_2Mo_2(CO)_4(N_2CAr_2) \quad (30)$$

	R	Ar
11a	H	Ph
11b	Me	Ph
11c	H	p-tolyl
11d	Me	p-tolyl

Ph_2CN_2 reacts with $Cp_2W_2(CO)_4$ only at elevated temperature to give still unidentified products.

A priori, the R_2CN_2 ligand may act either as a 4-electron or 2-electron donor giving complexes of type VI or VII, respectively. In VI, the CNN group is expected to be linear in accordance with the canonical form, $R_2C = N^+ = N:^-$. In fact, the actual structure is neither of the two forms, but is best described as a hybrid, VIIIa-c, with VIIIb predominating (Fig. 6, 7). The R_2CN_2 ligand is thus a 4-electron donor, but the CNN group is non-linear as in the mononuclear complexes, M≡N-N=CR_2 (39).

The intraligand bond distances in 11a are similar to those in several mononuclear tungsten diazoalkane complexes (39,40), and the N-N (1.35Å) and C=N (1.28Å) distances are compatible with the assignment of bond orders of 1.3 and 1.9, respectively, to these bonds. Similar bond length/bond order relations suggest bond orders of ca. 1.0 for Mo2-N1 and 1.7 for Mo1-N1. The Mo-Mo bond

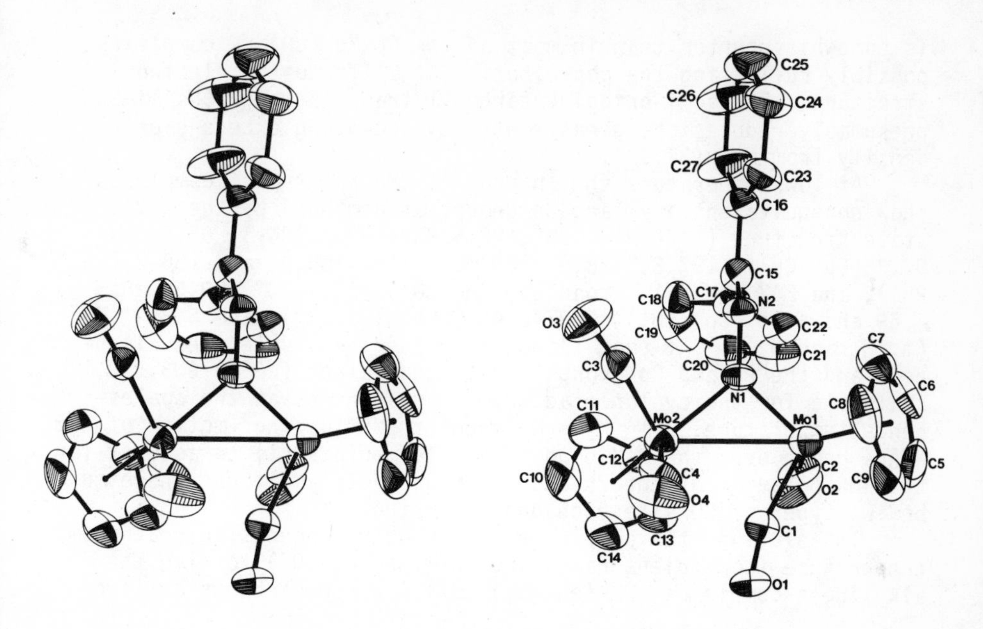

Figure 6. ORTEP stereoview plot of $Cp_2Mo_2(CO)_4(N_2CPh_2)$ (11a) (38)

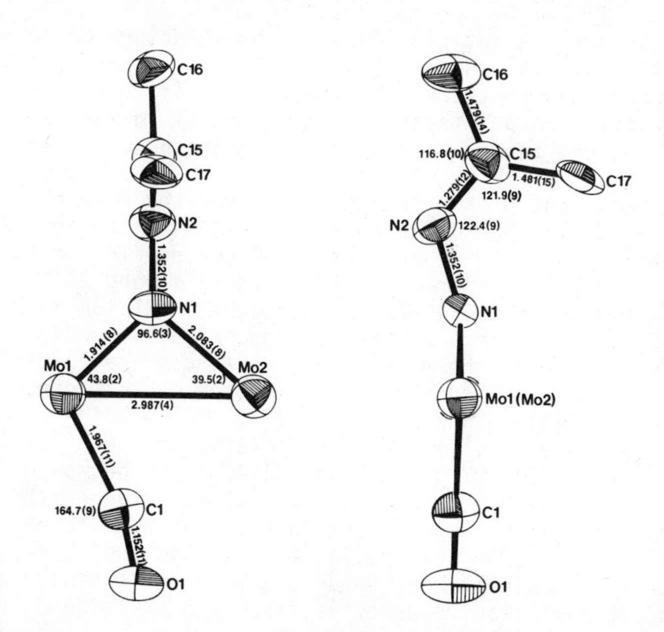

Journal of the American Chemical Society

Figure 7. ORTEP plot of portions of the bridging ligands in $Cp_2Mo_2(CO)_4(N_2CPh_2)$ (11a) (37)

is somewhat shorter than in most of the $Cp_2Mo_2(CO)_4L_2$ complexes, possibly reflecting the contribution of VIIIc to the electronic structure. The semi-bridging carbonyl <u>trans</u> to the R_2CN_2 ligand presumably reduces charge asymmetry by accepting excess electron density from Mo2 (<u>37</u>,<u>38</u>).

At low temperature the ^1H and ^{13}C NMR spectra of complexes 11 show nonequivalent aryl and Cp groups as demanded by the solid state structure ($^{13}C\{^1H\}$, -38°, 22.5 MHz: 93.7 ($\underline{C}N_2$), 95.5 and 97.3 (Cp, Cp'), 127.8, 128.5, 129.1, 131.5, 135.8, and 136.2 (Ph, Ph'), and 247 and 241 (broad, CO and CO'); ^1H, -70°, 89.56 MHz: 4.69 and 4.95 (Cp, Cp'), 6.37-6.45 (broad, d, H_o), and 6.85-7.78 (m, H_m and H_p)). At room temperature the inequivalencies disappear and the Ph and Cp groups appear equivalent ($\Delta G^{\ddagger}$ = 6.3 ± 0.1 kcal/mole for the cyclopentadienyl group protons at the coalescence temperature, 241 °K). We suggest a "flopping" motion of the $NCAr_2$ group over the M-M bond (via intermediate VI) to account for the NMR behavior of the Ph and Cp groups. In addition, there must be two types of CO site exchange occurring.

In solution, the complexes 11 lose dinitrogen slowly at room temperature (convenient rates are obtained at 60°) to give the alkylidene complexes, 12 (eq. 31) (<u>37</u>). Since carbenes are two-

$$11 \xrightarrow[-N_2]{60^\circ} \underset{(12)}{(C_5H_4R)_2Mo_2(CO)_4(CAr_2)} \qquad (31)$$

electron donors, we anticipated that the structure of 12 would be that depicted in IX. However, x-ray diffraction and solution NMR spectroscopy show that the actual structure is that shown in X, in which one aryl group interacts with one metal to form a π-allyl type ligand, one terminus of which is σ-bonded to the second metal.

Figures 8 and 9 show the structure for the Ar = p-tolyl complex (<u>37</u>,<u>38</u>). The Mo-Mo distance (3.087(2)Å) is in the range expected for a single bond, and the Mo-CB (CB = bridging carbon) distance of 2.22Å is appropriate for a Mo-C σ-bond (d(Mo-Et = 2.38Å in $CpMo(CO)_3Et$ (<u>41</u>) and d($Mo-CH_2SiMe_3$) = 2.13Å in $Mo_2(CH_2SiMe_3)_6$ (<u>42</u>)). In a large number of symmetrically-bonded π-allyl molybdenum complexes the Mo-C(allyl) distances are about 2.3Å for the terminal carbons and about 2.2Å for the central carbon of the π-allyl group (<u>43</u>). Bonding constraints of the bridging group and the fact that the "π-allyl" is incorporated in the p-tolyl ring distort the CB-C22-C27 group (Fig. 9) so that the Mo-CB distance is shorter and the Mo-C22 and Mo-C27 distances longer than in an unconstrained π-allyl coordinated to molybdenum. The Mo-C(allyl) distances in $CpMo(CO)_2(\eta^3\text{-benzyl})$ (<u>44</u>,<u>45</u>) also show a progression from a short "terminal" Mo-C bond to a long Mo-C(ortho) bond, although the variation reported is less than that found here. Very recently, the structure of complex 13 was reported with the indicated bond distances (<u>46</u>).

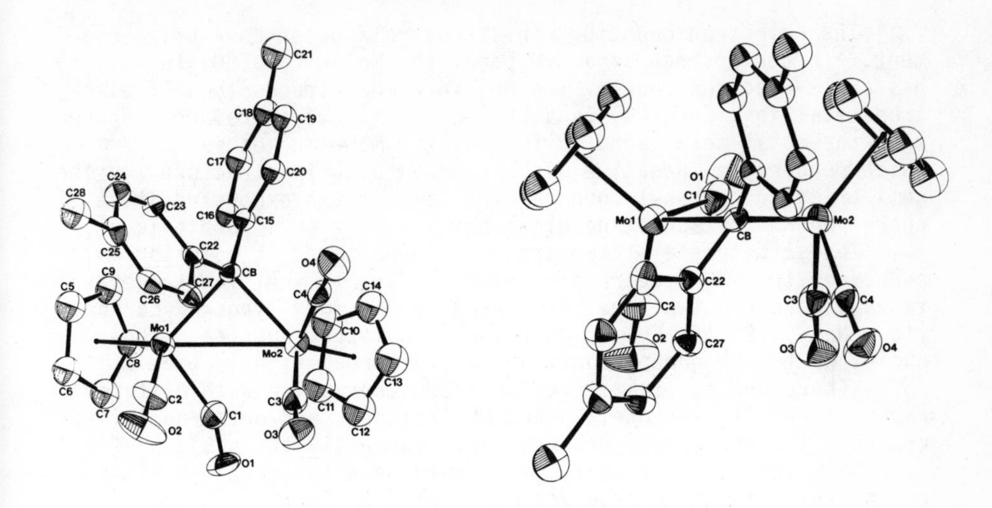

Figure 8. ORTEP plot of $Cp_2Mo_2(CO)_4(CAr_2)$ *(Ar =* p-tolyl*) (12c) (38)*

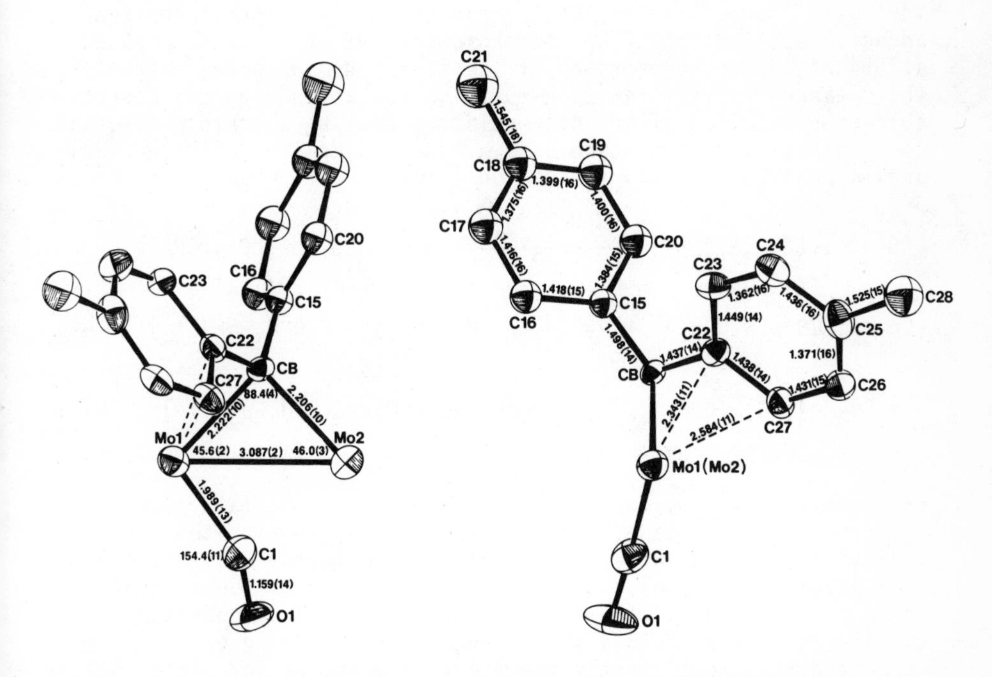

Journal of the American Chemical Society

Figure 9. ORTEP plot of portions of the bridging ligands in $Cp_2Mo_2(CO)_4(CAr_2)$ *(Ar =* p-tolyl*) (12) (37)*

The "electron counting" in structure X deserves a brief comment. In the absence of a M-M bond, the Mo in $CpMo(CO)_2(\eta^3\text{-allyl})$ has an 18-electron count. The Mo* in X corresponds to a 16-electron metal in a $CpMo(CO)_2(\sigma\text{-alkyl})$ complex. We therefore propose that the metal-metal bond is formed by a Mo⟶Mo* dative bond which renders Mo* negative with respect to Mo in structure X. The semi-bridging carbonyl then removes some of the excess negative charge on Mo* by accepting electrons into the CO π^*-orbitals.

The structure of 12 apparently is unaltered in solution as seen from the NMR spectra presented in Fig. 10. At low temperatures, there are two types of Cp and Ph and four types of carbonyls. Near -60° the Cp and Ph signals coalesce, but the ortho-nuclei of the Ph groups continue to give broad signals even at 25°. There are at least three processes operative with different rates: carbonyl exchange, Cp and Ph exchange, and ortho nuclei exchange. These exchange processes are currently the subject of a lineshape analysis. Finally, it should be noted that the ^{13}C signal of the bridging carbon occurs at δ176.0, in the same range found for C_α of several bridging diphenylmethylene complexes of rhodium (47).

Several experiments have provided information concerning the mechanism of formation of 12 from 11, and all evidence is consistent with intramolecular loss of dinitrogen from the diazoalkane adduct. Observation by spectrophotometry of an isosbestic point at 554 nm. during conversion of 11a to 12a demonstrated not only the cleanness of the reaction but also the absence of any appreciable concentration of an intermediate. Heating a benzene solution containing either a mixture of 11a and $(C_5H_4Me)_2Mo_2(CO)_4$ (eq. 32) or 11a and 11d (eq. 33) yielded only the direct products; no cross

$$\underset{(11a)}{Cp_2Mo_2(CO)_4(N_2CPh_2)} \xrightarrow{(C_5H_4Me)_2Mo_2(CO)_4} \underset{(12a)}{Cp_2Mo_2(CO)_4(CPh_2)} \quad (32)$$

$$\underset{(11a)}{Cp_2Mo_2(CO)_4(N_2CPh_2)} + \underset{(11d)}{(C_5H_4Me)_2Mo_2(CO)_4[N_2C(C_6H_4\text{-p-Me})_2]} \longrightarrow$$

$$\underset{(12a)}{Cp_2Mo_2(CO)_4(CPh_2)} + \underset{(12d)}{(C_5H_4Me)_2Mo_2(CO)_4[C(C_6H_4\text{-p-Me})_2]} \quad (33)$$

products (12b in the former and 12b and 12c in the latter) were observed in either case. These results, coupled with the observation that $(p\text{-MeC}_6H_4)_2CN_2$ itself showed no evidence of decomposition under the thermolysis conditions, ruled out a route involving prior dissociation of the diazoalkane from 11, decomposition of the diazoalkane to form a free carbene, and trapping of the free carbene by 1. We currently favor a cyclic transition state (XI) from which dinitrogen is evolved; similar 1,3-dipolar additions are observed in reactions of diazoalkanes with carbon-carbon unsaturated systems (48).

(VI) (VII)

(VIIIa) (b) (c)

(IX) (X)

13

a: 2.20 Å
b: 2.18
c: 2.31
d: 2.63
e: 3.19

(XI)

(XII)

$$\underset{\sim}{1} + RSSR \longrightarrow Cp(OC)_2Mo(\mu\text{-}SR)_2Mo(CO)_2Cp \quad \underset{\sim}{20} \qquad (45)$$

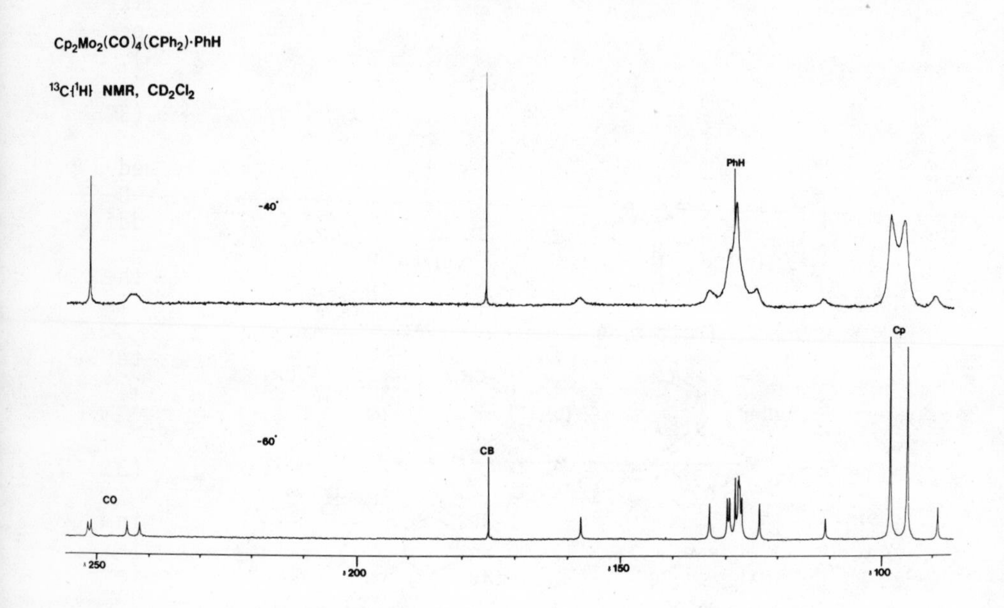

Figure 10. NMR spectra of $Cp_2Mo_2(CO)_4(CPh_2)$ (12a)

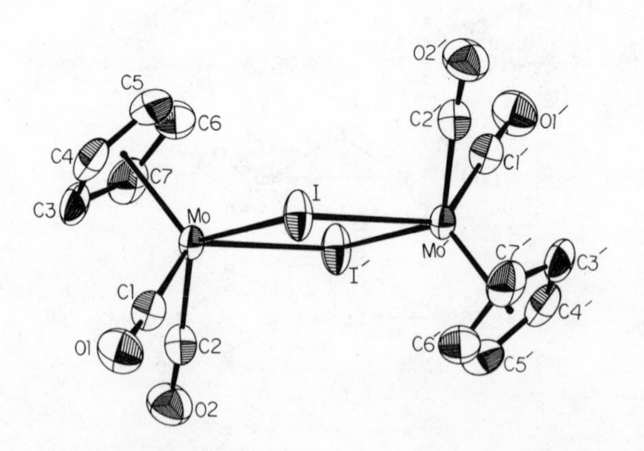

Figure 11. ORTEP plot of $Cp_2Mo_2(CO)_4I_2$ (14)

The bridging alkylidene ligand in 12 has been found to be a reactive species towards several small molecules. The addition of dihydrogen (45 psi, 50°) to 12a resulted in cleavage of the alkylidene from the metals; diphenylmethane (confirmed by ^1H NMR spectroscopy and GC/MS) and 1 were obtained in high yield (eq. 34).

$$Cp_2Mo_2(CO)_4(CPh_2) + H_2 \longrightarrow Cp_2Mo_2(CO)_4 + Ph_2CH_2 \quad (34)$$

Mass spectroscopy confirmed that only diphenylmethane-d_2 formed from the reaction of 12a and D_2. Interestingly, the use of a D_2/H_2 mixture in the hydrogenolysis reaction yielded Ph_2CHD in addition to Ph_2CD_2 and Ph_2CH_2. These labelling experiments demonstrate that the cleavage of Ph_2CH_2 is not concerted and that the ortho-hydrogens of the phenyl groups are not involved, but it is premature to speculate on the mechanism.

Carbon monoxide also cleaved off the bridging diphenylmethylene moiety under the same mild conditions to give $Cp_2Mo_2(CO)_6$, the product from carbonylation of 1, and diphenylketene (eq. 35).

$$Cp_2Mo_2(CO)_4(CPh_2) + 3CO \longrightarrow Cp_2Mo_2(CO)_4 + Ph_2C{=}C{=}O \quad (35)$$

Trimethylphosphine gave a still unidentified yellow, crystalline Mo-containing product <u>which lacks aryl groups</u>; preliminary evidence is consistent with liberation of the diarylmethylene moiety to form a free phosphorus ylide.

As this section amply shows, the reactivity of the Mo≡Mo triple bond toward nucleophiles shows a surprising diversity. This results from the fact that both the triple bond and the carbonyls are sites of potential attack, depending on the nucleophilic reagent in use. The next section shows a similar diversity of behavior toward electrophilic reagents.

Electrophilic Additions to M≡M Bonds

Reactions with Halogens. As a consequence of the high electron density in the M≡M bond, complex 1 is much more reactive toward electrophiles and mild oxidizing agents than the parent, singly-bonded $Cp_2Mo_2(CO)_6$. Thus, 1 reacts with I_2 even at -78° to give the iodide-bridged dinuclear complex, 14 (eq. 36), as the

$$1 + I_2 \longrightarrow Cp(OC)_2Mo(\mu\text{-}I)_2Mo(CO)_2Cp \quad (36)$$

(14)

kinetic product. The structure of 14 has been determined and is portrayed in Fig. 11 (49). In this binuclear oxidative addition, I_2 is a *net* 6-electron donor (2 X $\ddot{I}$:) thereby completely displacing all three M≡M bonds. The Mo····Mo separation in 14 is 4.441-(1)Å, well outside the range of bonding interactions. The large Mo-I-Mo angle of 102.24(2)° also attests to the lack of a Mo-Mo bond in 14.

Complex 14 is red-brown and stable at room temperature as a solid. In solution, however, 14 rapidly isomerizes to a violet isomer (6). This isomerization may also be followed by low temperature ^{1}H NMR spectroscopy which shows that a new Cp singlet at δ5.51 grows in as the δ5.57 signal of 14 diminishes. The isomerization may be simply trans→cis, but may also be the bridge→terminal transformation indicated in eq. 37. Repeated attempts to

$$M(\mu\text{-}I)_2M \longrightarrow I\text{—}M{=}M\text{—}I \qquad (37)$$

grow crystals of the violet isomer have invariably led to what appears to be intimately twinned crystals, frustrating all attempts to solve the structure by crystallography. The reaction of I_2 with $(C_5H_4Me)_2Mo_2(CO)_4$ gives the bridged form which does not rearrange cleanly.

Chlorine reacts rapidly with 1, but even at low temperature the product appears to be an intractable solid of approximate composition $(CpMoCl_2)_x$. However, a dichloride may be prepared from the reaction of 1 and HCl (vide infra) or of 1 and the halogen transfer reagent, $PhICl_2$ (50).

$$Cp_2Mo_2(CO)_4 \ + \ PhICl_2 \longrightarrow \underset{(15)}{[CpMo(CO)_2Cl]_n} \ + \ PhI \qquad (38)$$

The structure of 15, a lustrous black, crystalline solid, is not completely established. Its IR spectrum (ν_{CO}: 1995, 1940, 1900, and 1855 cm^{-1}) is quite different from the iodide, 14, which shows only three CO stretches at 1950, 1940, and 1860 cm^{-1}. The ^{1}H NMR spectrum of 15 also differs from 14 in that the former exhibits two Cp resonances of equal intensity separated by about 0.2 ppm and centered at δ4.9 whereas the latter has only a single Cp resonance at δ5.4.

The Cl(2p) and Mo(3d) x-ray photoelectron spectra (XPS) of 15 were kindly recorded by Prof. R.E. McCarley and Mr. M. Luly. The chlorine XPS, displayed in Fig. 12, clearly shows two types of Cl atoms (each peak is a doublet composed of the $^2P_{3/2}$, $^2P_{1/2}$ spin-orbit pair) separated by 1.2 eV. The higher energy peak (198.0 eV referred to C(1s) at 285.0 eV) may be attributed to a bridging Cl-atom and the lower energy peak (197.6 eV) to a terminal chlorine. Structure XII is assigned to 15, even though only one Mo(3d) peak could be resolved in the Mo XPS (the energy separation may be too small to resolve due to the broad peaks).

In XII one molybdenum is formally Mo(I) and the other Mo(III).

Structure XII is consistent with all the data in hand and gives each metal the 18-electron configuration. We are attempting to grow crystals of 15 suitable for crystallographic analysis.

The propensity of the M≡M bond to lead to unique products is especially evident when one considers that a dichloride, $Cp_2Mo_2(CO)_4Cl_2$ (16), may be prepared in low yield by photolysis (eq. 39) (51), and that 16 is dark red whereas 15 is black. Complex 16 presumably has the bridged structure analogous to 14 (Fig. 11) since the IR and ^{1}H NMR spectra of 16 (ν_{CO}: 1955, 1858 cm^{-1}; δ(Cp): 5.8) are similar to those of 14 (see above).

$$CpMo(CO)_3Cl \xrightarrow[-CO]{h\nu} [CpMo(CO)_2Cl] \longrightarrow \frac{1}{2}\, \underset{(16)}{Cp_2Mo_2(CO)_4Cl_2} \quad (39)$$

Reaction with Hydrogen Halides. Gaseous hydrogen halides react rapidly with the Mo≡Mo bond of 1 at and below room temperature. Hydrogen chloride reacts to give low and erratic yields of the HCl adduct, $Cp_2Mo_2(CO)_4(HCl)$ (6). The major reaction product is a brown, insoluble powder. On the basis of spectroscopic evidence, the HCl adduct was assigned a structure with bridging H and Cl atoms (6).

The reaction of 1 with HI is much cleaner and affords the adduct, 17, in moderate to good yields (eq. 40). The structure of

$$Cp_2Mo_2(CO)_4 + HI \xrightarrow[PhMe]{0^\circ} \underset{(17)}{Cp(OC)_2Mo(\mu\text{-}H)(\mu\text{-}I)Mo(CO)_2Cp} \quad (40)$$

17 has been determined (Fig. 13). The H atom was not located but undoubtedly occupies a bridging site *trans* to the bridging iodine. The Mo-Mo distance (3.310(2)Å) in 17 is 0.05Å longer than the corresponding distance in the isoelectronic $Cp_2Mo_2(CO)_4(\mu\text{-}H)(\mu\text{-}PMe_2)$ (18) (52). The Mo-I bonds are considerably shorter in 17 (2.759(1)Å) than in 14 (2.853(1)Å) and in other Mo-I complexes (2.836-2.858Å) (53). Doedens and Dahl (52) also noted that the Mo-P distances in 18 were short and proposed Mo-P π-bonding. We have discussed previously the intermingling of metal-ligand and metal-metal bonding orbitals in bridged M-M bonds and such interactions may lead to contraction of metal-ligand distances as well as the metal-metal distance (54,55). The presence of a Mo-Mo bond in 17 (as required by the 18-electron rule) is also manifested in the acute Mo-I-Mo angle (73.70(4)°).

The complex, 17, reacts with a further equivalent of HI to give $Cp_2Mo_2(CO)_4I_2$ (14) and presumably H_2 (eq. 41).

$$\underset{(17)}{Cp_2Mo_2(CO)_4(\mu\text{-}H)(\mu\text{-}I)} + HI \longrightarrow H_2 + \underset{(14)}{Cp_2Mo_2(CO)_4(\mu\text{-}I)_2} \quad (41)$$

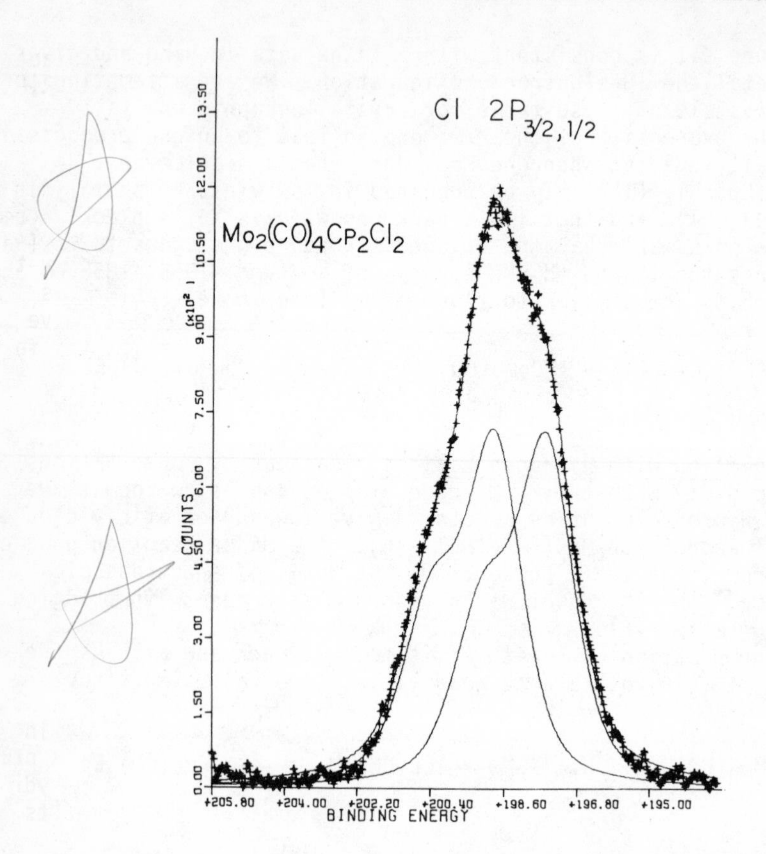

Figure 12. Cl(2p) x-ray photoelectron spectrum of $Cp_2Mo_2(CO)_4Cl_2$ (15)

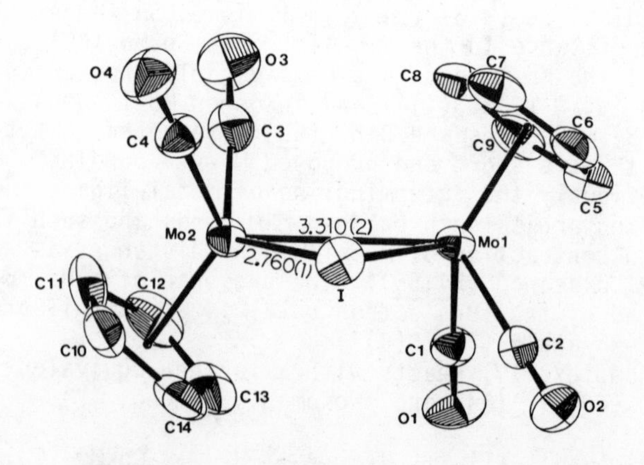

Figure 13. ORTEP plot of $Cp_2Mo_2(CO)_4(HI)$ (17)

The dichloride, 15, has also been isolated from the reaction products of HCl with 1. By following the reaction by ^1H NMR, the reaction sequence in eq. 42 was also established.

$$1 \xrightarrow{HCl} Cp_2Mo_2(CO)_4(\mu\text{-}H)(\mu\text{-}Cl) \longrightarrow \underset{(15)}{Cp_2Mo_2(CO)_4Cl_2} \quad (42)$$

Interestingly, hydride donors react with the diiodide, 14, to give the HI-adduct (17) which in turn may be reduced further to 1 and H_2 (eq. 43,44). Presumably the latter reaction proceeds through a di-hydride intermediate, 19, which loses H_2 to give 1. That the presumed dihydride loses H_2 is consistent with the fact that 1 does not react with H_2 under mild conditions.

$$\underset{(14)}{Cp(OC)_2Mo(\mu\text{-}I)_2Mo(CO)_2Cp} \xrightarrow[-I^-]{+H^-} \underset{(17)}{Cp(OC)_2Mo(\mu\text{-}H)(\mu\text{-}I)Mo(CO)_2Cp} \quad (43)$$

$$17 \xrightarrow[-I^-]{+H^-} \underset{(19)}{Cp(OC)_2Mo(\mu\text{-}H)_2Mo(CO)_2Cp} \longrightarrow 1 + H_2 \quad (44)$$

The sequence of reactions, 40-44, shows that the bridging "hydride" in 17 is amphoteric. In eq. 40, the "hydride" is produced from a proton and reacts further with a proton as a hydride to produce H_2 (eq. 41). In eq. 44, the same "hydride" reacts as a proton with H^- to produce H_2.

We are currently investigating the reactions of 1 with other protic acids and the interactions of the resulting products with olefins and acetylenes. The object of this work is to establish if dinuclear complexes can function as catalysts for the anti-Markovnikov addition of HX to terminal olefins or acetylenes. Preliminary results show that the HI adduct (17) reacts slowly with 1-hexene at 80° (products not yet definitely established). The reaction of 17 with ethylene (45 psi, 75°) yields 3-pentanone as the primary organic product. Acetylene (1 atm., 25°) reacts rapidly at room temperature with 17 to give as yet unidentified products.

Reactions with R_2S_2 and S_8. In their reaction with 1, dimethyl and diphenyl disulfide behave as pseudo-halogens giving the bridged complexes, 20 (eq. 45). These complexes are also obtained from the reaction of R_2S_2 with $Cp_2Mo_2(CO)_6$ under more forcing conditions (56,6).

The reaction of 1 with elemental sulfur is quite complex. The addition of solid sulfur to a solution of 1 or the dropwise addition of a solution of 1 to a solution of sulfur causes a rapid evolution of CO and the precipitation of dark red solids insoluble in all common solvents. These solids appear to be identical to the $[CpMoS_x]_n$ products isolated from the reaction of sulfur or episulfides with $Cp_2Mo_2(CO)_6$ (57).

However, when 1.0 to 1.5 equivalents of sulfur in acetone are added slowly to a solution of 1 in CH_2Cl_2 or acetone (i.e., under conditions where 1 is in excess during most of the reaction), a red-black solid precipitates. Recrystallization affords a crystalline solid (50% yield, dec. 200°) which displays carbonyl stretches at 1995, 1970, 1950, 1910(sh), 1892, 1880(sh), 1785, and 1760 cm^{-1} (KBr disc). The nature of this solid was established by x-ray crystallography to be a salt composed of the cluster cation, $Cp_3Mo_3(CO)_6S^+$ (21) and the anion, $CpMo(CO)_3^-$ (eq. 46) (58). The ν_{CO} bands at 1880, 1785, and 1760 cm^{-1} are thus attributed to the anion, and the remainder to the cation.

$$1 + \frac{1}{8} S_8 \longrightarrow \underset{(21)}{[Cp_3Mo_3(CO)_6S]^+}[CpMo(CO)_3]^- + \cdots \quad (46)$$

Cation 21 is the first homonuclear molybdenum carbonyl cluster to be characterized and is formed in a novel, sulfur-mediated disproportionation of two dimers into a trimer and a monomer with an accompanying electron transfer. Since there are nine CO groups per four Mo atoms in the product, but only 8 CO / 4 Mo in the starting dimers, a carbonyl-poor fraction must be produced. This fraction has been isolated but not characterized.

Figures 14 and 15 show the ORTEP plot of the cluster, 21 (58). In the cation, which has virtual C_3-symmetry, the carbonyls are arranged in two sets, axial and equatorial. The latter set may be classed as barely semi-bridging (the average $Mo\text{-}C\text{-}O_{eq}$ angle is 168(1)° *vs*. 176(2)° for $Mo\text{-}C\text{-}O_{ax}$) (14). The three Mo-Mo distances in 21 are 3.106(2), 3.085(2), and 3.064(2)Å (ave. 3.085(21)Å). The Mo-S distances are all equal and have an average value of 2.360(4) Å. Both the Mo-Mo and Mo-S distances are longer than the corresponding bond lengths in the related cluster, $Cp_3Mo_3S_4^+$ (22) viz. 2.812 (Mo-Mo), 2.293 ($Mo\text{-}\mu_3S$) (59).

In the tetrasulfide, each Mo(IV) has a d^2 configuration which is precisely the number required for three Mo-Mo electron pair σ-bonds. Each Mo atom in 21, however, has the d^4 configuration. A qualitative MO scheme based upon linear combinations of the four frontier orbitals of the $CpMo(CO)_2$ groups (19) suggests that the $Cp_3(CO)_6Mo_3$ triangle gives rise to five bonding MO's (3a+e) and seven antibonding MO's (a+3e). The 3p-orbitals (a+e) of the μ_3-S then interact to give seven bonding (3a+2e), two approximately nonbonding (e), and six antibonding (2a+2e) molecular orbitals. The 18 electrons in the Mo_3S core then completely fill all bonding plus the nonbonding (e) orbitals.

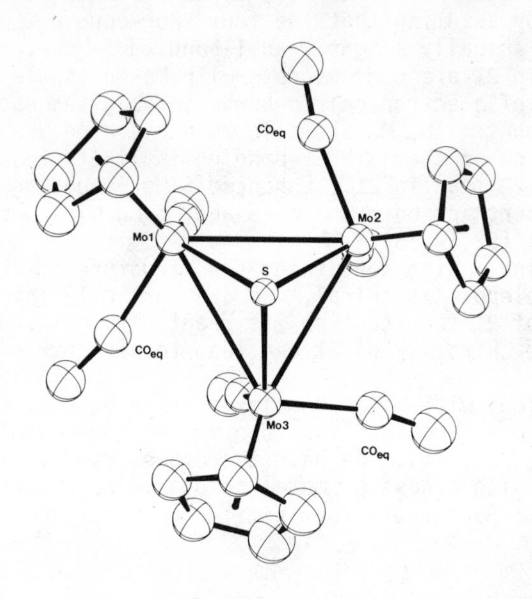

Figure 14. ORTEP plot of $Cp_3Mo_3(CO)_6S^+$ cation (21) viewed down the approximate C_3-axis (58)

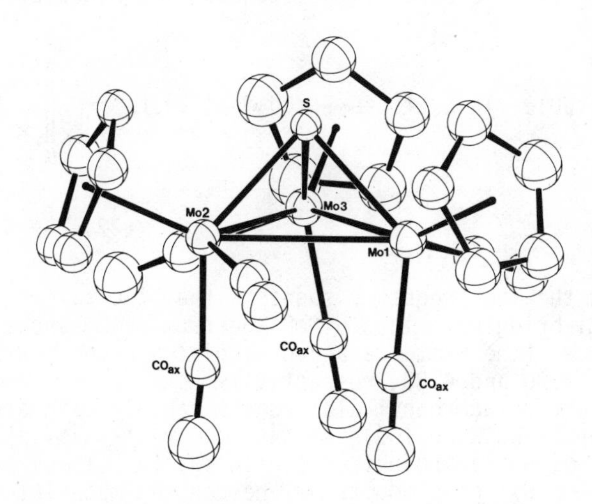

Figure 15. ORTEP plot of $Cp_3Mo_3(CO)_6S^+$ (21) cation viewed from slightly above the Mo_3-plane

The greater Mo-Mo and Mo-S distances in *21* vs. *22* could be explained by assuming that the four "non-bonding" electrons in the former are actually slightly anti-bonding. However, the Mo-Mo distances in *21* are commensurate with Mo-Mo single bonds, and the assumption of electron pair σ-bonds in *21* gives each Mo the 18-electron count. The Mo atoms in *22* achieve only a 16-electron configuration so that S→Mo π-bonding is likely to be much more important in *22* than in *21*. Enhanced S→Mo π-bonding would be expected to contract both the Mo-S and Mo-Mo distances (see discussion above) (52,54,55).

One final point concerning the structure of the cluster *21* is that the molecule is chiral. We are currently investigating the chemistry of *21* to establish its reactivity patterns, and to determine if the chirality of *21* can lead to asymmetric induction.

Reactions with Metal Complexes. The reactions of acetylenes with metal complexes or ions to give π-bonded adducts is well known (e.g., eq. 47). We have therefore investigated the reactions of *1* with a host of metal complexes to determine if the Mo≡Mo triple bond would form similar "π-complexes" to metals.

$$RC\equiv CR \; + \; L_4Pt \longrightarrow L_2Pt \leftarrow \overset{\displaystyle R}{\overset{|}{C}}\!\!\equiv\!\!\underset{|}{\underset{\displaystyle R}{C}} \longleftrightarrow L_2Pt\!\!\lhd\!\!\begin{matrix} CR \\ \| \\ CR \end{matrix} \qquad (47)$$

Complex *1* does react readily with L_4Pt ($L = Ph_3P$) according to eq. 48 (6). The carbonyl stretches of *23* occur at 1980 and

$$2\ Cp_2Mo_2(CO)_4 \; + \; L_4Pt \longrightarrow Cp_2Mo_2(CO)_4L_2 \; + \; \begin{matrix} M \\ \| \\ M \end{matrix}\!\!\rhd Pt\begin{matrix} L \\ \\ L \end{matrix} \qquad (48)$$

$(23,\ M = CpMo(CO)_2)$

1800 cm^{-1}; the low frequency suggests that the CO groups are bridging or semi-bridging. The ^1H NMR spectrum of *23* shows only Ph and Cp resonances (the latter a sharp singlet) in the correct area ratios at δ2.90 and 4.80, respectively.

Dicobalt octacarbonyl also reacts rapidly with acetylenes to give the $(\mu\text{-}RCCR)Co_2(CO)_6$ complexes. With *1*, $Co_2(CO)_8$ reacts rapidly to give the products shown in eq. 49. These products may be the decomposition products of the desired metallatetrahedrane complex (6).

$$1 + Co_2(CO)_8 \longrightarrow Co_4(CO)_{12} + CpMo(CO)_3Co(CO)_4 + Cp_2Mo_2(CO)_6 + \cdots\cdots \quad (49)$$

Diiron enneacarbonyl reacted with 1 to give a host of products, the characterization of which was severly hampered by their instability. However, the complex, 24, was obtained in an indirect way as shown in eq. 50 (6). The IR spectrum of 24 shows three

$$Cp_2Mo_2(CO)_4I_2 + Fe(CO)_4^{2-} \xrightarrow{-2I^-} M{\equiv}M{>}Fe(CO)_4 \quad (50)$$

(24, M = $CpMo(CO)_2$)

CO-stretches (2050, 2020, 2000 cm^{-1}) that are readily assigned to the C_{2v}-$Fe(CO)_4$ fragment and two (1900, 1875 cm^{-1}) to the $CpMo(CO)_2$ fragment.

The reactions of 1 with a variety of other transition metal carbonyl species have led to disappointing results. Typically, the reactants do not react at room temperature (if at all) or react at higher temperatures to give intractable products. Table III lists some of the results found to date (60). The dark residues may contain cluster species, but their lack of solubility in any solvent has precluded their analysis. Reaction 9 (Table III) gives a mixture with three Cp singlets in the 'H NMR and merits further investigation. The transfer of Cp from Mo to Re (example 6) bespeaks the extensive rearrangements possible under forceful conditions.

Summary and Prognosis

If the reactions of $Cp_2Mo_2(CO)_4$ can be taken as at all representative, then metal-metal multiple bonds indeed have promise as a versatile and useful functional group in inorganic chemistry. The new compounds obtainable directly from the M≡M multiple bonds have also been shown to possess some remarkable reactivity as a result of the proximity of the two metal centers and the ability of the M-M bond to transfer electrons via changes in the M-M and M-L bond orders.

Nucleophilic and electrophilic additions and oligomerization (cluster formation) have been demonstrated. Unusual coordination geometries and bonding may be adopted by the coordinated ligands. These in turn may lead to new reaction types for these ligands. As more information on the reactivity of multiply-bonded metals becomes available, new catalytic applications will undoubtedly

Table III. Reactions of $Cp_2Mo_2(CO)_4$ with Transition Metal Complexes

Rxn.	Reactant	Conditions[a]	Products
1	$Ru_3(CO)_{12}$	1	Starting materials, higher $Ru_n(CO)_x$ clusters
2	$H_4Ru_4(CO)_{12}$	1	$Ru_3(CO)_{12}$, insol. residue with $Mo_{11}Ru_7$ anal.
3	$Os_3(CO)_{12}$	1	No reaction
4	$H_2Os_3(CO)_{10}$	2	$Cp_2Mo_2(CO)_6$ + insol. residues
5	$Co_4(CO)_{12}$	1	No reaction
6	$HRe_3(CO)_{14}$	3	$CpRe(CO)_3$, unreacted 1, insol. residue (Mo_7Re_3).
7	Cp_2WH_2	4	No reaction
8	$L_2Ir(CO)Cl$[b]	1	1, $Cp_2Mo_2(CO)_6$, unreact. $L_2Ir(CO)Cl$, insol. residue.
9	$CpCoLMe_2$	1	Mixture, no crystalline products

a) Conditions: 1 = refluxing toluene; 2 = hexane, 25°; 3 = hexane/CH_2Cl_2, 25°; 4 = hexane or THF, $h\nu$;

b) L = Ph_3P.

become obvious. The inherent unsaturation of the metal centers, and the ability of the M≡M unit to accommodate a range of coordination types and electron counts offer some similarities to surface atoms on small metallic aggregates. New theoretical insights will hopefully follow observations on novel structure-bonding-reactivity relationships which will develop as more knowledge of these systems accrue.

In short, the chemistry of M≡M multiple bonds may be likened to an unearthed chest of buried treasure. We have opened the lid and gazed upon the surface, but precious treasures still lie below.

Acknowledgements

The authors thank the National Science Foundation (Grant CHE-7907748), the donors of the Petroleum Research Fund administered by the American Chemical Society, and the AMAX Foundation for support of this work.

Literature Cited

1. Cotton, F.A. Acc. Chem. Res., 1978, 11, 225 and references therein.

2. Trogler, W.C.; Gray, H.B. Acc. Chem. Res., 1978, 11, 232.

3. Chisholm, M.H. Acc. Chem. Res., 1978, 11, 356.

4. Chisholm, M.H. Trans. Met. Chem., 1978, 3, 321.

5. Chisholm, M.H. Adv. Chem. Ser., 1979, 173, 396.

6. Curtis, M.D.; Klingler, R.J. J. Organometal. Chem., 1978, 161, 23.

7. (a) Bino, A.; Cotton, F.A. Proc. of the Climax Third Internat. Conf. on Chem. and Uses of Molybdenum (H. F. Barry and P.C.H. Mitchell, Eds.), Climax Molydenum Co., Ann Arbor, MI, 1979, p. 1; (b) Templeton, J.L. Prog. Inorg. Chem., 1979, 26, 211.

8. McGinnis, R.N.; Ryan, T.R.; McCarley, R.E. J. Am. Chem. Soc., 1978, 100, 7900.

9. Chisholm, M.H.; Cotton, F.A.; Murillo, C.A.; Reichert, W.W. Inorg. Chem., 1977, 16, 1801.

10. Curtis, M.D.; Fotinos, N.A.; Sattelberger, A.P.; Messerle, L., to be published.

11. Curtis, M.D.; Butler, W.M. J. Organometal. Chem., 1978, 155, 131.

12. Klingler, R.J.; Butler, W.M.; Curtis, M.D. J. Am. Chem. Soc., 1978, 100, 5034.

13. Potenza, J.; Giordano, P.; Mastropaolo, D.; Efraty, A. Inorg. Chem., 1974, 13, 2540.

14. Cotton, F.A.; Wilkinson, G. "Advanced Inorganic Chemistry", John Wiley and Sons, New York, N.Y., 4th Ed., p. 1057f.

15. Curtis, M.D.; Han, K.R.; Butler, W.M. Inorg. Chem., 1980, 19, 2096.

16. Cotton, R.; Commons, C.J. Aust. J. Chem., 1975, 28, 1673; Commons, C.J.; Hoskins, B.F. ibid., 1975, 28, 1663.

17. Pasynskii, A.A.; Skripkin, Yu. V.; Eremenko, I.L.; Kalinnikov, V.T.; Aleksandrov, G.G.; Andrianov, V.G.; Struchkov, Yu. T. J. Organometal. Chem., 1979, 165, 49.

18. Cotton, F.A.; Kruczynski, L.; Frenz, B.A. J. Organometal. Chem., 1978, 160, 93.

19. Jemmis, E.D.; Pinhas, A.R.; Hoffmann, R. J. Am. Chem. Soc., 1980, 102, 2576.

20. Hackett, P.; O'Neil, P.S.; Manning, A.R. J. Chem. Soc., Dalton Trans., 1974, 1625.

21. Chisholm, M.H.; Cotton, F.A.; Extine, M.W.; Rankel, L.A. J. Am. Chem. Soc., 1978, 100, 807.

22. Alper, H.; Silavwe, N.D.; Birnbaum, G.I.; Ahmed, F.R. J. Am. Chem. Soc., 1979, 101, 6582.

23. Alper, H.; Hartzgerink, J. J. Organometal. Chem., 1980, 190, C25.

24. King, R.B.; Iqbal, M.Z.; King, Jr., A.D. J. Organometal. Chem., 1979, 171, 53.

25. Bailey, Jr., W.I.; Chisholm, M.H.; Cotton, F.A.; Rankel, L.A. J. Am. Chem. Soc., 1978, 100, 5764.

26. Curtis, M.D.; Fotinos, N.A., unpublished results.

27. Allmann, R., Chap. 2 in "The Chemistry of Hydrazo, Azo, and Azoxy Groups," S. Patai, Ed., John Wiley and Sons, New York, N.Y., 1975.

28. Slater, S.; Muetterties, E.L., submitted for publication.

29. Bradley, J.S. J. Organometal. Chem., 1978, 150, C1.

30. Knox, S.A.R.; Stansfield, R.F.D.; Stone, F.G.A.; Winter, M.J.; Woodward, P. J. Chem. Soc., Chem. Commun., 1978, 221.

31. Knox, S.A.R.; Stansfield, R.F.D.; Stone, F.G.A.; Winter, M.J.; Woodward, P. J. Chem. Soc., Chem. Commun., 1979, 934.

32. Bailey, Jr., W.I.; Chisholm, M.H.; Cotton, F.A.; Murillo, C.A.; Rankel, L.A. J. Am. Chem. Soc., 1978, 100, 802.

33. Curtis, M.D.; Han, K.R., unpublished results.

34. Casey, C.P.; Andrews, M.A.; McAlister, D.R.; Ring, J.E. J. Am. Chem. Soc., 1980, 102, 1927.

35. Masters, C. Adv. Organometal. Chem., 1979, 17, 61.

36. Herrmann, W.A. Angew. Chem., Int. Ed. Engl., 1978, 17, 800.

37. Messerle, L.; Curtis, M.D. J. Am. Chem. Soc., 1980, 102, 0000.

38. Messerle, L.; Curtis, M.D.; Butler, W.M., to be submitted.

39. Ben-Shoshan, R.; Chatt, J.; Leigh, G.J.; Hussein, W. J. Chem. Soc. Dalton Trans., 1980, 771 and references therein.

40. Hidai, M.; Mizobe, Y.; Sato, M.; Kodama, T.; Uchida, Y. J. Am. Chem. Soc., 1978, 100, 5740.

41. Bennett, M.J.; Mason, R. Proc. Chem. Soc. (London), 1963, 273.

42. Huq, F.; Mowat, W.; Shortland, A.; Shapski, A.C.; Wilkinson, G. J. Chem. Soc., Chem. Commun., 1971, 1079.

43. Brisdon, B.J.; Woolf, A.A. J. Chem. Soc., Dalton Trans., 1978, 291.

44. King, R.B.; Fronzaglia, A. J. Am. Chem. Soc., 1966, 88, 709.

45. Cotton, F.A.; LaPrade, M.D. J. Am. Chem. Soc., 1968, 90, 5418.

46. Levisalles, J.; Rudler, H.; Dahan, F.; Jeannin, Y. J. Organometal. Chem., 1980, 188, 193.

47. Yamamoto, T.; Garber, A.R.; Wilkinson, J.R.; Boss, C.B.; Streib, W.E.; Todd, L.J. J. Chem. Soc., Chem. Commun., 1974, 354.

48. Overberger, C.G.; Anselme, J-P.; Lombardino, J.G. "Organic Compounds with Nitrogen-Nitrogen Bonds", Ronald Press, New York, N.Y., 1966, p. 56.

49. Curtis, M.D.; Han, K.R.; Butler, W.M., to be published.

50. Fotinos, N.A.; Curtis, M.D., to be published.

51. Ali, L.H.; Cox, A.; Kemp, T.J. J. Chem. Soc., Dalton Trans., 1973, 1475.

52. Doedens, R.J.; Dahl, L.F. J. Am. Chem. Soc., 1965, 87, 2576.

53. Adams, R.D.; Chodosh, D.F. J. Am. Chem. Soc., 1977, 99, 6544.

54. Triplett, K.; Curtis, M.D. J. Am. Chem. Soc., 1975, 97, 5747.

55. Triplett, K.; Curtis, M. D. Inorg. Chem., 1976, 15, 431.

56. Cameron, T.S.; Prout, C.K.; Rees, G.V.; Green, M.L.H.; Joshi, K.K.; Davies, G.R.; Kilbourn, B.J.; Braterman, P.S.; Wilson, V.A. J. Chem. Soc. (D), 1971, 14.

57. Dubois, M.R.; Haltiwanger, R.C.; Miller, D.J.; Glatzmeier, G. J. Am. Chem. Soc., 1979, 101, 5245; Dubois, M.R.; Dubois, D.L.; Vanderveer, M.C.; Haltiwanger, R.C. Paper No. 131, Division of Inorganic Chemistry, American Chemical Society, Abstracts 179th Meeting, 1980.

58. Curtis, M.D.; Butler, W.M., submitted for publication.

59. Vergamini, P.J.; Vakrenkamp, H.; Dahl, L.F. J. Am. Chem. Soc., 1971, 93, 6327.

60. Gerlach, R.F.; Curtis, M.D., unpublished results.

RECEIVED December 11, 1980.

13

Reactivity of Dimetallocycles

ANDREW F. DYKE, STEPHEN R. FINNIMORE, SELBY A. R. KNOX, PAMELA J. NAISH, A. GUY ORPEN, GEOFFREY H. RIDING, and GRAHAM E. TAYLOR

Department of Inorganic Chemistry, The University, Bristol BS8 1TS, England

The involvement of metallocycles in the catalysis of organic reactions is well established, and is exemplified by the studies of Wilke and co-workers in the field of organo-nickel chemistry (1) and by the role of metallocyclobutanes in the alkene metathesis reaction. Until very recently attention was given almost exclusively to the chemistry of metallocycles containing a single metal atom. There is now, however, a growing body of evidence prompting speculation (1-3) that metallocycles based on a dinuclear metal centre ("dimetallocycles") may be equally important. In seeking to understand catalysis of organic reactions by metal surfaces or by metal clusters it is clearly important to consider the dinuclear metal centre (4) and particularly the nature and reactivity of species co-ordinated at (or bridging) the centre.

In this paper we describe studies on the synthesis and reactivity of di-iron and di-ruthenium metallocycles derived from alkynes. Their chemistry is marked by an ease of carbon-carbon bond-making and -breaking which provides access to a range of simple organic species (e.g. CCH_2, $CHCH_2$, CMe, CHMe, CMe_2, CHCHCHMe) bridging the dinuclear metal centre. Related studies with dimolybdenum and ditungsten complexes will be described briefly.

DI-IRON AND DIRUTHENIUM METALLOCYCLES.

Under u.v. irradiation the well-known and readily available dimer $[Fe_2(CO)_4(\eta\text{-}C_5H_5)_2]$ reacts with alkynes RC_2R' to give the dimetallocycles $[Fe_2(CO)(\mu\text{-}CO)\{\mu\text{-}C(O)CRCR'\}(\eta\text{-}C_5H_5)_2]$ (1a-1f) in good yield (e.g. 1a, 42 %; 1d, 86 %) (5). Only (1g) (35 %) may be obtained in an analogous manner from $[Ru_2(CO)_4(\eta\text{-}C_5H_5)_2]$, but on heating (1g) in toluene with other alkynes an unusual exchange occurs rapidly to provide the appropriate complex (1h-1k) in high yield. In the formation of (1) an alkyne and CO ligand have become linked and the exchange is notable for the breaking and

0097-6156/81/0155-0259$05.00/0

(1a) M = Fe, R = R′ = H
(1b) M = Fe, R = R′ = Me
(1c) M = Fe, R = R′ = Ph
(1d) M = Fe, R = R′ = CO_2Me
(1e) M = Fe, R = H, R′ = Me
(1f) M = Fe, R = H, R′ = Ph
(1g) M = Ru, R = R′ = Ph
(1h) M = Ru, R = R′ = H
(1i) M = Ru, R = R′ = Me
(1j) M = Ru, R = H, R′ = Me
(1k) M = Ru, R = H, R′ = Ph

regeneration of this carbon-carbon bond. This behaviour is matched by an unprecedented fluxionality of dimetallocycles (1) derived from alkynes for which R = R′. This has been characterised by variable temperature ^1H and ^{13}C n.m.r. spectroscopy as being of the form illustrated. It comprises a rapid breaking and re-forming of the 'alkyne'-CO link, involving both carbons of the 'alkyne', with CO effectively entering and leaving the dimetallocycle. Free energies of activation are in the range 67 (1g) to 85 (1a) kJ mol^{-1}. The structure of (1g) was determined by X-ray diffraction, revealing that relatively small atomic movements are associated with the fluxional rearrangement.

In addition to providing an indication that dimetallocycles may have unexpectedly high lability, the complexes (1) are an excellent source of a range of dinuclear metal species containing bridging organic ligands. A substantial organic chemistry of di-iron and di-ruthenium centres is in process of being established as a consequence. Scheme 1 summarises reaction sequences evolved from the ethyne-derived dimetallocycles (1a) and (1h) and will be described in some detail. Comparable systems exist for other complexes (1).

The sequence (1)→(2)→(3)→(4) is unique to ruthenium, and to alkynes RC_2H (R = H, Me, Ph) in that step (1)→(2) involves a hydrogen shift. Up to about 100 °C (1h) undergoes the fluxional breaking and regeneration of the 'alkyne'-CO link previously described, but in boiling toluene (111 °C) irreversible cleavage of that bond occurs, coupled with a hydrogen shift, giving the μ-vinylidene complex $[Ru_2(CO)_2(\mu\text{-}CO)(\mu\text{-}CCH_2)(\eta\text{-}C_5H_5)_2]$ (2) as cis- and trans-isomers in 65 % yield. On heating to the same temperature (1a) decomposes. Addition of dry HBF_4 to (2) results in protonation at the methylenic carbon and formation of the

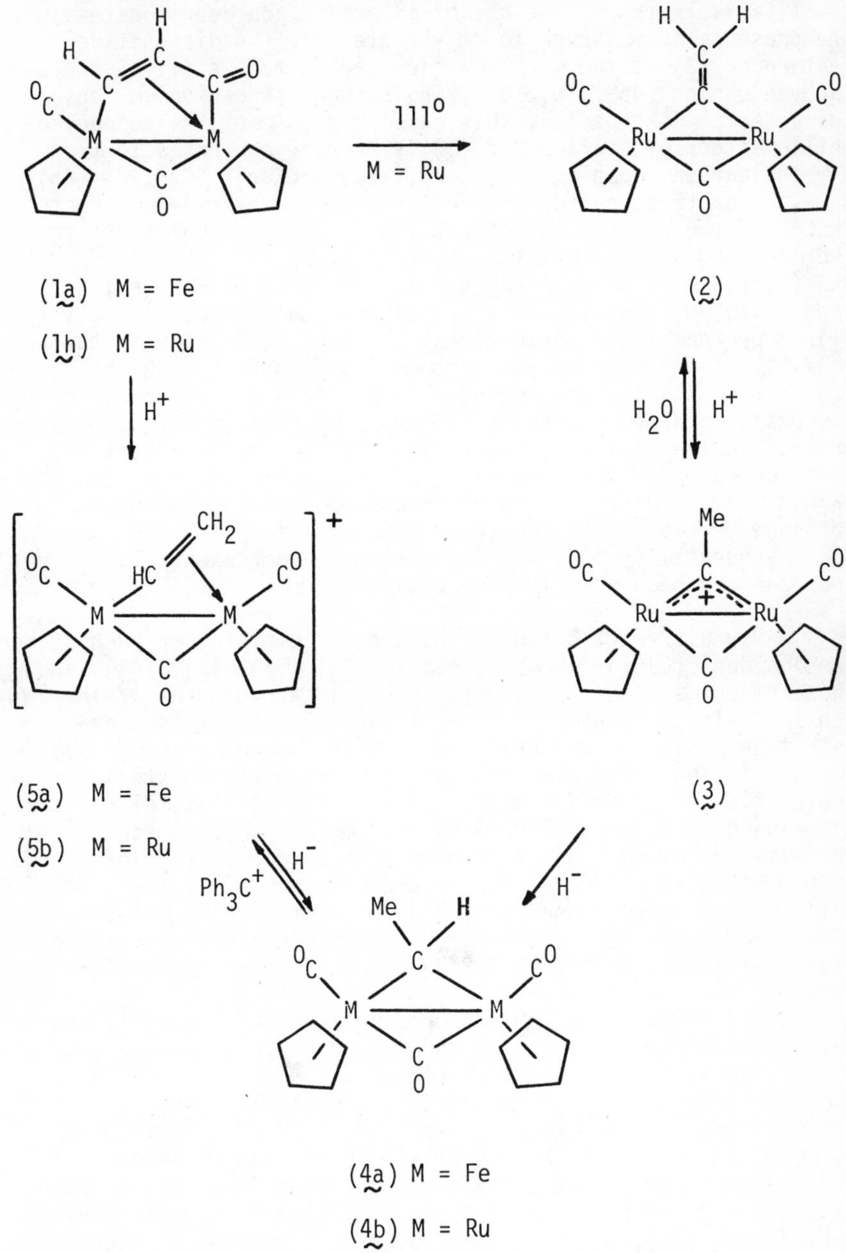

111°

M = Ru

(1a) M = Fe

(1h) M = Ru

H^+

(2)

H_2O H^+

(5a) M = Fe

(5b) M = Ru

(3)

Ph_3C^+ H^-

H^-

(4a) M = Fe

(4b) M = Ru

Scheme 1.

μ-methylcarbyne cation cis-$[Ru_2(CO)_2(\mu\text{-}CO)(\mu\text{-}CMe)(\eta\text{-}C_5H_5)_2]^+$ (3) quantitatively (6). The cation is acidic and deprotonates in the presence of moisture to regenerate (2). A distinctive feature of (3) is the very low field ^{13}C n.m.r. shift of the μ-carbyne carbon (469.7 p.p.m.), which suggests carbonium ion character. As expected, this carbon is susceptible to nucleophilic attack by hydride and (3) is thereby converted to the μ-methylcarbene complex $[Ru_2(CO)_2(\mu\text{-}CO)(\mu\text{-}CHMe)(\eta\text{-}C_5H_5)_2]$ (4b). However, on treatment of the latter with the hydride abstractor $Ph_3C^+BF_4^-$ the process is not reversed; abstraction occurs from the methyl group to form the μ-vinyl cation $[Ru_2(CO)_2(\mu\text{-}CO)(\mu\text{-}CHCH_2)(\eta\text{-}C_5H_5)_2]^+$ (5b). This complex, and its iron analogue, are best obtained directly and quantitatively from (1) by protonation, a transformation which again highlights the lability of the 'alkyne'-CO carbon-carbon bond. Both (5a) and (5b) are fluxional, undergoing a combination of cis-trans interconversion and reorientation of the μ-vinyl ligand (7). Nucleophilic hydride attack on the μ-vinyl ligand of (5) occurs preferentially at the μ-vinylic carbon, affording the μ-methylcarbene complexes (4). The iron analogue of (3) has been obtained by subjecting $[Fe_2(CO)_4(\eta\text{-}C_5H_5)_2]$ to methyl lithium and HBF_4 sequentially (8), while dimanganese complexes related to (2) and (3) have been described recently (9).

X-Ray diffraction studies have been completed on each of the species depicted in Scheme 1, namely (1g), (2), (3), (4a), and (5a). These reveal a clear preference for a cis arrangement of the cyclopentadienyls throughout. For (4a) and (5a) the methyl and methylene groups of the bridging organic ligands are anti with respect to the cis cyclopentadienyls, presumably on steric grounds. Protonation of (2) has little effect on the metal-metal distance (it lengthens by about 0.02 Å) but the metal-carbon distance decreases by ca. 0.1 Å, in keeping with delocalisation of charge to the metal atoms and partial M=C double bond character. Bridging μ-methylcarbene in (4a) is indicated to be a less effective π-acceptor than bridging CO by consideration of the Fe-C distances [1.986(3) c.f. 1.902(3) Å].

Hydride attack on (5b) occurs, unlike (5a), not only at the β-vinlynic carbon but also at the α-carbon of the μ-vinyl, giving a very low yield of the ethene complex $[Ru_2(CO)(C_2H_4)(\mu\text{-}CO)_2(\eta\text{-}C_5H_5)_2]$ (6). Variable temperature ^1H n.m.r. spectroscopy reveals the complex to be fluxional, undergoing both cis-trans interconversion and ethene rotation. Fortunately, this interesting species, whose chemistry is under study, may be obtained in 70 % yield by exploiting once more the lability of (1g), which suffers rapid displacement of diphenylacetylene by ethene in boiling toluene (10). The exchange is reversed when (6) is irradiated with u.v. light in the presence of diphenylacetylene. It may be noted that the formation of (6) from

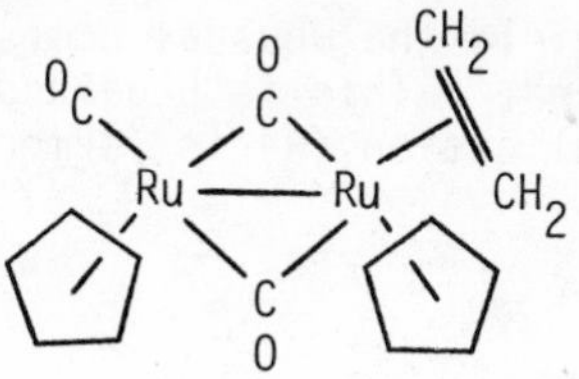

(6)

(1h) represents a conversion of ethyne to ethene at the diruthenium centre. In principle, addition may occur cis or trans and this feature is being investigated.

In essence, Scheme 1 describes two pathways by which ethyne may be converted to μ-methylcarbene at a dinuclear metal centre. In this connection it is interesting to note that studies of the chemisorption of ethyne on metal surfaces have led to suggestions that μ-methylcarbene or μ-methylcarbyne are formed as surface species, perhaps via μ-vinylidene (11-13). The sequence (1a)→(5a)→(4a) may be achieved for a variety of complexes (1), providing to date μ-carbenes as summarised in Scheme 2. Given the considerable choice of alkynes and nucleophiles available there is clearly scope for the preparation of many more.

We are currently developing an equally promising route to μ-carbenes, derived uniquely from allenes (10). For unsubstituted allene the synthesis is presented in Scheme 3. It is clear from results described earlier that (1g) is an efficient source of "$Ru_2(CO)_3(\eta\text{-}C_5H_5)_2$" and it again liberates diphenylacetylene when

(i) H^+
(ii) H^-

R = R´ = H
R = H, R´ = Me
R = Me, R´ = H
R = R´ = Me

μ-C(H)Me
μ-C(H)Et
μ-CMe_2
μ-C(Me)Et

Scheme 2.

heated with allene to yield the unusual complex (7), which has lost the metal-metal bond. This is regained on protonation, when the μ-1-methylvinyl cation (8) is formed. Hydride attack on

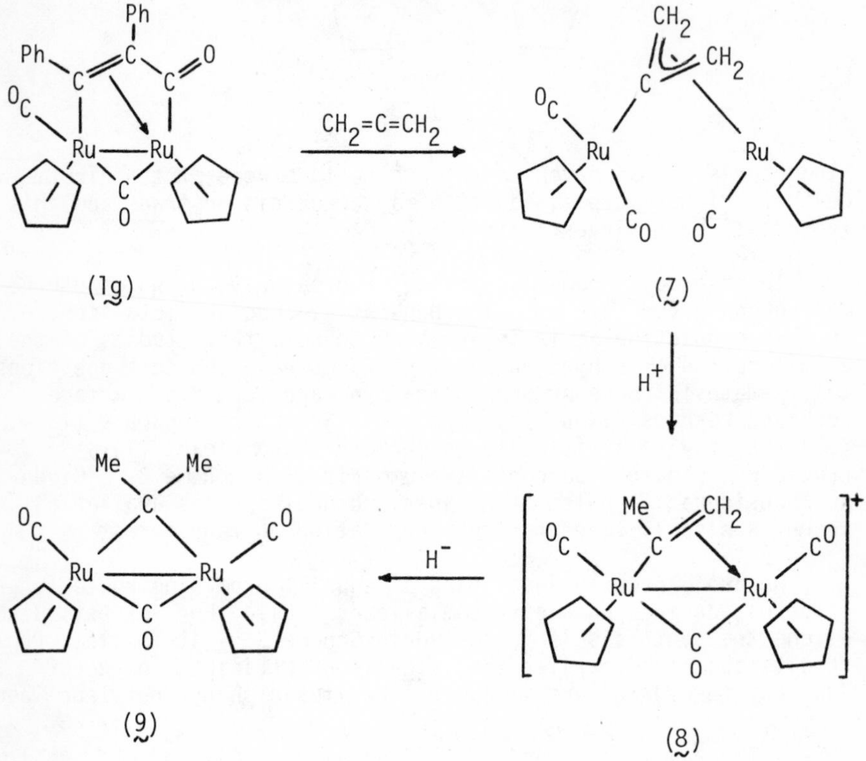

Scheme 3.

(8) then occurs specifically at the β-vinyl carbon, generating the μ-dimethylcarbene complex (9) in 60 % yield overall. Substituted allenes are available and we anticipate being able to convert these in the same manner (e.g. buta-1,2-diene to μ-C(Me)Et and penta-2,3-diene to μ-CEt_2).

The transition metal chemistry of μ-carbenes (14) is of interest because of the possible involvement of such species in Fischer-Tropsch synthesis (15) and alkene metathesis (16,17). However, apart from μ-CH_2 and one example of μ-CHMe, simple hydrocarbon species have, until the work described here, been generally unavailable and the reactivity of μ-carbenes is effectively unexplored. An opportunity was therefore presented for such study, in which carbon-carbon bond formation has taken precedence.

The μ-carbene complexes are structurally related to the carbonyl-bridged forms of the dimers $[M_2(CO)_4(\eta\text{-}C_5H_5)_2]$ and they react similarly with alkynes under u.v. irradiation. Reactions of the μ-methylcarbene complexes (4) are summarised in Scheme 4. The new complexes are derived by linking of the carbene and alkyne

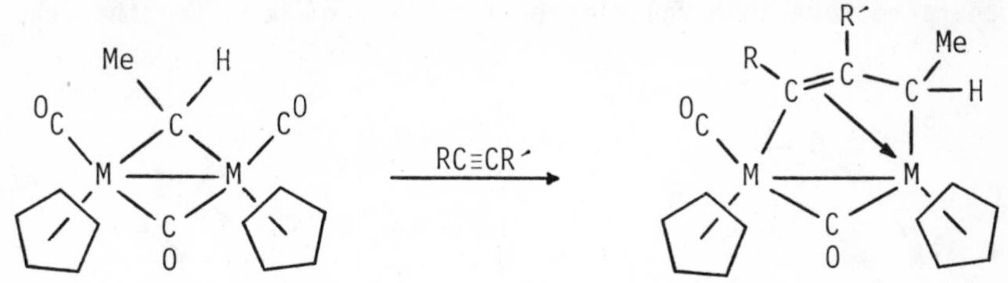

(4) $\xrightarrow{RC\equiv CR'}$ (9a)

R = R′ = H, Me, CO_2Me
R = H, R′ = Me
R = Me, R′ = Ph

Scheme 4.

and may be viewed as 'insertion' products. Obtained in 50 - 90 % yields, they were characterised structurally through an X-ray diffraction study of $[Fe_2(CO)(\mu\text{-}CO)\{\mu\text{-}CH(Me)C_2(CO_2Me)_2\}(\eta\text{-}C_5H_5)_2]$ (18). Bridging ligands of this type have been reported recently (17,19), but were not obtained in this way. The structure of the complexes (9) is strikingly related to that of the complexes (1). However, whereas the 'alkyne'-CO bond in (1) is very labile the 'alkyne'-carbene carbon-carbon bond in (9) is not. Each of (9) is stereochemically rigid and stable at temperatures up to 100 °C. Slow transformation of the X-ray compound does occur, in boiling toluene over several days to produce $[Fe_2(CO)\{C_2H_2C_2(CO_2Me)_2\}(\eta\text{-}C_5H_5)_2]$ (10). This extraordinary

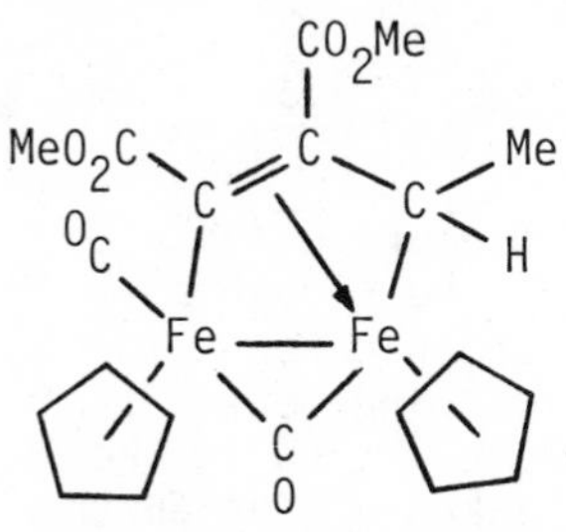

(9) 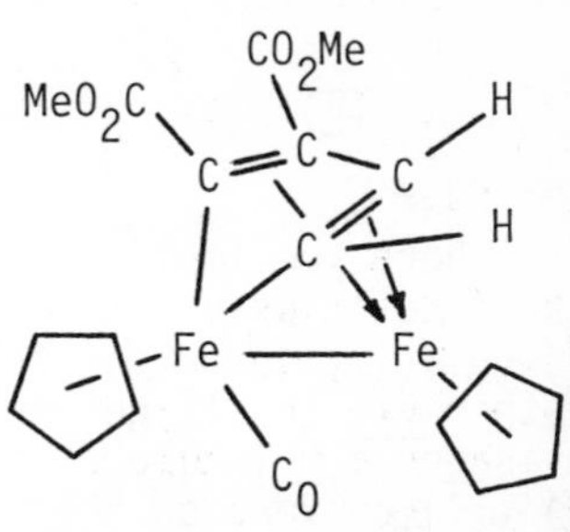 (10)

reaction comprises a double β-elimination from the CMe group, bringing the methyl carbon into co-ordination.

The co-ordination of the bridging ligand in (9) may be represented in three ways; that shown earlier (9a), a 'μ-allyl' mode (9b) and a 'μ-vinylcarbene' representation (9c). The latter appears to have some validity on the basis of bond lengths (the

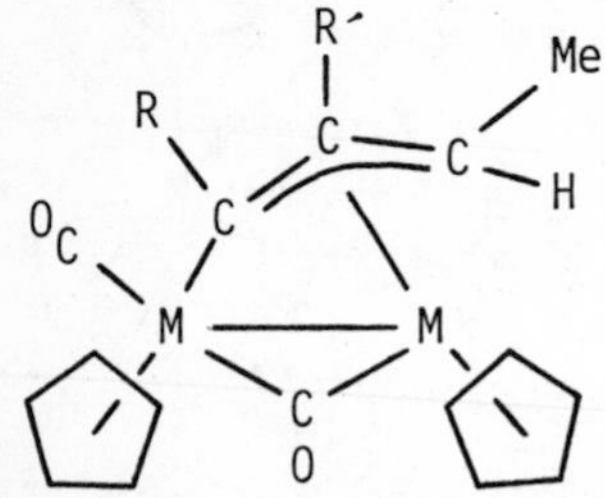

(9b)

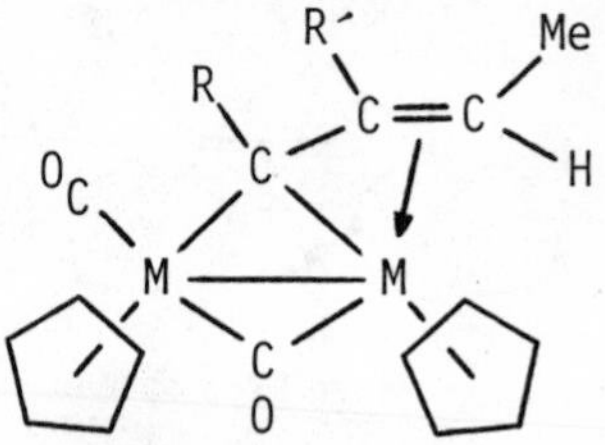

(9c)

μ-C is equidistant from both metals) and n.m.r. data, and suggested that the vinyl substituent of the carbene might be released from complexation. This was achieved when under 100 atm. of CO $[Fe_2(CO)_2(\mu\text{-}CO)\{\mu\text{-}CH(Me)C_2H_2\}(\eta\text{-}C_5H_5)_2]$ (11) was formed as shown, in high yield. Heating or irradiating (11) reverses the process.

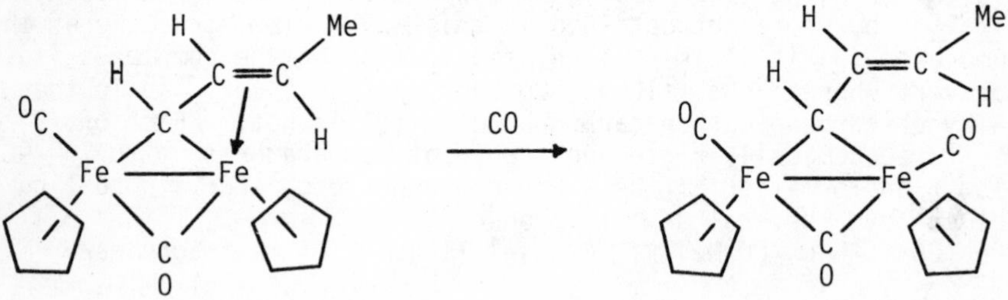

(9) $\xrightarrow{CO}$ (11)

There are implications for both alkyne polymerisation and alkene metathesis in the above observations. The sequence (4)→(9)→(11) comprises the transformation of one μ-carbene to another *via* an alkyne insertion followed by rearrangement. If one envisages a molecule of alkyne in the role played by CO in the (9)→(11) conversion then the sequence can be taken a step further through insertion of alkyne into the new μ-carbyne. Successive insertions and rearrangements of this type then provide a mechanism fo alkyne polymerisation at a dinuclear metal centre, initiated by a μ-carbene. This is laid out in Scheme 5. A scheme for alkyne polymerisation initiated by carbenes co-ordinated at a mononuclear metal centre has been postulated (20), and recently it was shown that such carbenes do initiate polymerisation (21).

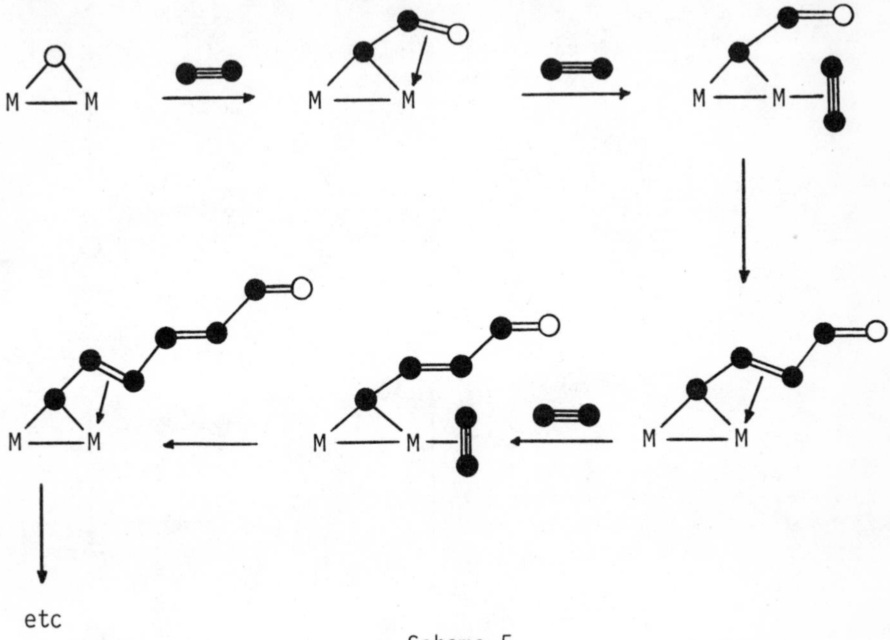

M—M M—M M—M— etc

Scheme 5.

The ability of a μ-carbene to react with an unsaturated hydrocarbon and form an enlarged dimetallocycle encourages speculation over their role in such processes as alkene metathesis and Fischer-Tropsch synthesis. In Scheme 6 a possible mechanism for metathesis initiated by a μ-carbene is presented, owing much to other workers (17,22). Reactions of μ-carbenes with alkenes are under investigation in our laboratory. Recently Pettit has observed that the μ-methylene complex $[Fe_2(CO)_8(\mu\text{-}CH_2)]$ generates propene when subjected to a pressure of ethene and has also suggested the intermediacy of a three-carbon dimetallocycle (23).

DIMOLYBDENUM AND DITUNGSTEN METALLOCYCLES.

Most metathesis catalysts involve molybdenum or tungsten and in view of our speculation over a possible role for μ-carbenes in the process it was of interest to attempt the preparation of such complexes of these metals. In an adaptation of the successful route described earlier, the μ-ethyne complex (12) (and analogous

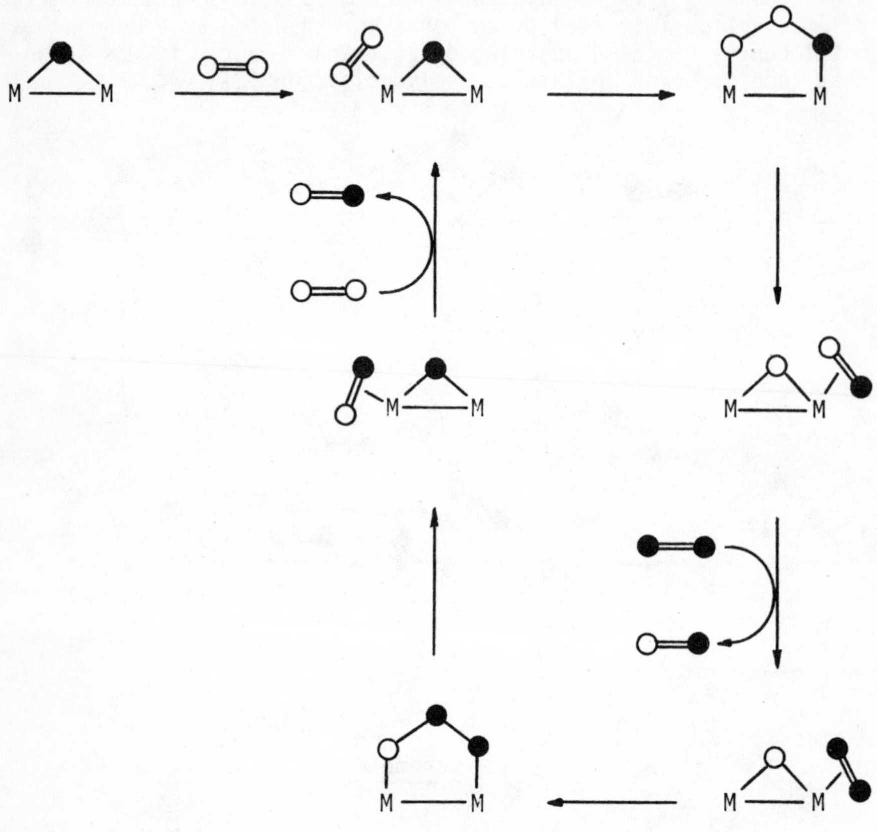

Scheme 6.

complexes of other alkynes) was treated with HBF_4 and $NaBH_4$ in sequence (24). The results are summarised in Scheme 7. Protonation does yield a μ-vinyl cation, but addition of $NaBH_4$ regenerates (12) rather than form the desired μ-methylcarbene complex (14). Other nucleophiles such as chloride, acetate, and trifluoroacetate attacked molybdenum, giving the complexes (15) with the μ-vinyl retained. Addition of the appropriate acid HX to (12) provides (15) directly. An X-ray diffraction study has been completed on the trifluoroacetate of (15).

U.v. irradiation of $[W_2(CO)_6(\mu\text{-}C_5H_5)_2]$ in the presence of dimethylacetylene dicarboxylate effects alkyne-CO linking as in the di-iron and diruthenium systems, to provide

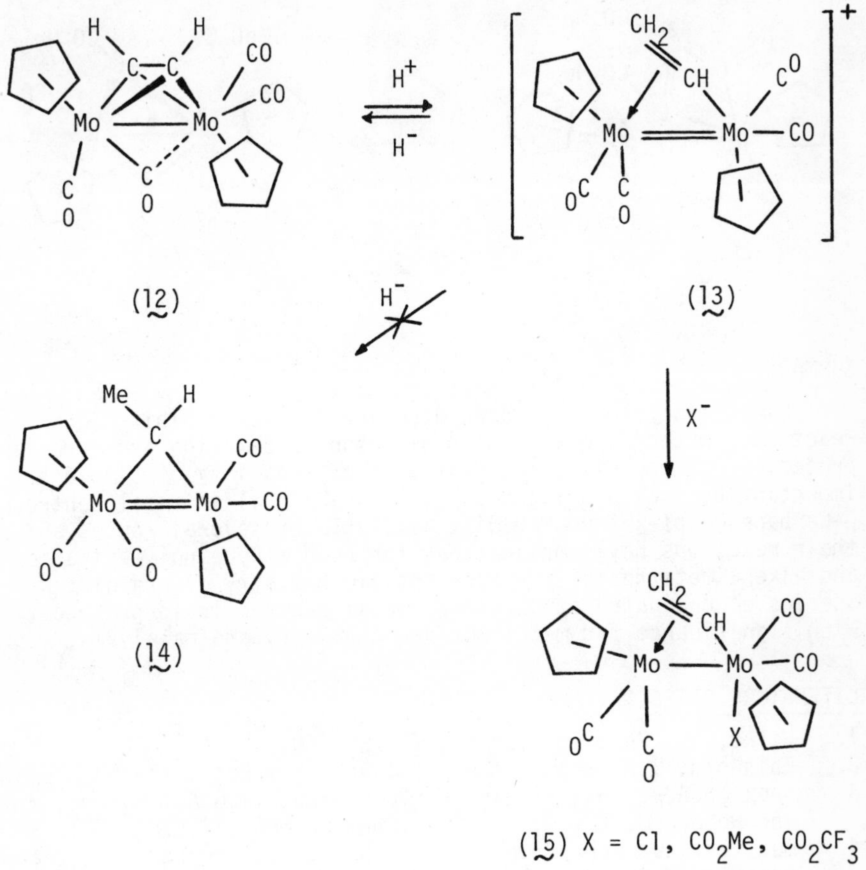

(12) (13) (14) (15) X = Cl, CO_2Me, CO_2CF_3

Scheme 7.

$[W_2(CO)_4\{\mu\text{-}C(O)C_2(CO_2Me)_2\}(\eta\text{-}C_5H_5)_2]$ (16) whose structure has been established by X-ray diffraction (25). The bridging unit in (16) is subtly different from that in (1), being an $\eta^2:\eta^2$ ligand rather than $\eta^1:\eta^3$. The new carbon-carbon bond is easily broken on warming to 50 °C, when CO is ejected to produce the μ-alkyne complex (17). Other alkynes react with $[W_2(CO)_6(\eta\text{-}C_5H_5)_2]$ to afford analogues of (17) directly, probably via thermally very unstable species of type (16). In contrast to the complexes (1), protonation of (16) occurs with retention of the 'alkyne'-CO link, so that the development of chemistry like that from (1) appears unlikely.

(16) -CO → (17)

SUMMARY

Dimetallocycles have been discovered which exhibit high reactivity with respect to carbon-carbon bond-making and -breaking processes. They allow the synthesis of a variety of simple but important hydrocarbon ligands bridging a dinuclear metal centre. μ-Carbene complexes are readily available by several routes and their reactions have implications for both alkyne polymerisation and alkene metathesis. A substantial chemistry of organic species co-ordinated at dinuclear metal centres is in prospect, with significance for metal surface chemistry and catalysis.

LITERATURE CITED.

1. Wilke, G., Pure Appl.Chem., (1978), 50, 677.
2. Chisholm, M.H. and Cotton F.A., Acc.Chem.Res., (1978), 11, 356.
3. Knox, S.A.R., Stansfield, R.F.D., Stone, F.G.A., Winter, M.J., and Woodward, P., J.Chem.Soc.,Chem.Commun., (1978), 221.
4. Muetterties, E.L., Rhodin, R.N., Band, E., Brucker, C.F., and Pretzer, W.R., Chem.Rev., (1979), 79, 91.
5. Dyke, A.F., Knox, S.A.R., Naish, P.J., and Taylor, G.E., J.Chem.Soc.,Chem.Commun., (1980), 409.
6. Davies, D.L., Dyke, A.F., Endesfelder, A., Knox, S.A.R., Naish, P.J., Orpen, A.G., Plaas, D., and Taylor, G.E., J.Organometallic Chem., (1980), 198, C43.
7. Dyke, A.F., Knox, S.A.R., Naish, P.J., and Orpen, A.G., J.Chem.Soc.,Chem.Commun., (1980), 441.
8. Nitay, M., Priester, W., and Rosenblum, M., J.Amer.Chem.Soc., (1978), 100, 3620.
9. Lewis, L.N., Huffman, J.C., and Caulton, K.G., J.Amer.Chem. Soc., (1980), 102, 403.
10. Dyke, A.F., Knox, S.A.R., and Naish, P.J. J.Organometallic Chem., (1980), in press.
11. Ibach, H., Hopster, H., and Sexton, B., Appl.Surf.,Sci., (1977), 1, 1.
12. Kesmodel, L.L., Dubois, L.H., and Somorjai, G.A., Chem.Phys. Lett., (1978), 56, 267.

13. Hemminger, J.C., Muetterties, E.L., and Somorjai, G.A., J.Amer.Chem.Soc., (1979), 101, 62.
14. Herrmann, W.A., Angew.Chem.Int.Ed., (1978), 17, 800.
15. Muetterties, E.L., and Stein, J., Chem.Rev., (1979), 79, 479.
16. Garnier, F., Krausz, P., and Dubois, J.E., J.Organometallic Chem., (1979), 170, 195.
17. Levisalles, J., Rudler, H., Dahan, F., and Jeannin, Y., J.Organometallic Chem., (1980), 188, 193.
18. Dyke, A.F., Knox, S.A.R., Naish, P.J., and Taylor, G.E., J.Chem.Soc.,Chem.Commun., (1980), 803.
19. Johnson B.F.G., Kelland, J.W., Lewis, J., Mann, A.L., and Raithby, P.R., J.Chem.Soc.Chem.Commun., (1980), 547.
20. Matsuda, T., Sasaki, N., and Higashimura, T., Macromolecules, (1975), 8, 717.
21. Katz, T.J., and Lee, S.J., J.Amer.Chem.Soc., (1980), 102, 422.
22. Grubbs, R.H., Progr.Inorg.Chem., (1978), 24, 1.
23. Sumner, C.E. Riley, P.E., Davis, R.E., and Pettit, R., J.Amer.Chem.Soc., (1980), 102, 1752.
24. Beck, J.A., Knox, S.A.R., Riding, G.H., Taylor, G.E., and Winter, M.J., J.Organometallic Chem., (1980), in press.
25. Finnimore, S.R., Knox, S.A.R., and Taylor, G.E., J.Chem.Soc.,Chem.Commun., (1980), 411.

RECEIVED December 4, 1980.

14

The Coordination Chemistry of Metal Surfaces

EARL L. MUETTERTIES

Department of Chemistry, University of California, Berkeley, CA 94720

A relatively common interaction in molecular coordination or organometallic compounds that nominally are coordinately unsaturated is the formation of a three-center two-electron bond between a metal center in the compound and a C-H bond of a hydrocarbon, a hydrocarbon fragment, or a hydrocarbon derivative that is a ligand in the complex. This interaction can be the prelude, the intermediate or transition state, to a subsequent reaction in which the CH hydrogen atom is transferred to the metal center and a direct σ bond is formed between the carbon atom and the metal atom especially if the C-H bond is an activated bond. Internal oxidative addition of CH is a term often applied to this subsequent reaction step. The overall sequence is schematically outlined in 1. Factors that materially

1

affect the forward and back rate constants in step 2 of 1 have not been adequately elucidated except that step two is especially favored if the formal oxidation state in the oxidized product is easily accessible. Hence, favorable and established reaction couples are Fe(0) ⇌ Fe(II), Ru(0) ⇌ Ru(II), Co(I) ⇌ Co(III), Nb(III) ⇌ Nb(V), Ta(III) ⇌ Ta(V), and Ti(II) ⇌ Ti(IV) with the $d^8 \rightleftharpoons d^6$ couple the most common one.

The first example of the internal CH oxidative addition came from organic research with diazobenzene and $PdCl_4^{2-}$. Cope and Siekman (1) discovered the palladation reaction of diazobenzene

0097-6156/81/0155-0273$06.25/0

in 1965. The palladation reaction (1,2) is general to a variety of organic nitrogen compounds such as benzyldialkylamines (equation 1) whereby the nitrogen compound is converted to a five-

$$(1)\ PdCl_4{}^{2-} + C_6H_5CH_2N(CH_3)_2 \xrightarrow{-HCl} \left[Cl_2Pd(C_6H_4CH_2N(CH_3)_2) \right]^- + Cl^-$$

membered ring metallocycle ostensibly through a first step shown in equation 2 of amine displacement of the chloride ligand.

$$(2)\ PdCl_4{}^{2-} + C_6H_5CH_2N(CH_3)_2 \rightleftharpoons PdCl_3[N(CH_3)_2C_6H_5]^- + Cl^-$$

Later steps can be envisioned as in 2. Presently, the list of

$$\left[Cl_3Pd\cdots H\text{–}C_6H_4CH_2N(CH_3)_2 \right]^- \longrightarrow \left[Cl_3Pd(H)(C_6H_4CH_2N(CH_3)_2) \right]^- \xrightarrow{-HCl} \text{Products}$$

2

such internal C-H oxidative addition reactions is large and involves such ligands as triaryl phosphites and phosphines, trialkylphosphines, cyclopentadienyl, benzene and olefins (olefin ⇌ allyl) (3,4). In addition, there is a large number of coordination compounds in which the intermediate state, discussed above, with the multicenter C-H-M bond has been established by crystallographic studies—the H-metal bond distance has varied from the very short, 1.75Å, to the long but apparently still bonding range of ∿1.9-2.5Å (5,6).

This established C-H chemistry in molecular coordination compounds should have its formal analog in metal surface chemistry. Carbon-hydrogen bond-cleavage is a common reaction at metal surfaces but it does not comprise mechanistically the simple collision of a hydrocarbon such as methane with a metal surface (*vide infra*). Methane neither reacts with nor is chemisorbed on a metal surface under moderate conditions. However, propylene does chemisorb, and then C-H bonds may be broken, depending upon the nature of the metal surface. Generally, the sequence at the surface comprises first the formation of a coordinate bond between a surface metal atom(s) and some functionality in the organic molecule. Then, if C-H hydrogen atoms of the chemisorbed molecule can closely approach the surface metal atoms, the three center C-H-M bond should develop (the surface is typically coordinatly unsaturated unless overladen with contaminant elements like carbon or sulfur) leading to an activation of the C-H bond and possible subsequent C-H bond cleavage particularly if the C-H bond is activated in the parent organic molecule.

Saturated hydrocarbons other than methane do react with metal surfaces at 20°C. Clean metal surfaces such as ruthenium absorb saturated hydrocarbons and effect dehydrogenation of the hydrocarbon, e.g., cyclohexane is converted to benzene. In these cases, perhaps the initial chemisorption process comprises the formation of several C-H-metal three-center two-electron bonds.

I present data from our recent experimental and theoretical studies of metal surfaces that implicate the very same sequential steps described above for C-H bond activation in molecular coordination and organometallic complexes.

Experimental

Nearly all the experiments described were performed in an ultra high vacuum chamber at pressures of about 10^{-10} torr. The specific equipment and experimental procedures used have been described elsewhere (7-9). Experimental protocol for the thermal desorption experiments and for the chemical displacement reactions is presented below. All these experiments were repeated with a control, blank experiment with a metal crystal that had the front and exposed face covered with gold; the sides and back of the crystal were exposed (8,9). These blank experiments were performed to ensure that all thermal desorption and chemical displacement experiments monitored only the surface chemistry of the front exposed face of the metal crystal under study.

Gas exposures were performed with a variable leak valve equipped with a dosing "needle" such that the gases could be introduced in close proximity to the surface, thus minimizing background contamination. Two separate valve-needle assemblies mounted symmetrically with respect to the mass spectrometer were used to introduce the different gases in displacement reactions. This avoided contamination of the displacing gas in the leak valve. During displacement reactions, the crystal face was directed 45° away from the line of sight of the mass spectrometer ionizer (so as to face the second valve assembly). This configuration decreased the mass spectrometer signal of gases evolved from the crystal during a displacement reaction as compared to a thermal desorption. The time interval between the two gas exposures in a displacement reaction was between 1-5 minutes. Oxide formation on nickel (111) and (100) was effected by a prolonged exposure (5-10 min.) of the crystal to 5×10^{-8} torr of O_2 with a crystal temperature of ∿350°C. The oxide was ordered with a c(2x2) low energy electron diffraction pattern.

Carbon contaminated surfaces were prepared by thermally decomposing benzene on the nickel surface. The carbon overlayer was ordered but the diffraction pattern was complex. It did not correspond to a graphitic ring structure. Approximate carbon coverages were estimated using Auger calibration curves based on thermally decomposed benzene (10).

All chemicals used were free by mass spectrometric criterion of any significant impurities. Isotopically labeled compounds were purchased or prepared; see references (8) and (9) for their isotopic characterization.

Chemisorption state modeling by extended Huckel molecular orbital calculations were made in a collaborative study with R. M. Gavin, and full details of these calculations will be published elsewhere.

Discussion

Surface Crystallography and Composition. Platinum (11) and nickel (8,9,12) have been the metal surfaces examined in our surface science studies to date. The surface coordination chemistry has been examined as a function of surface crystallography and surface composition. Surfaces specifically chosen for an assay of metal coordination number and of geometric effects were the three low Miller index planes (111), (110) and (100) as well as the stepped 9(111)x(111) and stepped-kinked 7(111)x(310) surfaces (both platinum and nickel are face centered cubic). These surfaces are depicted in Figures 1-5. The coordination

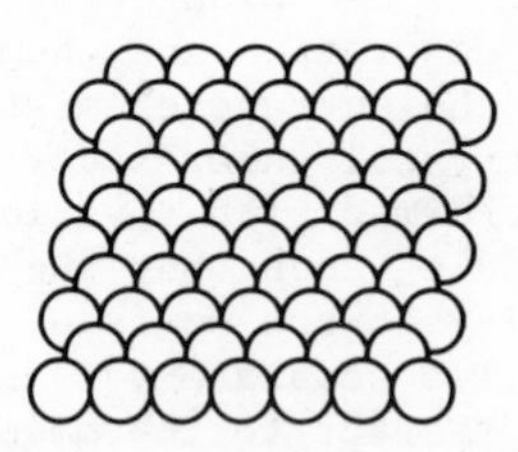

Figure 1. A perspective of the thermodynamically most stable crystal plane for face-centered cubic, the (111) low Miller index plane. All surface atoms have nine nearest neighbors compared to twelve for the bulk atoms.

chemistry generally was established for these five clean surfaces and for their carbon, sulfur and oxygen modified surfaces.

Chemistry of Benzene and Toluene. One of our best defined chemical systems for C-H bond activation and C-H bond scission is that of benzene and toluene on nickel surfaces (9). The chemistry was a sensitive function of surface crystallography and of surface composition (9).

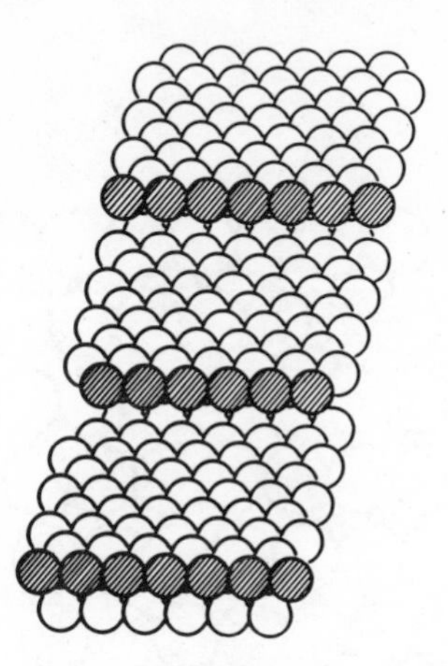

Figure 2. The most common type of imperfection site on a (111) crystal plane is steps. This illustration shows the perspective of regularly placed steps on the (111) surface. Every ninth row is a step (the step atoms are shaded) and this may be indexed as 9(111) × (111). The coordination number of step atoms is seven.

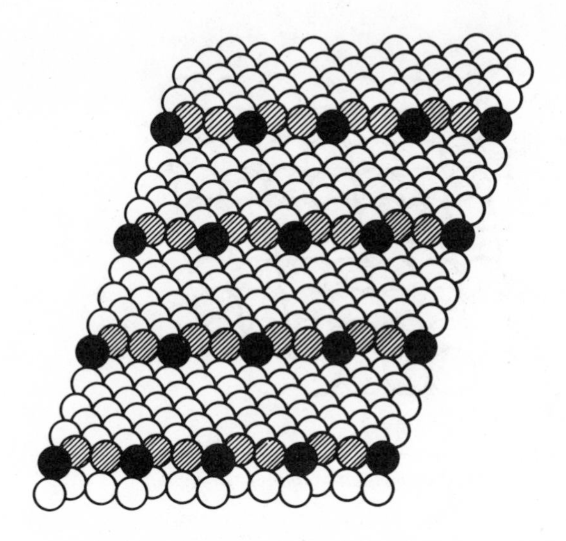

Figure 3. The other common imperfection site on a (111) surface is a kink. This 7(111) × (310) surface shown in perspective is a regularly stepped and kinked surface with a step sequence beginning every seventh row and with a kink site at every third atom in the row. The atoms in kink sites have a coordination number of six (black spheres).

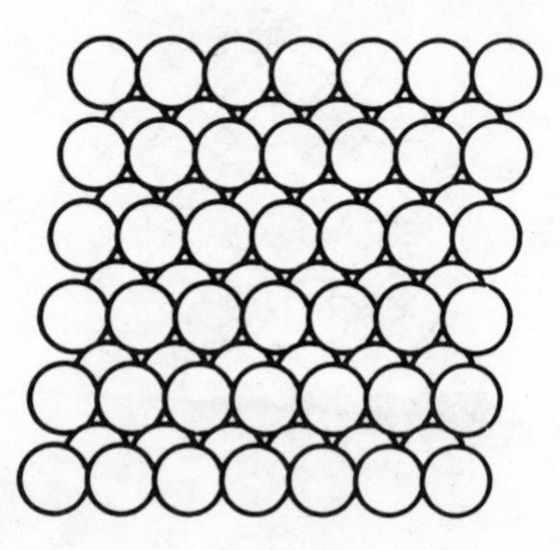

Figure 4. This (110) low Miller index plane is like a super-stepped surface—every other row is a step. The coordination numbers of surface metal atoms in the crests and troughs of this stepped or ruffled surface are seven and eleven, respectively.

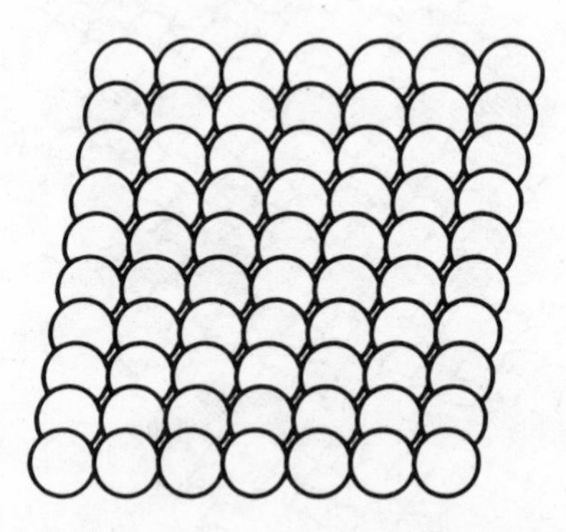

Figure 5. This representation is for the flat (100) plane where the surface atoms have a coordination number of eight.

Chemisorption of benzene on the flat (111) and (100) surfaces was fully molecular (associative) below 100°C and ∿150-200°C, respectively. For each surface, the benzene was quantitatively displaced from the crystal surface by trimethylphosphine below these temperatures. Furthermore, when these experiments were effected with C_6H_6+C_6D_6 mixtures, only C_6H_6 and C_6D_6 molecules were displaced. Thus, no reversible C-H (or C-D) bond breaking occurs on these clean surfaces within the time scale of the experiments (minutes). However, irreversible bond breaking does occur at temperatures above 100°C and 150°C for the (111) and (100) surfaces, respectively, as shown by thermal desorption experiments where only a fraction of the benzene molecules reversibly desorbed: ∿20% on Ni(111) with a maximum at 115 - 125°C and ∿40% on Ni(100) with a maximum at ∿220°C. In each case thermal decomposition of the chemisorbed benzene to H_2 + Ni-C was a competitive process to thermal desorption. Thermal desorption experiments with C_6H_6 + C_6D_6 mixtures on these two surfaces yielded in the fraction that reversibly desorbed only C_6H_6 and C_6D_6 molecules. No H-D exchange was observed. *The following qualifications must be noted. In the realm of ultra high vacuum experiments, the concentration or activity of surface species is relatively low at least with respect to those present on surfaces at near normal pressures. Accordingly, carbon-hydrogen bond scission reactions are typically irreversible; the hydrogen surface activity is apparently too low for the reverse reaction to proceed at measurable rates. However, with extension of the ultra high vacuum studies to real surfaces at normal pressures, reversible C-H bond scission may be and has been observed. Also of note is the fact that desorption of hydrogen atoms as H_2 on these surfaces is an activated process and minimally requires temperatures of about 130°C (since the desorption process is second order in hydrogen atoms, the desorption temperature is dependent upon surface coverage by hydrogen atoms). In the decomposition experiments, there was no detectable H-D isotope effect upon the H_2 (or D_2) desorption temperature maxima.*

The stepped nickel 9(111)x(111) and stepped-kinked Ni 7(111)x(310) surfaces displayed a benzene coordination chemistry that was quantitatively and qualitatively identical with that of the Ni(111) surface except that not all the benzene was displaced by trimethylphosphine indicating that either a small percentage (∿10%) of the benzene on these surfaces either was present in different environments or was dissociatively (9) chemisorbed; see later discussion of stereochemistry. Benzene chemisorption behavior on Ni(110) was similar to that on Ni(111) except that the thermal desorption maximum was lower, ∿100°C, and that trimethylphosphine did not quantitatively displace benzene from the Ni(110)-C_6H_6 surface. In these experiments, no H-D exchange was observed with C_6H_6 + C_6D_6 mixtures.

The benzene surface chemistry was not qualitatively altered by the presence of carbon or sulfur impurities (up to ∿0.2-0.3

monolayer); only the benzene sticking coefficient was lowered by the presence of such impurity atoms. Surface oxygen atoms present as oxide oxygen reduced the benzene sticking coefficient to near zero values at 20°C.

Toluene surface coordination chemistry was quite different from that of benzene. Toluene chemisorption on all the clean surfaces was thermally irreversible. In addition, toluene was not displaced from these surfaces by trimethylphosphine nor by any other potentially strong field ligand examined to date, e.g., carbon monoxide or methyl isocyanide. In the thermal decomposition of toluene on these surfaces (attempted thermal desorption experiments), there were *two* thermal desorption maxima for H_2 (or D_2 from perdeuterotoluene) with the exception of the Ni(110) surface. This is illustrated in Figure 6 for Ni(111)-C_7D_8.

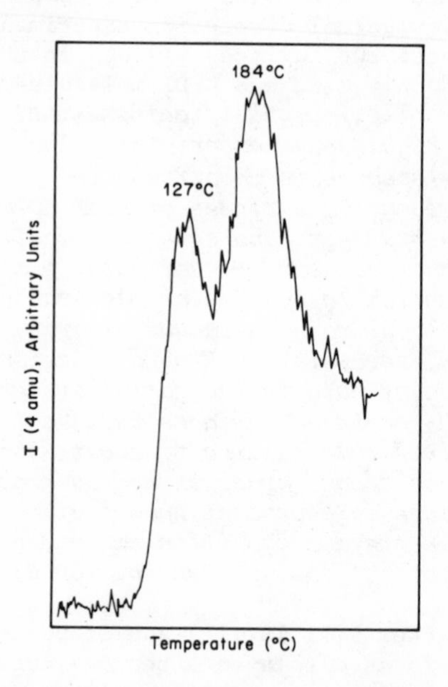

Figure 6. Desorption curve for D_2 from the decomposition of $C_6D_5CD_3$ on the (111) surface. D_2 was monitored by mass spectrometry as the crystal was heated at a rate of 25°/s.

Experiments with the deuterium labeled molecules, $C_6H_5CD_3$ and $C_6D_5CH_3$ incisively delineated the molecular details of toluene chemisorption and decomposition on the nickel (111), stepped, stepped-kinked, and (100) surfaces. Following the D_2 formation from Ni(111)-$C_6H_5CD_3$, there was a *single* desorption maximum for D_2 at ∿130°C; none was detected in the higher temperature region. For Ni(111)-$C_6D_5CH_3$, no D_2 appeared in the low temperature range (∿130°C) and there was a *single* desorption maximum at ∿185°C. One experiment is illustrated in Figure 7 (compare with

Figure 6 for the Ni(111)-C_7D_8 decomposition). Thus, no aromatic C-H (or C-D) bonds in toluene chemisorbed on Ni(111) are broken below temperatures of ∿150°C whereas all aliphatic C-H (or C-D) bonds are broken and the hydrogen atoms desorb as H_2 (or D_2) below 150°C. The experiments do not establish the temperatures at which aliphatic C-H bonds are first cleaved but all these processes are completed by 130-150°C. Essentially the same

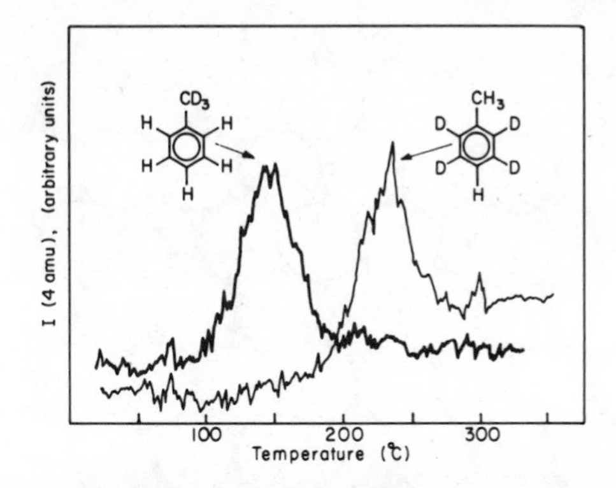

Figure 7. In these thermal desorption experiments, D_2 was monitored as a function of crystal temperature in the decomposition of $C_6H_5CD_3$ and $C_6D_5CH_3$ on the Ni(111) surface.

behavior and the same desorption maxima were observed for the $C_6H_5CD_3$ and $C_6D_5CH_3$ decompositions on the stepped 9(111)x(111) and stepped-kinked 7(111)x(310) surfaces. The same behavior was observed for the two labeled toluenes on the flat (100) surface: Single D_2 desorption maxima were observed with the temperature maximum of 110°C for $C_6H_5CD_3$ and of 230°C for $C_6D_5CH_3$. In sharp contrast, toluene thermal decomposition on the super-stepped (110) surface (Figure 4) gave only a single H_2 (or D_2 for C_7D_8) desorption maximum at 150°C.

Impurity atoms do perturb the character of the toluene thermal decomposition on the flat surfaces. At carbon or sulfur coverage levels of 20% or higher on Ni(100), where an ordered c(2x2) low energy difraction pattern was obtained and where the impurity atoms reside over four-fold sites, the differentiation between aliphatic and aromatic C-H bond cleavage rates was lost. Under these conditions, only a single hydrogen desorption maximum was observed.

Stereochemical Features of Benzene and Toluene Coordination Chemistry. Benzene forms an ordered chemisorption state on the flat Ni(111) and Ni(100) surfaces at 20°C with unit cells of $(2\sqrt{3}x2\sqrt{3})R$ 30° and c(4x4), respectively (13). The symmetry data do not fix the registry of the benzene with respect to the metal atoms nor the orientation of the ring plane to the surface plane. However, basic coordination principles would suggest that the benzene ring plane should be parallel to the surface plane. In Figures 8 and 9, possible registries of the benzene with respect

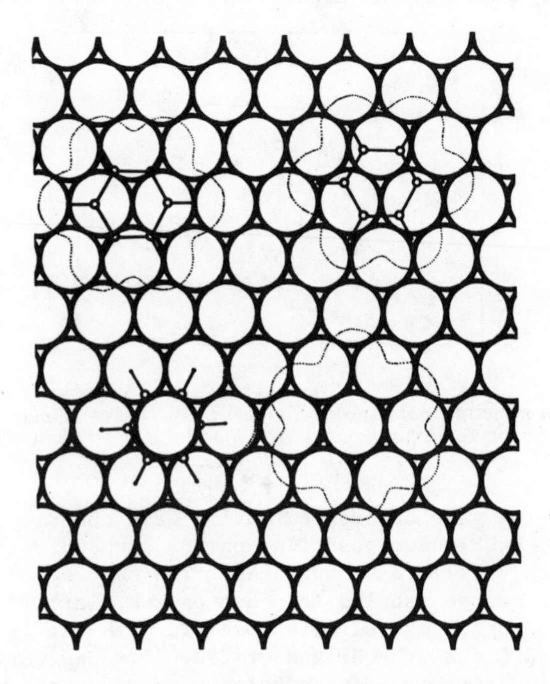

Figure 8. Illustration of possible symmetric registries between chemisorbed benzene and the metal atoms in the (111) nickel surface.

Shown are registries with the C_6 centroid over a single nickel atom, over a two-fold site, and over a three-fold site. Because the distance between hydrogen atoms at para positions in the benzene molecule is virtually identical to twice the nickel atom diameter, the registry with the C_6 centroid atop a single nickel atom is precise (lower left and right)*. Overlaps with the π and π^* benzene orbitals and metal surface orbitals should be excellent. In one rotational form* (lower left) *the hydrogen atoms of the benzene lie directly over single nickel atoms so as to generate multicenter C—H—Ni bonds. The van der Waals extension of the benzene molecule (· · ·) is virtually identical with the space occupied by the central nickel atoms and the surrounding six nickel atoms. Another form generated from the above by a rotation of 30° has the C—H hydrogen atoms lying nearly above the two-fold sites* (lower right)*. Either of these may be the most stable registry for benzene on this surface.*

to the metal atoms are presented. For the Ni(111) surface, one of the more favorable registries has the C_6 centroid directly over a single metal atom. This registry provides for optimal overlap of not only π and π^* orbitals with metal surface orbitals but also for the development of three-center two-electron bonds between the C-H hydrogen atoms and surface metal atoms; one registry places each hydrogen atom directly over a nickel atom and rotation by 30°

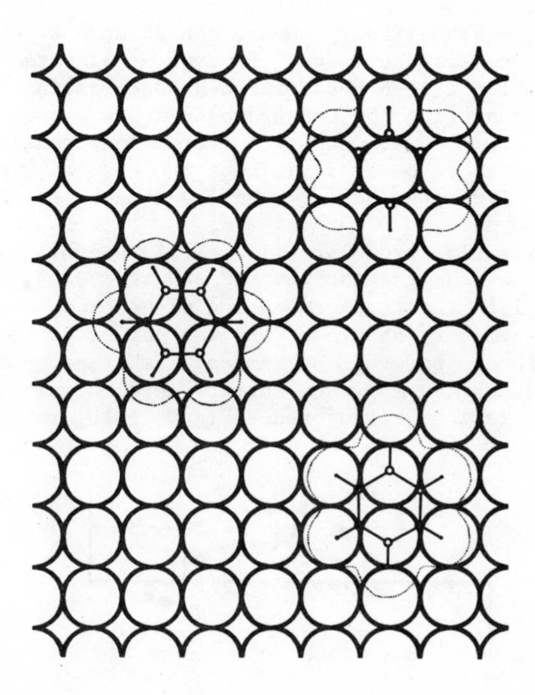

Figure 9. Some of the more symmetrical registries of the benzene molecule with respect to the Ni(100) lattice plane. Although none of these registries can have as close a 1:1 correspondence of H—Ni atom positions as on the (111) surface (Figure 8), benzene appears to be more strongly bound (9) on this surface than on the (111) surface.

places the hydrogen atoms nearly directly over 2-fold bridging (Ni-Ni) sites. Electronic (14) and vibrational (13,15) spectroscopic data are consistent with these basic formulations of chemisorbed benzenes in planes parallel to these two flat nickel surfaces. Importantly, this configuration does not allow close approach of hydrogen atoms to surface metal atoms in the 20°C chemisorption state *(That is, close in a relative sense. Actually the C-H hydrogen atoms probably are slightly out of the C_6 plane towards the metal surface so as to develop better C_6H_6 π and π^* overlaps and to achieve stronger three-center two-electron C-H-M bonds. In $C_6H_6Cr(CO)_3$, such a displacement of hydrogen atoms has been established from an X-ray and neutron crystallographic study (16))*. Hence the absence of C-H bond breaking on these flat surfaces. However at elevated temperatures, thermal excitation of C-H bonds for C-H bond bending and of C_6 plane tipping can bring the C-H hydrogen atoms closer to the surface metal atoms - with consequent C-H bond breaking that, in fact, was observed to be a significant process, at high temperatures, competitive with the (reversible) benzene thermal desorption process.

On the slightly irregular 9(111)x(111) and 7(111)x(310) surfaces, most of the chemisorbed benzene molecules should be in planes parallel to the terrace section planes and, in this context, the chemistry should be and was analogous to that on the (111) surface. However, because of the steps some of the C-H hydrogen atoms can approach surface metal atoms very closely and C-H bond scission should ensue (Figure 10). This stereo-

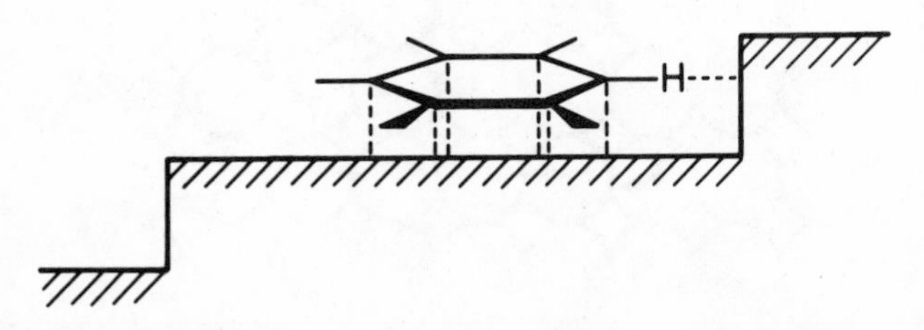

Figure 10. An exaggerated representation of benzene molecules chemisorbed near stepped or stepped-kinked sites where C—H hydrogen atoms may closely approach metal atoms in the next plane. This can account for the higher reactivity (C—H bond breaking) observed for benzene on such irregular stepped or stepped-kinked surfaces.

chemical feature was clearly evident in the slightly less than quantitative displacement of benzene from these surfaces by trimethylphosphine in contrast to the quantitative displacement on the fully flat (111) and (100) surfaces. For the fully ruffled, or super-stepped, (110) surface, close approach of C-H hydrogen and facile C-H bond scission should be more extensive - and apparently was, based on the thermal desorption and chemical displacement experiments. The specific stereochemical features of the Ni(110)-C_6H_6 chemisorption state(s) cannot be characterized with our available data.

Toluene should coordinate initially in a stereochemical fashion analogous to benzene on the flat surfaces. This orientation will necessarily bring an aliphatic C-H hydrogen atom of the methyl group very close to the surface. Again C-H bond cleavage should be facile. Cleavage would then generate a benzyl species in which, if all carbon atoms lie in a plane parallel to the surface, all carbon atoms can interact with the surface metal atoms as can the seven hydrogen atoms in three-center two-electron C-H-M bonds. This is a probable stereochemistry for the toluene chemisorption state (dissociative) at 20°C on the flat surface. It is a representation that is fully consistent with all of our experimental data. When the surface was substantially perturbed by the presence of impurity atoms, like C or S, the characteristic toluene surface chemistry established with the labeled molecules, $C_6H_5CD_3$ and $C_6D_5CH_3$ was lost.

Our data suggest a basis for selective H-D exchange between toluene and D_2 centered at the methyl site. Actually, Crawford and Kemball (17) earlier had established such selectivity for toluene and D_2 on evaporated nickel films with sintered films showing higher selectivity, by one order of magnitude, than unsintered films (ostensibly the sintered films had a higher percentage of metal atoms in (111) or (100) type environments than did the unsintered films). We have also established that irregular nickel surfaces show essentially no selectivity (18). All these results, with "irregular" surfaces under normal pressure conditions, are fully consistent with our ultra high vacuum studies.

Ethylene and Acetylene. On nickel (111), both ethylene and acetylene are irreversibly chemisorbed; neither can be thermally desorbed. We also find that trimethylphosphine cannot displace ethylene or acetylene from these surfaces. There have been suggestions that ethylene and acetylene are not present on the surface as molecules but as molecular fragments. Many ultra-high vacuum studies of ethylene and of acetylene chemisorption on nickel crystal planes have been reported. Most of these studies seem to implicate states in which C-H bond cleavage reactions have accompanied the basic chemisorption process (19).

For a flat surface, we would expect that both acetylene and ethylene would chemisorb initially such that the C-C bond vectors would be parallel to the surface plane. The hydrogen atoms could be significantly farther from the surface than for the benzene case discussed above. *(They may be farther away as in coordinately saturated metal olefin complexes where the acetylenic or ethylenic C-H bonds are bent from the C-C bond vector away from the metal center. However, generation of C-H-M multicenter interactions for the surface case may be substantial. The typically coordinately saturated molecular olefin or alkyne complex may in this case be quite inadequate as a reference model.)* Hence on ground state considerations and the benzene surface data, C-H bond scission for ethylene and for acetylene should not be a facile process at least at 20°C. Torsional twisting of these molecules, bound in the conventional π fashion, about the C-C bond may be a lower energy process than for ring tipping or ring deformation for benzene chemisorbed on a flat surface. If these barriers are quite low, then C-H bond scission should be facile at 20°C. At this point, the basic molecular details of ethylene and of acetylene chemisorption on nickel surfaces are not definitively established and further experimental data are required. We do anticipate that propyne or propylene chemisorption would be dissociative in character because of the easier close approach of methyl hydrogen atoms to the surface (by formal analogy to the benzene and toluene systems discussed above).

Our studies of olefin and acetylene chemisorption states on platinum surfaces is presently incomplete. Ethylene and acetylene chemisorption on platinum (111) are complicated by the apparent presence of more than one chemisorption state (indicated by thermal desorption studies). When C_2H_4 and C_2D_4 are chemisorbed on Pt(111), the small fraction of ethylene thermally desorbed as ethylene comprises nearly a statistical mixture of all possible $C_2H_xD_{4-x}$ molecules. Thus we see here reversible C-H (and C-D) bond breaking on this flat platinum surface. In an analogous experiment with C_2H_2 and C_2D_2, only a small extent of H-D exchange was observed for the small fraction of acetylene molecules that reversibly desorb from this surface (11).

Acetonitrile and Methyl Isocyanide (8). Acetonitrile forms an ordered chemisorption state on the fully flat nickel surfaces, a p(2x2) and a c(2x2) on Ni(111) and Ni(100), respectively. Acetonitrile thermal desorption from these two surfaces was nearly quantitative (a small amount of acetonitrile decomposed at the temperatures characteristic of the reversible thermal desorption from these surfaces). Importantly from an interpretive context, acetonitrile was *quantitatively* displaced from these two flat low Miller index planes by trimethylphosphine (8). However, the displacement was not quantitative (only 90-95% complete) from the stepped and stepped-kinked surfaces. For the super-stepped (110) surface, chemisorption was nearly *irreversible* and no acetonitrile could be displaced from this surface by trimethylphosphine.

For the ordered p(2x2) and c(2x2) states on Ni(111) and Ni(100), respectively, the acetonitrile must be bound solely through the CN nitrogen atom and the CN vector must be largely normal to the surface planes. As shown in Figure 11, such stereochemistry maintains the methyl hydrogen atoms well removed from the surface metal atoms even with nominal departures from a surface-N-C angle of 90° and with CH_3 group bending at the nitrile carbon atom. Near step or kink sites, a close approach of methyl hydrogen atoms can occur (Figure 11) and this probably accounts for the slightly different behavior of acetonitrile on the stepped

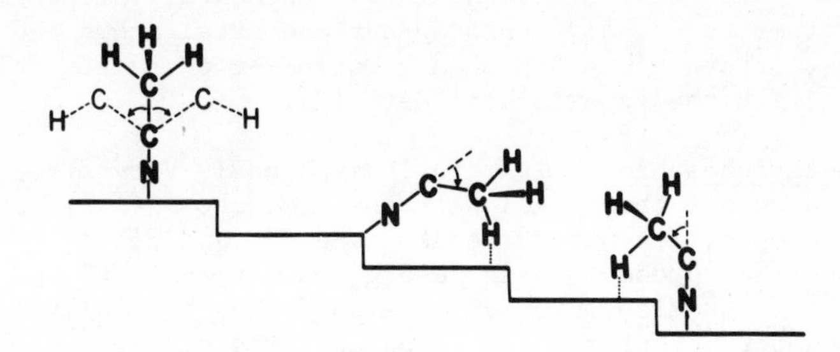

Figure 11. On the flat (111) and (100) surfaces acetonitrile is bonded to surface metal atoms only through the nitrogen atom, and the CN bond vector is more or less normal to the surface.

Thus, methyl hydrogen atoms cannot easily closely approach surface metal atoms even through vibrationally excited states as shown schematically at the left in the figure. However, acetonitrile molecules bound near step or kink sites can have methyl hydrogen atoms in positions from which there can be a facile close approach of these hydrogen atoms to the surface metal atoms. This geometric or stereochemical feature explains the reactivity (irreversible C—H bond–breaking processes) observed for acetonitrile on the stepped, stepped-kinked, and super-stepped (110) nickel surfaces.

or stepped-kinked surface relative to that for the (111) surface: probably some acetonitrile molecules initially chemisorbed near such surface irregularities as steps undergo irreversible C-H bond cleavage reactions and hence are not displaced by the phosphine. Fully consistent with this stereochemical interpretation was the thermal reactivity of acetonitrile on the super-stepped (110) surface and the nondisplacement of the nitrile from this surface by the phosphine.

The precise registry of the nitrile nitrogen atom with respect to surface metal atoms has not been established. Normally acetonitrile is bound in coordinately saturated mononuclear metal compounds or clusters through the nitrogen atom to a *single* metal atom (8). However, the ordered chemisorption states for CH_3CN on Ni(111) and Ni(100) are coordinately unsaturated. Possibly, the nitrogen atom may lie at bridging sites, e.g., the three-fold site on the (111) and four-fold site on the (100) surface. We plan vibrational studies of these surface states and these may be informative about this stereochemical feature.

Methyl isocyanide was irreversibly chemisorbed on all the clean nickel surfaces. Based on the tendency of isocyanides to bind through both the carbon and the nitrogen atoms in coordinately unsaturated metal clusters (20), we would expect the N-C bond vector of methyl isocyanide chemisorbed on the flat nickel (111) and (100) surfaces to be more or less parallel to the surface with both the isocyanide carbon and nitrogen atoms bonded to surface metal atoms (8). This stereochemistry necessarily allows methyl hydrogen atoms to closely approach surface metal atoms and allows for facile, irreversible C-H bond scission—a postulate fully consistent with the experimental data (8).

Trimethylphosphine. Trimethylphosphine is very strongly chemisorbed on all the nickel surfaces. On the Ni(111) surface, thermal decomposition occurs readily and CH_4 and H_2 are desorbed as decomposition products with desorption maxima at 90 and 98°C respectively. Chemisorption of this phosphine initially must involve a donor-acceptor interaction centered at the phosphorus atom. Models show that the methyl hydrogen atoms can then closely approach the surface metal atoms. Cleavage of C-H bonds probably occurs at or near 25°C, and P-C-Ni bonds are then irreversibly formed. This surface chemistry qualitatively mirrors that of trimethylphosphine in the coordinately unsaturated complex, $Fe[P(CH_3)_3]_4$, which is primarily $HFe[\eta^2\text{-}\underline{C}H_2\underline{P}(CH_3)_2][P(CH_3)_3]_3$ in the solution state (21).

Hydrocarbon Fragments - Modeling by Molecular Orbital and Cluster Chemistry. A basic guideline for metal surface coordination chemistry with respect to hydrocarbon or hydrocarbon derivatives may be formulated as follows: If the stereochemistry of the chemisorption state allows C-H hydrogen atoms to closely approach surface metal atoms then the chemisorption state should be further stabilized by the formation of a three-center two-electron C-H-metal bond. This effect should be more pronounced the more electron deficient the metal surface. There should be an activation of the C-H bond and the hydrogen atom should become more protonic in character. If the C-H bond is sufficiently weakened by this interaction then C-H bond cleavage should result. Certainly, all our metal surface studies point to the importance of this stereochemical feature in determining the chemistry of chemisorbed molecules. One further illustrative example that is incompletely defined as yet is that for pyridine on the Ni(100) and Pt(111) surfaces where the pyridine is irreversibly chemisorbed and three hydrogen thermal desorption maxima of relative intensities 1:2:2 were observed (11,12). These data suggest that pyridine initially chemisorbs in a plane normal to the surface bonding through the nitrogen atom. This places the α - CH hydrogen atoms close to surface metal atoms and C-H bond cleavage should occur very readily (at 20°C or slightly higher temperatures) to form a pyridyl species (11,12). At high temperatures, ring tipping so as to ultimately lie in a plane parallel to the surface may then occur in two stages as shown in 3. Studies with deuterium labeled pyridines are in progress to confirm or refute this hypothesis.

If the above generalization is correct, then simple extended Hückel molecular orbital calculations for the surface chemisorption states of simple hydrocarbons or hydrocarbon fragments should sense this tendency for multicenter C-H-metal bonds to form (these

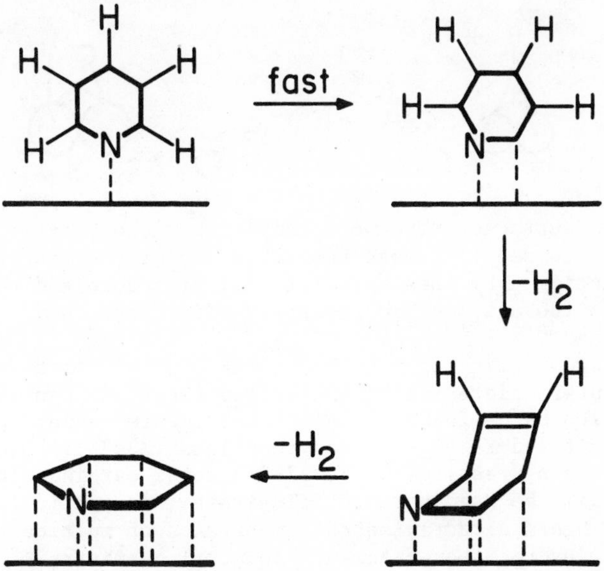

H H H H H N fast H H H H N -H2 H H N -H2 N

3

particular calculations are inappropriate for modeling features of the C-H bond cleavage step). We (22) have examined this aspect for the nickel (111) surface and find that multi-center C-H-metal bond formation does appear to be important. For example, in the case of the CH_2 fragment the lowest energy site for bonding was found to be the two-fold site as would be expected from coordination principles but that stereochemistry 4 was of higher energy

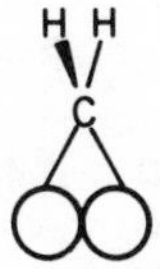

H H C H C H

4 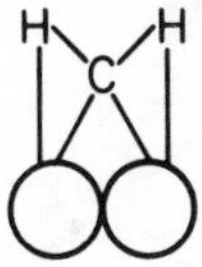 5

than 5 simply because of the added stabilization in 5 of the multicenter C-H-M interactions (all known dinuclear and cluster complexes with CH_2 ligands have form 4 because these complexes are coordinately saturated). For the methylidyne fragment, CH, the three center site, 6, is explicably more stable than the two-center and sitting atop a single metal atom site when the C-H vector was fixed normal to the surface plane. However, the most stable stereochemistry was 7 where the carbon is off the three-fold site toward a two-fold site and the C-H vector is tipped so as to generate a multicenter C-H-M bond. In fact, every system tested so far by these calculations exhibits a stabilization as C-H-metal interactions are generated for these coordinately

unsaturated surfaces. The more complex benzene system is presently under study to see if a most favorable registry (Figures 8 and 9) can be identified by this calculational procedure and how it conforms to those symmetric registries discussed above and in Figures 8 and 9.

Cluster modeling of possible chemisorption states and of possible intermediate states in surface reactions can to a first approximation be useful in guiding experiments or interpretations of experimental data for surface reactions (23-25). One important and enlightening result (6, 26, 27) in metal carbide cluster chemistry will be used here to illustrate this particular point because it bears directly on the importance of multicenter C-H-M bonding for hydrocarbon fragments in metal chemistry.

Oxidation of the iron carbide cluster anion, $Fe_4C(CO)_{12}{}^{2-}$ (Figure 12), yielded the coordinately unsaturated and thermally very reactive $Fe_4C(CO)_{12}$ cluster; both reactant and product have a four coordinate carbide carbon atom. When the oxidation is carried out at ∿0°C in the presence of hydrogen, hydrogen addition occurred to give a new cluster with both a C-H and a Fe-H-Fe hydrogen site. The methylidyne or CH ligand was not simply bound to iron atoms through the carbon atom—a type of bonding (see 6 above) conventionally found in clusters such as $HCCo_3(CO)_9$ (28).

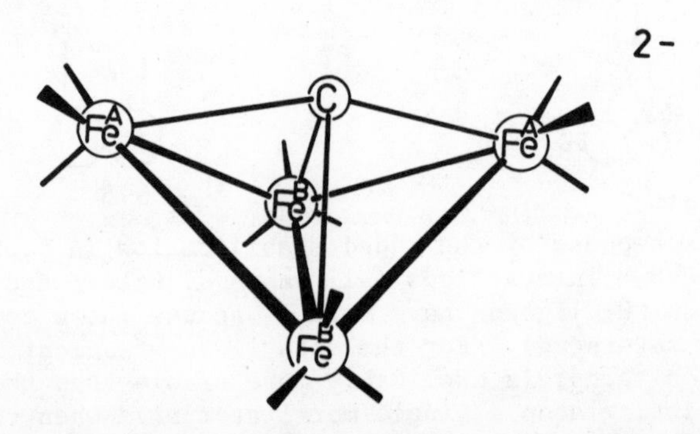

Figure 12. In $[Fe_4C(CO)_{12}{}^{2-}]$, the iron atoms form a butterfly array with the carbide carbon atom centered above the wings (26). The carbide carbon atom has the very low coordination number of four, a featured showed only by this cluster and the related cluster $[HFe_4C(CO)_{12}{}^{-}]$ (26). Each iron atom has three terminally bound CO ligands.

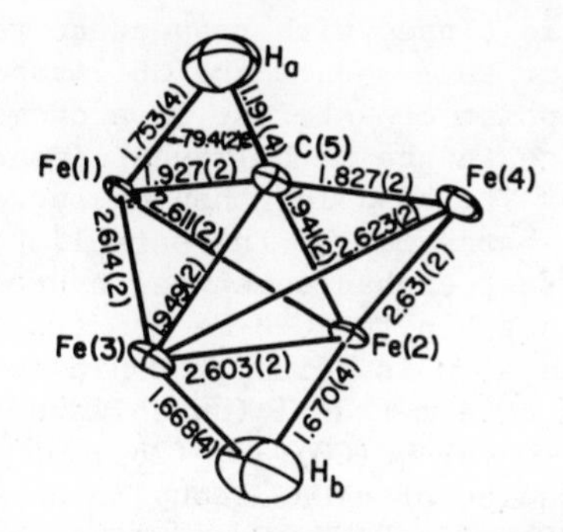

Figure 13. Structure of $HFe_4(\eta^2\text{-}CH)(CO)_{12}$, minus the twelve carbonyl ligand atoms, as determined by a neutron crystallographic study (27).

The most notable features of this structure center on the methylidyne, CH, ligand which forms a strong, closed three-center, two-electron bond with the apical iron atom Fe(1). There are two molecules in the asymmetric unit. The distances shown are for molecule (1); no experimentally significant differences exist between the two molecules. Each iron atom in this cluster has three terminally bound carbonyl ligands.

Such a conventional bonding would have given, for the open butterfly array of four iron atoms, a coordinately unsaturated cluster. Instead, both the carbon and the hydrogen atoms of the methylidyne ligand were bound to iron atoms as shown in Figure 13—with the additional C-H-Fe multicenter bonding a coordinately saturated cluster obtains. The stereochemical feature of the CH ligand in this cluster is analogous in form to the molecular orbital prediction for CH on a metal surface. Furthermore, there are spectroscopic data (29) for CH chemisorbed on Ni(111) that suggest

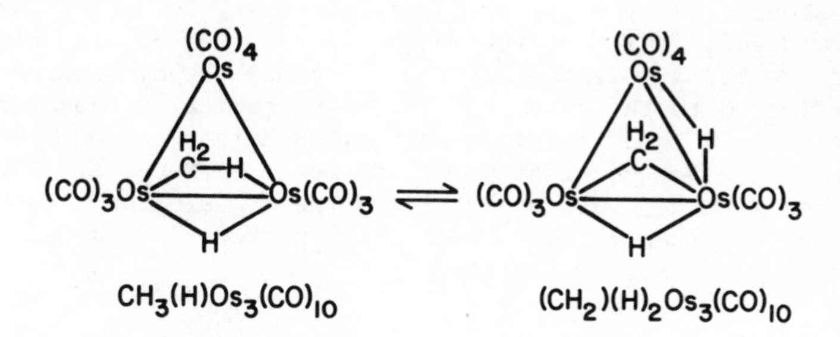

Figure 14. Representation of the solution state of two clusters, a methylene and a methyl species, which rapidly interconvert (30).

The methylene structure has been established by x-ray analysis, but the methyl derivative has not been so defined, although NMR spectroscopic data have shown that the methyl group unsymmetrically bridges two osmium atoms with a three-center, two-electron C—H—Os interaction.

that the CH vector is tipped with respect to the surface normal and is ostensibly similar to 7 and to the CH ligand in the iron cluster. Another structural point of interest in a chemical context is that the C-H bond distance in the methylidyne ligand in the cluster is rather long, 1.18Å. The bond lengthening suggests C-H bond activation—a point borne out by the established cluster chemistry where (A) there is fast exchange between hydrogen atom sites in the cluster, i.e., between the C-H-Fe and the Fe-H-Fe sites and (B) the C-H hydrogen atom is quite protonic in character and is removed by the weak base methanol (unpublished observation from our research) to form $[HFe_4C(CO)_{12}^-]$ (26).

Multicenter bonding of a CH_3 group in a metal cluster has also been established (30). In $H(CH_3)Os_3(CO)_{10}$, the CH_3 group is sigma bonded to one osmium atom in the cluster and is also bonded through a multicenter Os-C-H-Os bond to a second. Without the latter interaction the cluster would have been coordinately unsaturated. The CH bond in this cluster is activated: The methyl derivative in solution is in equilibrium with a second cluster that has a methylene ligand, $H_2(CH_2)Os_3(CO)_{10}$. The methylene ligand bridges two osmium atoms in the cluster (Figure 14). This cluster chemistry with the facile and reversible C-H bond cleavage reaction illustrates the two steps postulated in the beginning of this article: the formation of a multicenter C-H-M bond which effects C-H bond activation and the subsequent (and reversible) step of C-H bond activation.

We seek in our cluster chemistry further examples of facile C-H bond activation and bond scission. Other established examples of such hydrocarbon chemistry had been reviewed earlier (23). In the following section the chemisorption of saturated hydrocarbons is considered.

Saturated Hydrocarbons. Dehydrogenation is *the* reaction of hydrocarbons at many metal surfaces. However, for this reaction to proceed under mild conditions the hydrocarbon must be bound to the surface for a reasonable period of time—the hydrocarbon must be coordinated to the surface. The most chemically reasonable bonding mode is through a multicenter C-H-M interaction. Methane neither reacts nor chemisorbs on clean metal surfaces at 20°C. In principle, methane could bond through the *aegis* of one, two, or three such multicenter bonds on a close packed clean metal surface

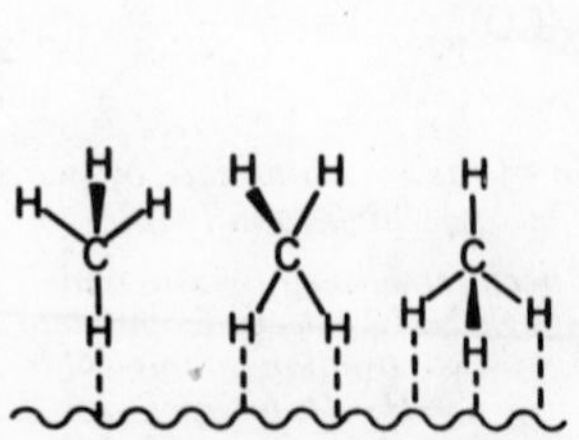

8

as shown in 8 although energetically very little should be gained in the binding in going from one to two or three C-H-M interactions per methane molecule (or per carbon atom in longer chain hydrocarbons). In fact, the binding of methane to clean metal surfaces is very weak, and methane thermally desorbs at temperatures of ∿-110°C. On the other hand, the binding potential for a hydrocarbon molecule should increase as the chain length increases; the number of single C-H-M interactions per molecule, 9, increases as

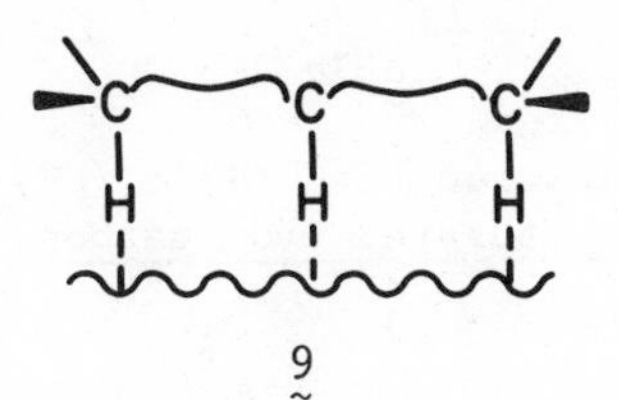

9

does the ionization potential decrease (Table I). Entropically, at least among the smaller hydrocarbons, cyclopentane and cyclohexane coordination, and the subsequent dehydrogenation reaction, at flat metal surfaces should be relatively favored. Cyclohexane could initially coordinate to a flat metal surface with up to six three-center C-H-M bonds, although three or four such bonds should be sufficient and would be a far more favorable representation in terms of energetics since minimization of ring strain could be realized (see Figure 15).

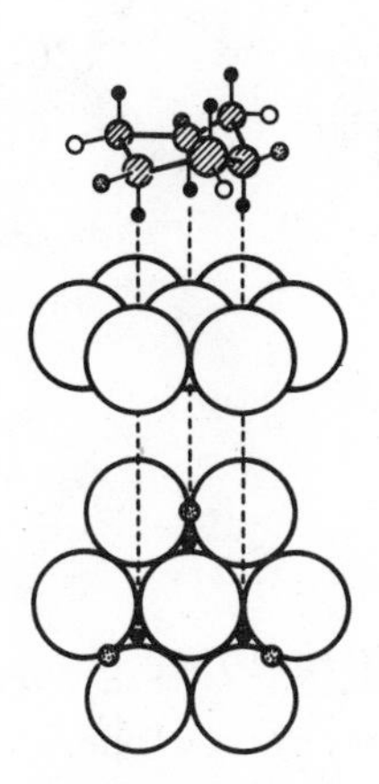

Figure 15. Representation of the interaction of cyclohexane (chair form) with Ru(0001) as suggested by Madey and Yates (31) in their classic study of hydrocarbon chemisorption on this basal plane of ruthenium. Three Ru—H—C three-center bonds are formed with three of the axial hydrogen atoms on one side of the chair form of cyclohexane. Weaker three-center Ru—H—C bonds may also be extant with equatorial C—H hydrogen atoms.

Table I

Ionization Potentials (eV) For Saturated Hydrocarbons[a]

Straight Chain Hydrocarbon	I.P.	Branched Hydrocarbon	I.P.	Cyclic or Polycyclic Hydrocarbons	I.P.
Methane	12.6				
Ethane	11.5				
Propane	11.1			*c*-C_3H_6	10.1
Butane	10.6	*i*-butane	10.6	*c*-C_4H_8	10.1
Pentane	10.4	*i*-pentane	10.3	*c*-C_5H_{10}	10.5
		neo-pentane	10.4	*c*-$C_5H_9CH_3$	10.5
Hexane	10.2	*i*-hexane	10.1	*c*-C_6H_{12}	9.8
		all other C_6H_{14}	∿10.1		
Heptane	10.1			*c*-$C_6H_{11}CH_3$	9.9
Octane	10.2			*c*-$C_6H_{10}(CH_3)_2$[b]	10.1
				Bicycloheptane	9.8
				Adamantane	9.3
				Cubane	8.7

a. J. Phys. Chem. Ref. Data 1977, 6 Supplement 1.

b. *cis* and *trans* 1,2-dimethylcyclohexane.

There is an elegant study (31) of hydrocarbon chemisorption by Madey and Yates for the basal plane, (0001), of ruthenium. They found that the activation energy for thermal desorption was a near linear function of the carbon chain length—at least for a limited set of hydrocarbons as shown in Table II. These results

Table II

Thermal Desorption Temperature Maxima and Thermal Desorption Activation Energies for Saturated Hydrocarbons on Ru (0001)

Hydrocarbon	Temperature for Reversible Thermal Desorption Maxima*	Activation Energy for Thermal Desorption
Ethane	91 K	5.0 kcal/mole
Cyclopropane	145 K	7.9 kcal/mole
Cyclohexane	227 K	14.2 kcal/mole
Cyclooctane	257 K	16.2 kcal/mole

*Temperature at which desorption rate is a maximum for mono- or sub-monolayer coverages.

suggest that to a first approximation there is a maximum of one M-H-C multicenter bond per carbon atom and that the bond strengths of these multicenters, although they must vary depending upon overall stereochemistry in the chemisorbed state (see Figure 15) are the same order of magnitude. It should be noted that the activation energies for thermal desorption are not a simple function of the ionization potential of the hydrocarbon.

We are presently exploring the chemistry of saturated hydrocarbons on nickel and platinum surfaces. We have found that cyclohexane does not react with Ni(111) at temperatures up to 70°C (the thermal desorption temperature is ∿-100°C), whereas the unsaturated derivatives cyclohexene and cyclohexadiene (either isomer) react with this surface at 20 - 70°C and undergo dehydrogenation to give chemisorbed benzene (11,12). These studies are still in progress.

Acknowledgment

This surface science research was supported by the Division of Chemical Sciences, Office of Basic Energy Sciences, U. S. Department of Energy under contract No. W-7405-Eng.-48 and the cluster chemistry by the National Science Foundation. The nickel and platinum surface studies were done by Cynthia Friend and Min-Chi Tsai, respectively and the cluster chemistry by Dr. Mamoru Tachikawa. The molecular orbital studies are collaborative with

Professor Robert M. Gavin, Department of Chemistry, Haverford College.

Abstract

In coordinately unsaturated molecular metal complexes, carbon-hydrogen bonds of the peripheral ligands may, if the stereochemistry allows, closely approach a metal center so as to develop a three-center two-electron bond between the carbon, the hydrogen, and the metal atoms, C-H-M. In some instances, the interaction is followed by a scission of the C-H bond whereby the metal is effectively oxidized and discrete M-H and M-C σ bonds are formed. This class of metal-hydrogen-carbon interactions and reactions is shown to be a common phenomenon in metal surface chemistry. Ultra high vacuum studies of nickel and platinum with simple organic molecules like olefins and arenes are described. These surface chemistry studies were done as a function of surface crystallography and surface composition. The discussion is limited to the chemistry of methyl isocyanide, acetonitrile, benzene and toluene, pyridine, trimethylphosphine, ethylene, acetylene and saturated hydrocarbons. Molecular orbital calculations are presented that support the experimental identification of the importance of C-H-M metal bonding for metal surfaces.

Literature Cited

1. Cope, A. C. and Siekman, R. W. J. Am. Chem. Soc. 1965, 87, 3272.
2. Cope, A. C. and Friedrich, E. C. J. Am. Chem. Soc. 1968, 90, 909.
3. Parshall, G. W. Accounts Chem. Res. 1970, 3, 139; 1975, 8, 113.
4. Parshall, G. W. "Catalysis" C. Kemball, Ed., Chemical Society Specialist Periodical Report, 1, 1977, 335.
5. Brown, R. K.; Williams, J. M.; Schultz, A. J.; Stucky, G. D.; Ittel, S. D. and Harlow, R. L. J. Am. Chem. Soc. 1980, 102, 981.
6. Beno, M. A.; Williams, J. M.; Tachikawa, M. and Muetterties, E. L. J. Am. Chem. Soc. 1980, 102, 4542.
7. Friend, C. M.; Gavin, R. M.; Muetterties, E. L. and Tsai, M.-C. J. Am. Chem. Soc. 1980, 102, 1717.
8. Friend, C. M.; Stein, J. and Muetterties, E. L. J. Am. Chem. Soc. 1981, 103, 0000.
9. Friend, C. M. and Muetterties, E. L. J. Am. Chem. Soc. 1981, 103, 0000.
10. Biberian, J. P. and Somorjai, G. A. Applications of Surface Science 1979, 2, 352.
11. Tsai, M.-C. and Muetterties, E. L., to be published.
12. Friend, C. M. and Muetterties, E. L., to be published.
13. Bertolini, J. C. and Rousseau, J. Surf. Sci. 1979, 89, 467.

14. Demuth, J. E. and Eastman, D. E. Phys. Rev. 1976, 13, 1523.
15. Lehwald, S.; Ibach, H.; Demuth, J. E. Surf. Sci. 1978, 78, 577.
16. Rees, B. and Coppens, P. Acta Cryst. 1973, B29, 2515.
17. Crawford, E. and Kemball, C. Trans. Faraday Soc. 1962, 58, 2452.
18. Stein, J. and Muetterties, E. L., unpublished data.
19. (a) Cattania, M. G., Simonetta, M. and Tescari, M. Surf. Sci. 1979, 82, L615; (b) Demuth, J. E. Surf. Sci. 1978, 76, L603; (c) Klimesch, P. and Henzler, M. Surf. Sci. 1979, 90, 57.
20. Day, V. W.; Day, R. O.; Kristoff, J. S.; Hirsekorn, F. J. and Muetterties, E. L. J. Am. Chem. Soc. 1975, 97, 2571.
21. Harris, T. V.; Rathke, J. W. and Muetterties, E. L. J. Am. Chem. Soc. 1978, 100, 6966.
22. Gavin, R. M. and Muetterties, E. L., unpublished data.
23. Muetterties, E. L.; Rhodin, T. N.; Band, E.; Brucker, C. F.; Pretzer, W. R. Chem. Rev. 1979, 79, 91.
24. Muetterties, E. L. and Stein, J. Chem. Rev. 1979, 79, 479.
25. Muetterties, E. L. Israeli J. Chem. 1980, 20, 84.
26. Tachikawa, M. and Muetterties, E. L. J. Am. Chem. Soc. 1980, 102, 4541.
27. Beno, M.; Williams, J. M.; Tachikawa, M. and Muetterties, E. L. J. Am. Chem Soc., to be published.
28. Seyferth, D. Adv. Organomet. Chem. 1976, 14, 98.
29. Demuth, J. E. and Ibach, H. Surf. Sci. 1978, 78, L238.
30. Calvert, R. B. and Shapley, J. R. J. Am. Chem. Soc. 1977, 99, 5225; 1978, 100, 6544; 1978, 100, 7726.
31. Madey, T. E. and Yates, J. T., Jr. Surf. Sci. 1978, 76, 397.

RECEIVED December 1, 1980.

New Approaches to the Chemistry of Di- and Trimetal Complexes

Synthesis, Chemical Reactivity, and Structural Studies

TERENCE V. ASHWORTH, MICHAEL J. CHETCUTI, LOUIS J. FARRUGIA, JUDITH A. K. HOWARD, JOHN C. JEFFERY, RONA MILLS, GEOFFREY N. PAIN, F. GORDON A. STONE, and PETER WOODWARD

Department of Inorganic Chemistry, The University, Bristol, BS8 1TS England

For some years we have been studying di- and tri-metal complexes in which organic ligands bridge the metal centres. These ligands often display unusual chemical behaviour not found, or rarely found, in mononuclear metal compounds. For example in $[Ru_3(CO)_8(C_8H_6)]$ the triruthenium cluster stabilises pentalene(1), an organic molecule having only a transient existence under normal conditions, while in the complex $[Ni_3(CO)_3\{\mu_3\text{-}(\eta^2\text{-}CF_3C_2CF_3)\}\{\mu_3\text{-}(\eta^8\text{-}C_8H_8)\}]$ the trinickel system bonds a planar cyclo-octatetraene ring(2). There are several known examples of situations where a ligand bridging a metal — metal bond displays unusual reactivity. Thus we have shown that alkynedimolybdenum complexes $[Mo_2(\mu\text{-}RC_2R)(CO)_4(\eta\text{-}C_5H_5)_2]$ undergo stepwise insertion reactions with alkyne molecules in a process which appears to depend on the dimolybdenum group alternating between Mo — Mo, Mo = Mo and Mo ≡ Mo bonding modes(3). Moreover, the acetylene molecule in the complex $[Mo_2(\mu\text{-}HC_2H)(CO)_4(\eta\text{-}C_5H_5)_2]$ shows Diels-Alder reactivity(4), unusual behaviour for this alkyne.

In spite of the evident indications that ligands bridging di- or tri-metal groups have a rich chemistry, there are few known rational synthetic procedures for preparing dimetal compounds, and virtually none for tri- or tetra-metal clusters. Most of the known compounds in these categories have not been obtained by designed syntheses(5). Yet logical synthetic pathways to di-, tri-, and tetra-nuclear metal complexes are required if the chemical reactivity of the co-ordinated ligands is to be fully exploited. In this paper we shall describe simple preparative routes to complexes having heteronuclear metal — metal bonds, and report some reactions of these species which depend on their multinuclear character. The new syntheses owe their success to two simple ideas. The first depends on the concept that C = M and C ≡ M bonds would react with low valent metal complexes as do C = C and C ≡ C linkages. The second follows from the premise that since Pt^0 readily react with molecules containing C = C and

0097-6156/81/0155-0299$05.00/0

C ═ M groups, then Pt^0 compounds and other low valent metal species would also react with M ═ M groups.

COMPOUNDS WITH BRIDGING ALKYLIDENE AND ALKYLIDYNE LIGANDS.

A serendipitous discovery(6) that $[Pt(cod)_2]$ (cod = cyclo-octa-1,5-diene) reacted with perfluoropropene to give the diplatinum complex $[Pt_2\{\mu\text{-}C(CF_3)_2\}(cod)_2]$, rather than the simple η^2 adduct $[Pt(CF_2{:}CFCF_3)(cod)]$, led to the idea(7) that mononuclear metal-carbene or -carbyne complexes would combine with nucleophilic and co-ordinatively unsaturated metal species (M′):

$$R^1R^2C{=}M \;+\; M' \longrightarrow M{-}C(R^1R^2){-}M' \text{ (three-membered ring, M–M′ bonded)} \qquad (A)$$

$$RC{\equiv}M \;+\; M' \longrightarrow M{=}C(R){-}M' \text{ (three-membered ring, M–M′ bonded)} \qquad (B)$$

(Ligands on M and M′ omitted for clarity)

In practice this approach to the synthesis of compounds with dimetallacyclo-propane or -propene ring systems has been very successful. Complexes (1) - (12) are representative of those which have been readily prepared by reacting various low-valent metal compounds either with carbene-metal species, or with the carbyne complexes $[W{\equiv}CR(CO)_2(\eta\text{-}C_5H_5)]$ or $[W{\equiv}CR(Br)(CO)_4]$ ($R = C_6H_4Me\text{-}4$) (8,9,10,11).

Complexes of type (A) or (B) can be readily identified by $^{13}C\{^1H\text{-decoupled}\}$ n.m.r. spectroscopy since the bridging $\underline{C}R^1R^2$ and $\underline{C}R$ groups give characteristic deshielded signals for the ligated carbon atoms. For the bridging alkylidene complexes the resonance is less deshielded than in the mononuclear metal precursor, e.g. for $[MnPt\{\mu\text{-}C(OMe)C_6H_4Me\text{-}4\}(CO)_2(PMe_3)_2(\eta\text{-}C_5H_5)]$ (13), δ for μ-C = 193.6 p.p.m., whereas for $[Mn{=}C(OMe)(C_6H_4Me\text{-}4)(CO)_2(\eta\text{-}C_5H_5)]$ the resonance for the ligated carbon of the carbene ligand is 334.8 p.p.m. In contrast, for the alkylidyne dimetal complexes the signal for the bridging carbyne-carbon atom is more deshielded than in the parent compound, e.g. for $[PtW(\mu\text{-}CC_6H_4Me\text{-}4)(CO)_2(PMe_2Ph)_2(\eta\text{-}C_5H_5)]$ (14) δ for $\mu\text{-}\underline{C}$ = 336 p.p.m., whereas in the spectrum of $[W{\equiv}CC_6H_4Me\text{-}4(CO)_2(\eta\text{-}C_5H_5)]$ the carbyne-carbon atom resonance is at 300 p.p.m. If the complexes are sufficiently soluble, the ^{13}C n.m.r. spectra of the platinum compounds reveal characteristic ^{195}Pt - ^{13}C satellites on the bridging $\mu\text{-}\underline{C}R^1R^2$ or $\mu\text{-}\underline{C}R$ resonances. Thus for (1) δ ($\mu\text{-}\underline{C}$) = 203 p.p.m. with J(PtC) = 880 Hz, while for $[PtW(\mu\text{-}CC_6H_4Me\text{-}4)(CO)_2(PMe_3)_2(\eta\text{-}C_5H_5)]$ (15) δ ($\mu\text{-}\underline{C}$) = 338 p.p.m. with J(PtC) = 732 Hz.

The molecular structure of several complexes of types (A) and (B) have been established by single-crystal X-ray diffraction studies (8,10,11). The results have revealed interesting features.

In compounds of type (A) containing alkylidene ligands, the latter do not necessarily symmetrically bridge the metal — metal bonds; indeed asymmetric bridging is probably the norm. In (1) the μ-C — Pt distance of 2.04(1) Å is at the short end of the range [1.99(3) - 2.15(2) Å] generally found for carbon — platinum σ bonds, whereas the much longer μ-C — W distance of 2.48(1) Å contrasts with the distances (2.1 - 2.3 Å) generally found for C (sp^3) — W σ bonds (12)(13), and the 2.34 Å reported(14) for the W — CH(OMe)Ph bond in the anion $[(OC)_5W — CH(OMe)Ph]^-$ of W^{-I}.

The dimensions of the dimetallacyclopropane rings, however, are susceptible to changes in the nature of the peripheral ligands. Thus in the complex $[(Me_3P)(OC)_4W\{\mu\text{-}C(OMe)C_6H_4Me\text{-}4\}Pt(PMe_3)_2]$ (16) the μ-C — W distance of 2.37(1) Å is shorter than that in (1). The Pt — W distance of 2.825(1) Å, compares with 2.861(1) Å in (1). This tightening of the ring system by substitution of PMe_3 for a CO ligand on tungsten is reflected in enhanced thermal and oxidative stability. Moreover, whereas (4) is a relatively unstable compound, the complex $[(Me_3P)(OC)_4W(\mu\text{-}CPh_2)Pt(PMe_3)_2]$(17) is stable. The compound $[CrPt\{\mu\text{-}C(OMe)Ph\}(CO)_5(PMe_3)_2]$ (18) rapidly decomposes in organic solvents above 80 °C into $[Cr(CO)_5(PMe_3)]$, $[Pt_3\{\mu\text{-}C(OMe)Ph\}_2(\mu\text{-}CO)(PMe_3)_3]$ (19), and $[Pt_3\{\mu\text{-}C(OMe)Ph\}_3(PMe_3)_3]$ (20). However, UV irradiation of (18) in the presence of PMe_3 affords a more stable complex $[(Me_3P)(OC)_4Cr\{\mu\text{-}C(OMe)Ph\}Pt(PMe_3)_2]$ (21), formed as a single isomer.

The decomposition of (18) into triplatinum species obviously reflects migration of the μ-C(OMe)Ph ligand across the Cr — Pt bond with simultaneous transfer of PMe_3 to chromium, and concomitant rupture of the metal — metal bond. The process is reminiscent of the earlier observation (15) that $[Mo\{C(OMe)Ph\}(CO)(NO)(\eta\text{-}C_5H_5)]$ reacts with $[Ni(CO)_4]$ to give $[Ni_3\{\mu\text{-}C(OMe)Ph\}_3(CO)_3]$ and $[Mo(CO)_2(NO)(\eta\text{-}C_5H_5)]$. It seems likely that this reaction proceeds via a dimetal species $[(\eta\text{-}C_5H_5)(ON)(OC)Mo\{\mu\text{-}C(OMe)Ph\}Ni(CO)_2]$ which was not isolated.

Transfer of the $\mu\text{-}CR^1R^2$ ligand in a structure of type (A) across the M — M′ bond resembles CO site exchange from one metal centre to another in polynuclear metal carbonyl chemistry(16). In this context it is interesting that the mononuclear manganese carbene compound $[MnI(COCH_2CH_2CH_2)(CO)_4]$ reacts with $[Pt(C_2H_4)_2(PBu^t{}_2Me)]$ to afford complex (22) in which the 2-oxacyclopentylidene ligand is terminally bonded to the platinum, the metal — metal bond being preserved albeit supported by the bridging iodo ligand(17).

Ph OMe
C
$(OC)_5W$ — $Pt(PMe_3)_2$

(1)

Ph OMe
C
$(OC)_5Mo$ — $Pt(PMe_2Ph)_2$

(2)

O
C
$(OC)_5Cr$ — $Pt(PMe_2Ph)_2$

(3)

Ph Ph
C
$(OC)_5W$ — $Pt(PMe_3)_2$

(4)

Ph OMe
C
$(\eta\text{-}C_5H_5)(OC)_2Mn$ — $Ni(PMe_3)_2$

(5)

Ph OMe
C
$(\eta\text{-}C_5H_5)(OC)_2Mn$ — $Pd(PMe_3)_2$

(6)

MeO R
C
$(\eta\text{-}C_5H_5)(OC)_2Re$ — $Pt(PMe_3)_2$

(7)*

* Throughout this paper R represents the group C_6H_4Me-4 in structural formulae.

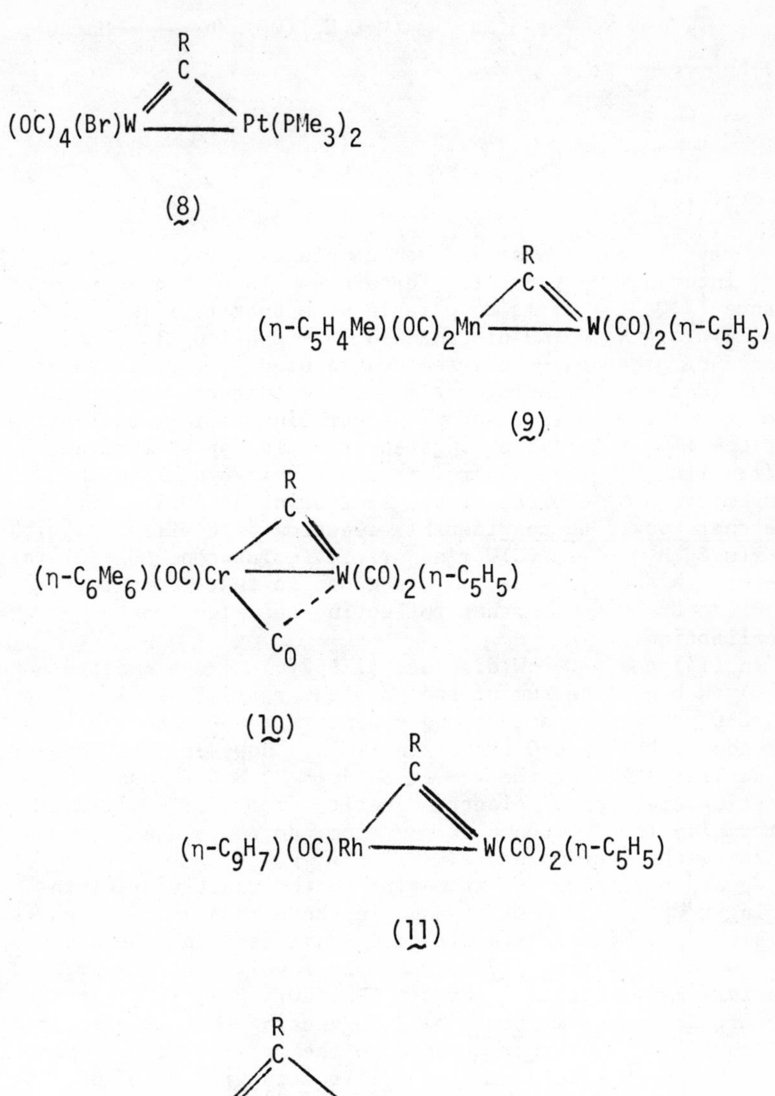
R
C
(OC)4(Br)W
Pt(PMe3)2
(8)
R
C
(η-C5H4Me)(OC)2Mn
W(CO)2(η-C5H5)
(9)
R
C
(η-C6Me6)(OC)Cr
W(CO)2(η-C5H5)
CO
(10)
R
C
(η-C9H7)(OC)Rh
W(CO)2(η-C5H5)
(11)
R
C
(η-C5H5)Co
W(CO)2(η-C5H5)
C
O
(12)

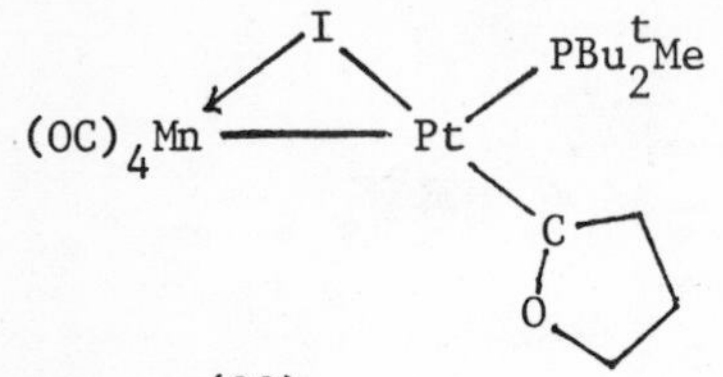

$(OC)_4Mn$ — Pt (μ-I), PBu^t_2Me, C(OCH$_2$CH$_2$CH$_2$)

(22)

$(\eta-C_5H_5)(OC)_2Mn$ — $Mn(CO)_4$, μ-C(CO)R

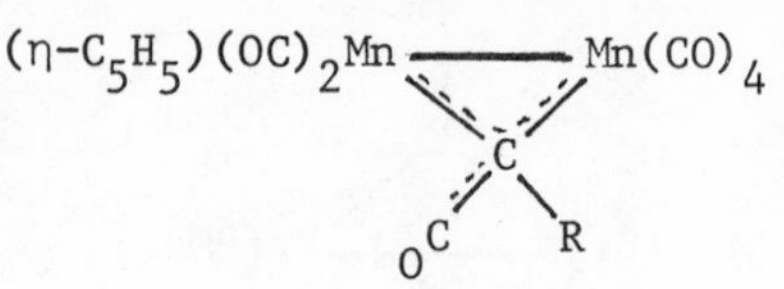

(26)

X-ray diffraction studies on complexes (10)(11) and (14)(10) reveal interesting structural features. In (10) the μ-C—W distance [2.025(6) Å] is comparable with that [2.14(2) Å] found(14) for the C=W distance in [W=$CPh_2(CO)_5$]. The presence of a carbon—tungsten double bond in (10) is in accord with an 18-electron configuration at the tungsten atom. However, since one of the CO ligands on the chromium atom is semi-bridging [< Cr·C·O 163(1)°] to the tungsten it would appear that an electron rich tungsten centre transfers charge back to the chromium *via* a π* orbital of the semi-bridging CO ligand(16). Hence there would be considerable electron delocalization within the Cr(μ-CC_6H_4Me-4)(μ-CO)W ring system. In compound (10) the μ-C—Cr distance [1.928(6) Å] is close to that found for Ph(MeO)C=Cr bonds, further reflecting the electron delocalization.

In (14) the μ-C—W distance [1.967(6) Å] again corresponds to a C=W bond (the sum of the covalent radii is 1.91 Å) but the two W·C·O groups are not strongly bent [< W·C·O mean 173(1)°], as is one of the Cr·C·O groups in (10). However, the carbonyl ligands lie back over the W—Pt bond [< Pt·W·C 72 and 67°] suggesting donation of electron density from the C≡O ligands to the formally 16-electron platinum atom, so as to equalize charge transfer within the molecule.

We are directing our attention to the reactivity of the bridging μ-CR^1R^2 and μ-CR ligands in the compounds of types (A) and (B). Some initial results are summarised in Scheme 1 (18). Formation of (23) from (18) presumably involves attack by $C_6H_4Me^-$ on an intermediate cation [CrPt(μ-CPh)$(CO)_5(PMe_3)_2$]$^+$, but the yield of (23) was low (*circa* 6 %) suggesting that side reactions had occurred. We referred above to the fragile Cr—Pt bond in (18). The more robust complex (21) reacts with Me_3O^+ BF_4^- to give a salt [CrPt(μ-CPh)$(CO)_4(PMe_3)_3$] [BF_4] (24) which on treatment with NaOMe afforded compound (25), the structure of which was confirmed by an X-ray diffraction study(17). Formation of (25) from (21) is non-stoichiometric but occurs in *circa* 60 % yield. Evidently a CO group migrates to the bridging μ-CPh group in (24), by a mechanism as yet unclear, and is attacked by OMe^-. However, it has been shown that a CO ligand can transfer to a carbyne group in the formation of complex (26) by reacting [Mn≡CC_6H_4Me-4$(CO)_2(\eta-C_5H_5)$] [BCl_4] with [PPN] [$Mn(CO)_5$] (19).

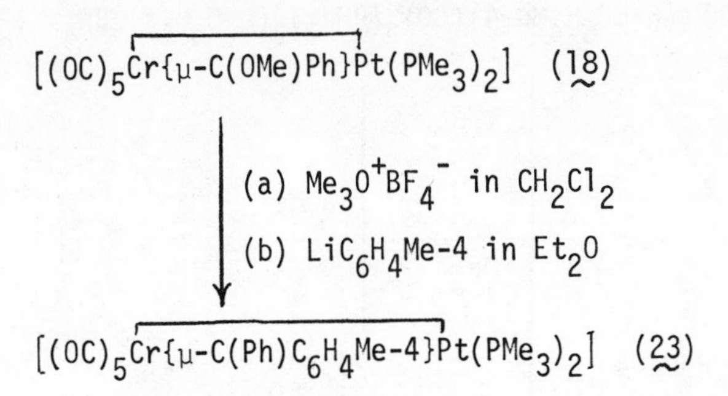

$[(Me_3P)(OC)_4Cr\{\mu\text{-}C(OMe)Ph\}Pt(PMe_3)_2]$ (21)

(a) $Me_3O^+BF_4^-$ in CH_2Cl_2

(b) NaOMe in MeOH

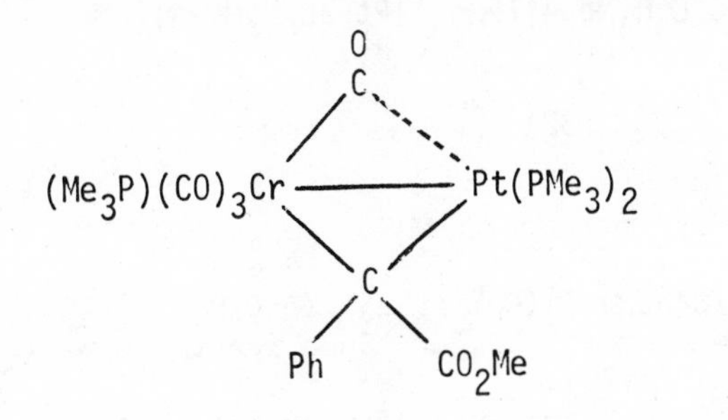

(25)

Scheme 1.

Clearly the transformation of complexes containing the ring system $[M\{\mu\text{-}C(OMe)R\}M']$ into the alkylidyne cation species $[M(\mu\text{-}CR)M']^+$ is of considerable importance since the latter would be expected to afford interesting products on treatment with a variety of nucleophiles. For this reason the salts $[MPt(\mu\text{-}CC_6H_4Me\text{-}4)(CO)_2(PMe_3)_2(\eta\text{-}C_5H_5)]$ $[BF_4]$ [(27), M = Mn; (28), M = Re] were prepared in 80 - 90 % yield by treating the neutral

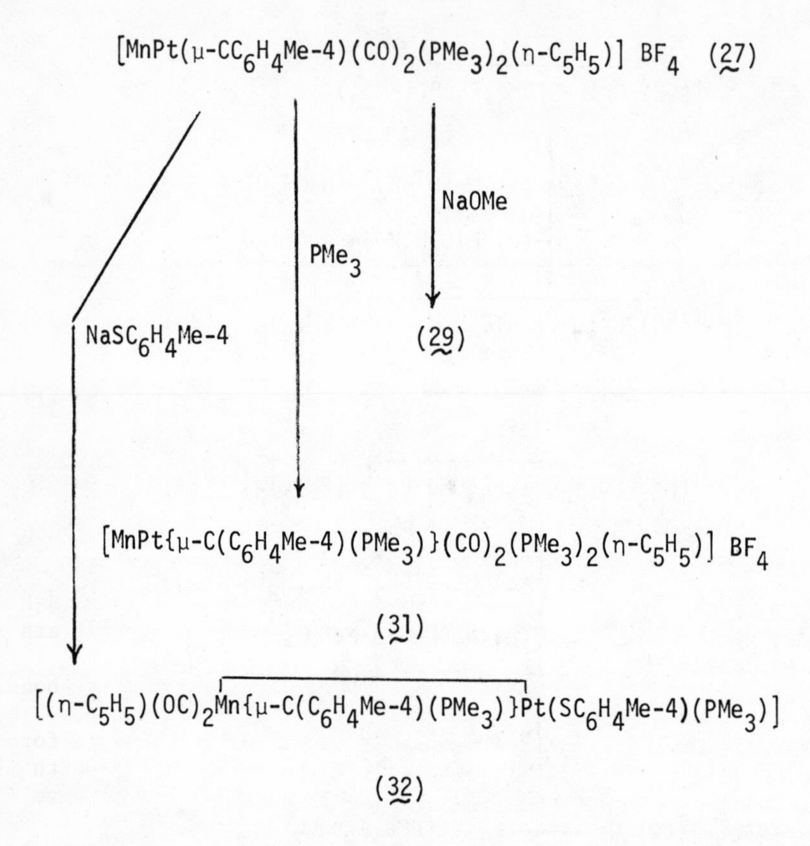

Scheme 2.

compounds $[MPt\{\mu\text{-}C(OMe)C_6H_4Me\text{-}4\}(CO)_2(PMe_3)_2(\eta\text{-}C_5H_5)]$ [(29), M = Mn; (30), M = Re] with $Me_3O^+ BF_4^-$. Some reactions of (27) are summarised in Scheme 2 (20).

The bridging alkylidyne complexes of type B can be used as intermediates to prepare triangular trimetal complexes with capping μ_3-CR ligands. Thus reaction of compounds (11) and $[PtW(\mu\text{-}CC_6H_4Me\text{-}4)(CO)_2(PEt_3)_2(\eta\text{-}C_5H_5)]$ (33) with $[Fe_2(CO)_9]$ gives compounds (34) and (35) respectively, characterised by X-ray diffraction studies(21). Carbon-13 n.m.r. studies reveal the μ-C resonances in (34) and (35) at δ (p.p.m.) 287 [d, J(RhC) 44 Hz] and 323 [d, J(PC) 9, J(PtC) 404, and J(WC) 110 Hz], respectively. Reaction of (14) with $[Fe_2(CO)_9]$ gives in 70 % yield purple crystals of $[FePtW(\mu_3\text{-}CC_6H_4Me\text{-}4)(\mu\text{-}CO)(CO)_5(PMe_2Ph)_2(\eta\text{-}C_5H_5)]$ (36) {^{13}C n.m.r. resonance for μ_3-C at 295 p.p.m. [t, J(PC) 11, J(PtC) 472, J(WC) 111 Hz]}. The chiral nature of the cluster reveals

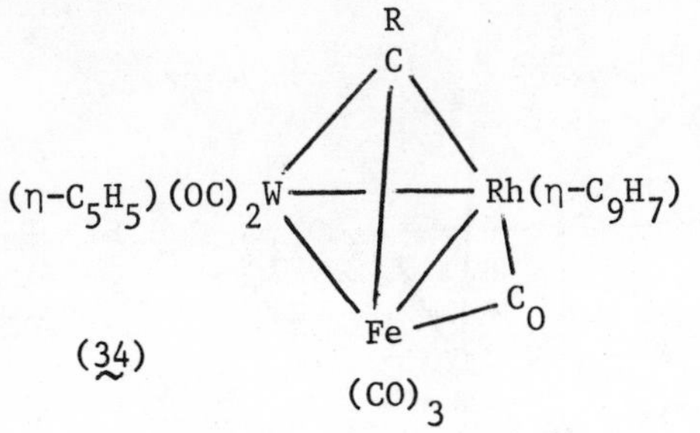

(34)

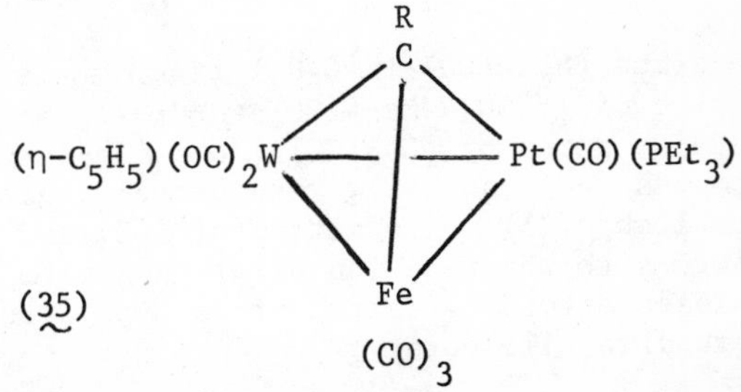

(35)

itself in the ^{13}C n.m.r. spectrum by the existence of four $\underline{C}H_3P$ resonances at 18, 16, 10.5 and 7.0 p.p.m. appearing as doublets [J(PC) 27 - 32 Hz].

This synthetic approach to heteronuclear metal—metal bonds can be extended to more complex systems. Thus reaction of $[Ni(cod)_2]$ or $[Pt(C_2H_4)_3]$ with $[W{\equiv}CC_6H_4Me\text{-}4(CO)_2(\eta\text{-}C_5H_5)]$ affords complexes (37) and (38), respectively. Reaction of (38) with $[Fe_2(CO)_9]$ affords compound (39). Compounds (37) - (39) have been fully characterised by X-ray diffraction studies(22). Complex (38) shows in its ^{13}C n.m.r. spectrum a resonance for the $\mu_2\text{-}\underline{C}R$ groups at 307 p.p.m. [J(PtC) 830, J(WC) 157 Hz]. In the spectrum of (39) the signals for the $\mu_3\text{-}\underline{C}R$ and $\mu_2\text{-}\underline{C}R$ groups occur at 297 [J(PtC) 547, J(WC) 117 Hz] and 314 [J(PtC) 807, J(WC) 158 Hz] p.p.m., respectively.

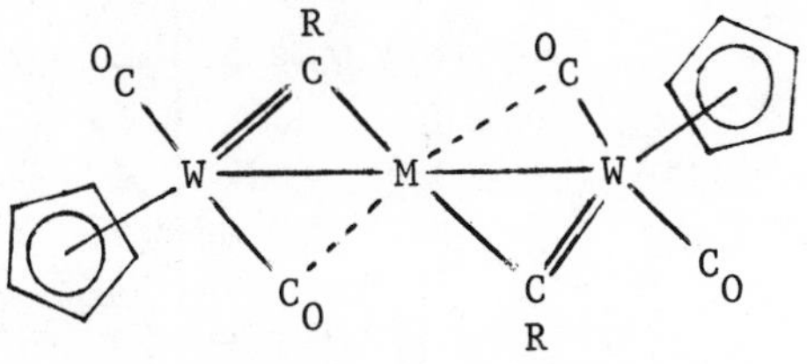

(37, M = Ni)

(38, M = Pt)

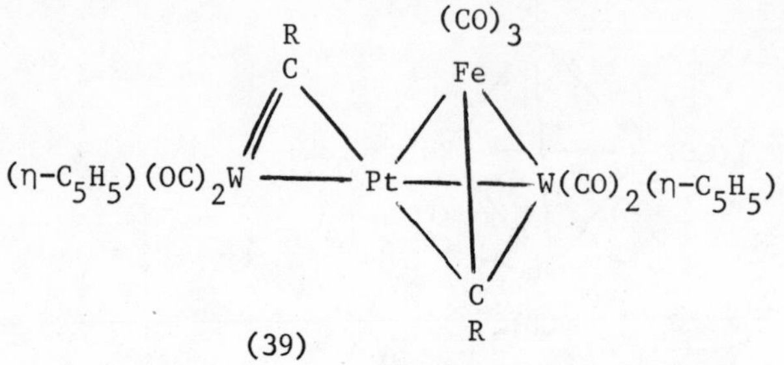

R
C (CO)$_3$
Fe
(η-C_5H_5)(OC)$_2$W — Pt — W(CO)$_2$(η-C_5H_5)
C
R

(39)

Since the $W(CO)_2(\eta\text{-}C_5H_5)$ fragment is isolobal with RC$\equiv$ (23), addition of $[W{\equiv}CC_6H_4Me\text{-}4(CO)_2(\eta\text{-}C_5H_5)]$ across the Co_2 and Rh_2 dimetal centres of $[Co_2(CO)_8]$ and $[Rh_2(\mu\text{-}CO)_2(\eta\text{-}C_5Me_5)_2]$ to give compounds (40) and (41), respectively, has also been successfully accomplished(21). Formation of (40) and (41) in this manner is analogous to the reaction of alkynes with these and other dimetal complexes affording tetrahedrane-type structures with a dicarbadimetalla core.

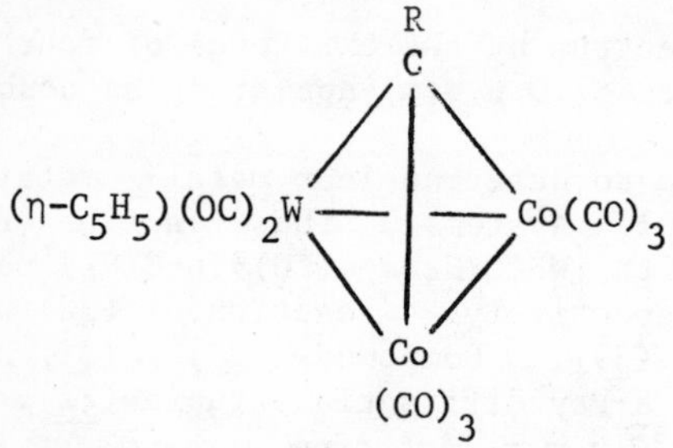

R
C
(η-C_5H_5)(OC)$_2$W — Co(CO)$_3$ (40) 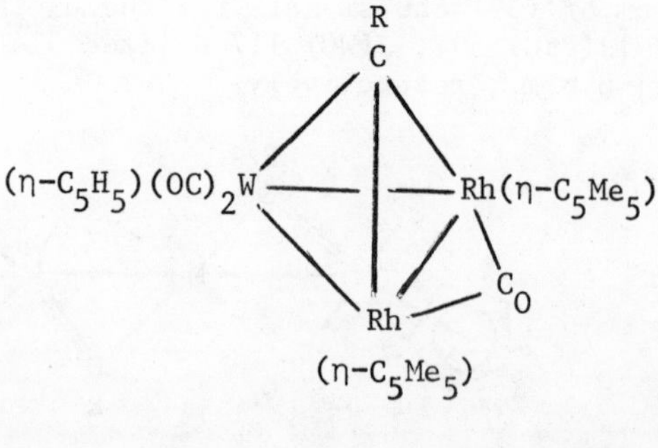
Co
(CO)$_3$

R
C
(η-C_5H_5)(OC)$_2$W — Rh(η-C_5Me_5)
C
O
(41) Rh
(η-C_5Me_5)

REACTIONS OF LOW VALENT METAL COMPOUNDS WITH COMPLEXES CONTAINING METAL — METAL MULTIPLE BONDS.

The processes represented by (C) and (D) are formally similar to those described earlier which yield the ring systems (A) and (B).

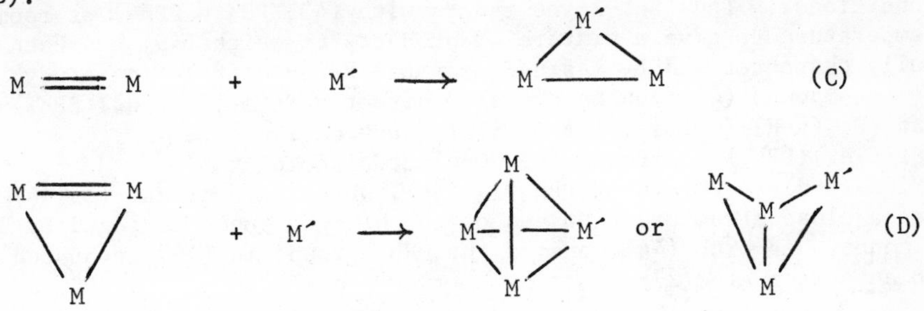

(Ligands on M and M´ omitted for clarity).

We have investigated the reactivity of the three compounds $[Re_2(\mu\text{-}H)_2(CO)_8]$ (24), $[Rh_2(\mu\text{-}CO)_2(\eta\text{-}C_5Me_5)_2]$ (25) and $[Os_3(\mu\text{-}H)_2(CO)_{10}]$ (26), containing metal — metal double bonds, towards low valent metal complexes in order to test the validity of the concept described above. Compound (42) is produced by reacting the dihydridodirhenium compound with either $[Pt(C_2H_4)_2(PPh_3)]$ or $[Pt(C_2H_4)(PPh_3)_2]$ (27). The structure has been confirmed by X-ray crystallography. Transfer of a CO ligand to the platinum is a common feature of reactions of this type; further examples are described below.

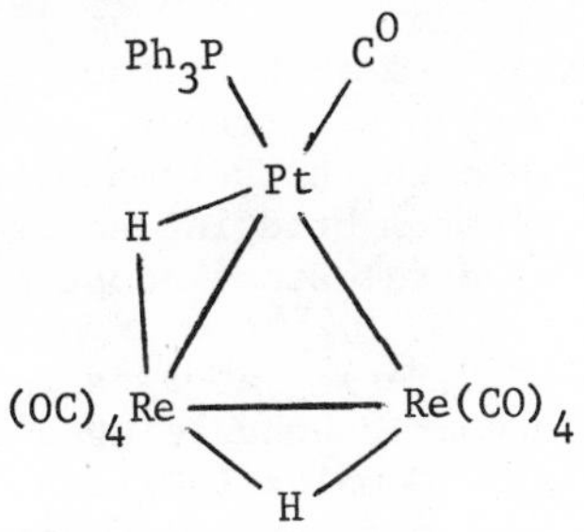

(42)

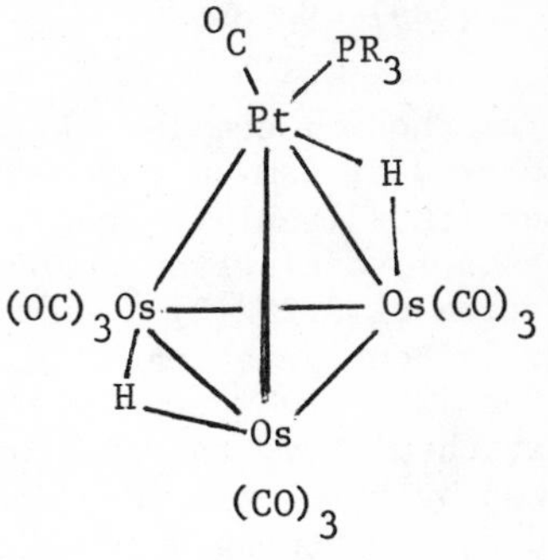

(43)

The osmium compound $[Os_3(\mu\text{-}H)_2(CO)_{10}]$ is a rich source of heteronuclear cluster complexes by virtue of the reactivity of the osmium — osmium double bond. The compounds $[Pt(C_2H_4)_2(PR_3)]$ $[PR_3 = P(cyclo\text{-}C_6H_{11})_3$, PPh_3 or $PBu^t{}_2Me]$ afford the '58-electron' clusters (43) which undergo dynamic behaviour in solution with site-exchange of the hydrido ligands (28). Compounds (43) react with CO, PPh_3 or $AsPh_3$ to give '60-electron' complexes, the process with CO being reversible. An X-ray crystallographic study on the PPh_3 adduct reveals a butterfly structure (44). Reactions of the 58-electron cluster (43) with alkynes results in rupture of some of the metal — metal bonds even under mild conditions. Thus but-2-yne reacts with (43, PR_3 = PPh_3) at room temperature to give a mixture of products of which (45) has been fully characterised by a single crystal X-ray diffraction study.

Compound (44) can be prepared directly from $[Pt(C_2H_4)(PPh_3)_2]$ and $[Os_3(\mu\text{-}H)_2(CO)_{10}]$. A similar reaction employing $[Ni(C_2H_4)(PPh_3)_2]$ affords the 60-electron complex $[NiOs_3(\mu\text{-}H)_2(\mu\text{-}CO)_2(CO)_8(PPh_3)_2]$ (46), which in contrast to (44) has a closo structure with two of the Ni — Os bonds bridged by CO ligands. As with (44), one of the PPh_3 groups in (46) is bonded to an osmium atom(27).

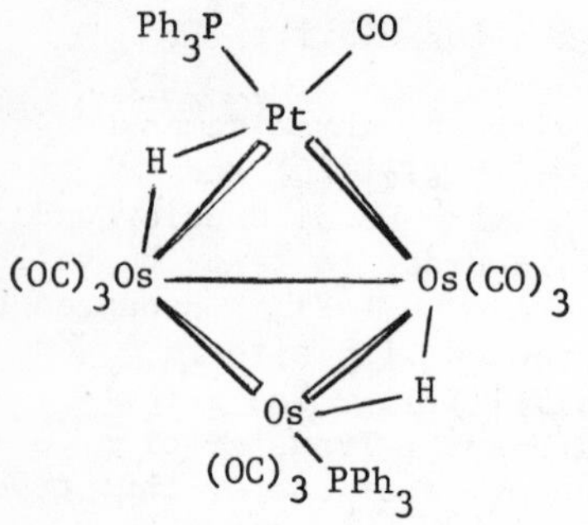

(44)

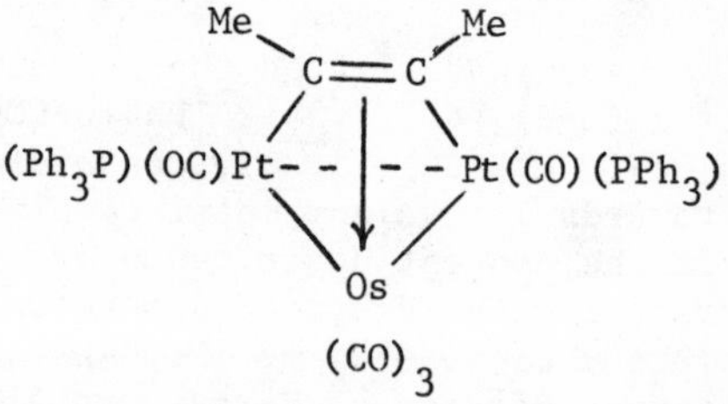

(45)

The rhodium complex $[Rh(acac)(C_2H_4)_2]$ reacts with the unsaturated triosmium compound to give the 60-electron cluster species $[RhOs_3(\mu\text{-}H)_2(\mu\text{-}acac)(CO)_{10}]$ (47) in which the acac ligand acts as a five-electron donor, one oxygen atom bridging an Os — Rh separation of 3.292(2) Å so that the four metal atoms adopt a butterfly configuration.

The dirhodium compound $[Rh_2(\mu\text{-}CO)_2(\eta\text{-}C_5Me_5)_2]$ engages in similar chemistry, the rhodium — rhodium double bond being readily attacked by Pt(0) species (Scheme 3)(29). Complex (50) can be viewed as comprising a 14- electron Pt(cod) group co-ordinated to a Rh ═ Rh bond, much as the anion $[Rh_3(\mu_3\text{-}CO)_2(CO)_2(\eta\text{-}C_5H_5)]^-$ can be visualised as arising through complexation of 14-electron $[Rh(CO)_2]^-$ with the Rh ═ Rh bond of $[Rh_2(\mu\text{-}CO)_2(\eta\text{-}C_5H_5)_2]$(30).

Not surprisingly, $[Rh_2(\mu\text{-}CO)_2(\eta\text{-}C_5Me_5)_2]$ reacts readily with the diazo compounds $RR'CN_2$ (R = H, R′ = H; R = CF_3, R′ = CF_3; R = H, R′ = $CH{:}CH_2$; R = H, R′ = CO_2Et) to give bridging alkylidene complexes (51). The latter undergo dynamic behaviour in solution. Herrmann et al(31) have prepared a number of related dirhodium complexes with bridging alkylidene groups by reacting either

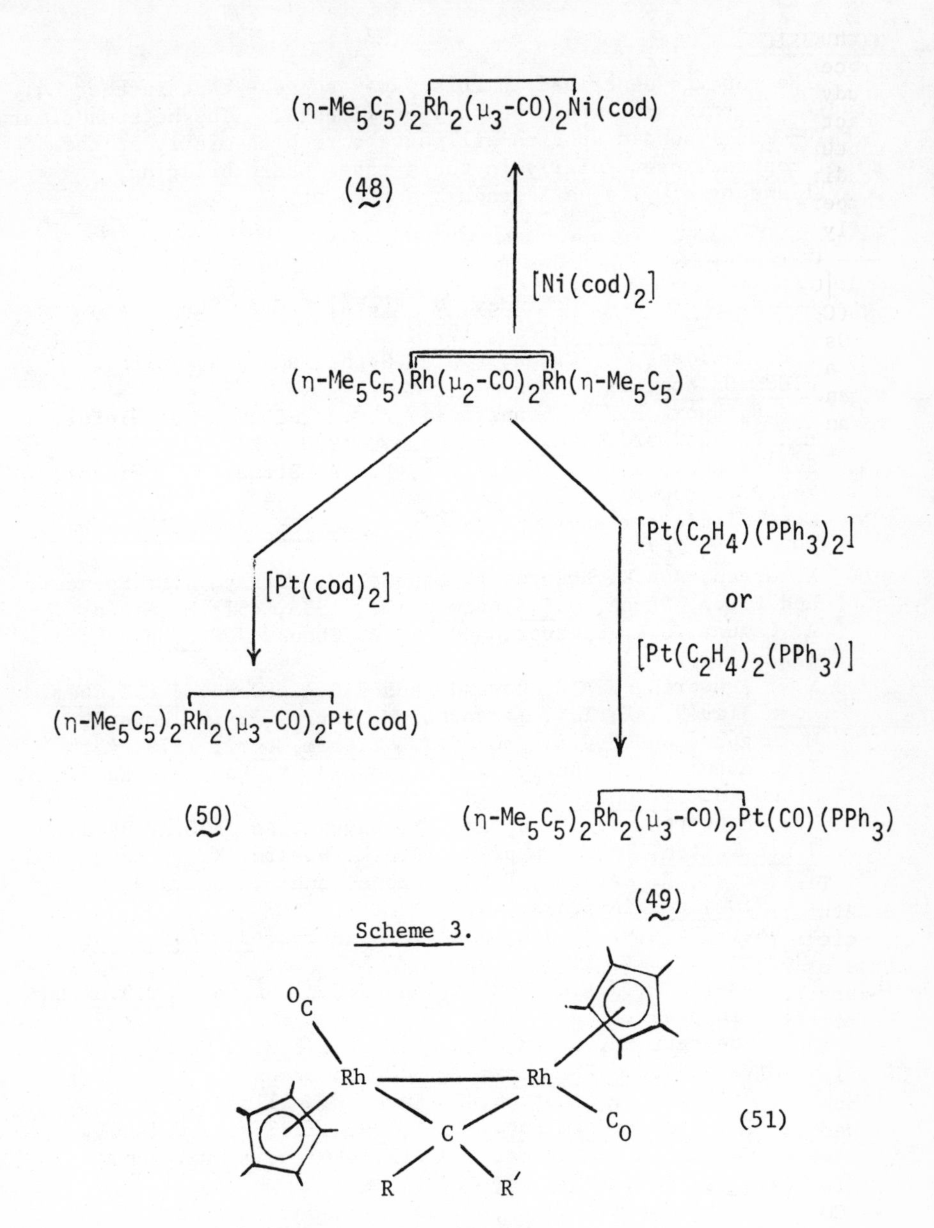

Scheme 3.

$[Rh(CO)_2(\eta-C_5H_5)]$ with diazo compounds, or $[Rh_2(\mu-CO)(CO)_2(\eta-C_5H_5)_2]$ with N-alkyl-N-nitrosoureas.

CONCLUSION.

The results described in this paper suggest that further studies in this area would be very profitable. The heteronuclear metal — metal bonded species will have a rich chemistry at the metal centres, particularly in those cases where bridging alkylidene or alkylidyne ligands are present.

LITERATURE CITED.

1. S.A.R. Knox, R.J. McKinney, V. Riera, F.G.A. Stone, and A.C. Szary, J.C.S.Dalton, 1979, 1801.
2. J.L. Davidson, M. Green, F.G.A. Stone, and A.J. Welch, J.C.S.Dalton, 1979, 506.
3. S.A.R. Knox, R.F.D. Stansfield, F.G.A. Stone, M.J. Winter, and P. Woodward, J.C.S.Chem.Comm., 1978, 221.
4. S.A.R. Knox, R.F.D. Stansfield, F.G.A. Stone, M.J. Winter, and P. Woodward, J.C.S.Chem.Comm., 1979, 434.
5. W.L. Gladfelter and G.L. Geoffroy, Adv.Organometallic Chem., 1980, 18, 207.
6. M. Green, J.A.K. Howard, A. Laguna, M. Murray, J.L. Spencer, and F.G.A. Stone, J.C.S.Chem.Comm., 1975, 451; M. Green, A. Laguna, J.L. Spencer, and F.G.A. Stone, J.C.S.Dalton, 1977, 1010.
7. T.V. Ashworth, J.A.K. Howard, and F.G.A. Stone, J.C.S.Chem. Comm., 1979, 42; T.V. Ashworth, M. Berry, J.A.K. Howard, M. Laguna, and F.G.A. Stone, J.C.S.Chem.Comm., 1979, 43; T.V. Ashworth, M. Berry, J.A.K. howard, M. Laguna, and F.G.A. Stone, J.C.S.Chem.Comm., 1979, 45.
8. T.V. Ashworth, J.A.K. Howard, M. Laguna, and F.G.A. Stone, J.C.S.Dalton, 1980, in press; J.A.K. Howard, K.A. Mead, J.R. Moss, R. Navarro, F.G.A. Stone, and P. Woodward, J.C.S.Dalton, in press.
9. M. Berry, J.A.K. Howard, and F.G.A. Stone, J.C.S.Dalton, 1980, in press.
10. T.V. Ashworth, J.A.K. Howard, and F.G.A. Stone, J.C.S.Dalton, 1980, in press.
11. M.J. Chetcuti, M. Green, J.C. Jeffery, F.G.A. Stone, and A. Wilson, J.C.S.Chem.Comm., 1980, in press.
12. D.M. Collins, F.A. Cotton, S. Koch, M. Millar, and C.A. Murillo, J.Amer.Chem.Soc., 1977, 99, 1259; M.H. Chisholm, F.A. Cotton, M.W. Extine, and B.R. Stults, Inorg.Chem., 1976, 15, 2252; 1977, 16, 603.
13. I. Levisalles, H. Rudler, T. Jeannin, and F. Dahan, J.Organometallic Chem., 1979, 178, C8.
14. C.P. Casey, T.J. Burkhardt, C.A. Bunnell, and J.C. Calabrese, J.Amer.Chem.Soc., 1977, 99, 2127.
15. E.O. Fischer and H.J. Beck, Angew.Chem.Internat.Edn., 1970, 9, 72.
16. F.A. Cotton, Progr.Inorg.Chem., 1976, 21, 1.

17. M. Berry, J. Martin-Gil, J.A.K. Howard, and F.G.A. Stone, J.C.S.Dalton, 1980, in press.
18. J.A.K. Howard, J.C. Jeffery, M. Laguna, R. Navarro, and F.G.A. Stone, J.C.S.Chem.Comm., 1979, 1170; idem. J.C.S.Dalton, in press.
19. J. Martin-Gil, J.A.K. Howard, R. Navarro, and F.G.A. Stone, J.C.S.Chem.Comm., 1979, 1168.
20. J.C. Jeffery, R. Navarro, and F.G.A. Stone, to be published.
21. M.J. Chetcuti, M. Green, J.A.K. Howard, J.C. Jeffery, R.M. Mills, G.N. Pain, S.J. Porter, F.G.A. Stone, A.A. Wilson, and P. Woodward, J.C.S.Chem.Comm., 1980 in press.
22. T.V. Ashworth, M.J. Chetcuti, J.A.K. Howard, F.G.A. Stone, and P. Woodward, J.C.S.Dalton, in press; R. Mills, J.A.K. Howard, and P. Woodward, to be published.
23. B.E.R. Schilling, R. Hoffmann, and D.L. Lichtenberger, J.Amer.Chem.Soc., 1979, 101, 585.
24. M.J. Bennett, W.A.G. Graham, J.K. Hoyano, and W.L. Hutcheon, J.Amer.Chem.Soc., 1972, 94, 6232.
25. A. Nutton and P.M. Maitlis, J.Organometallic Chem., 1979, 166, C21.
26. B.F.G. Johnson, J. Lewis, and P.A. Kilty, J.Chem.Soc.(A), 1968, 2859; R.W. Broach and J.M. Williams, Inorg.Chem., 1979, 18, 314.
27. L.J. Farrugia, J.A.K. Howard, P. Mitrprachachon, F.G.A. Stone, and P. Woodward, to be published.
28. L.J. Farrugia, J.A.K. Howard, P. Mitrprachachon, F.G.A. Stone, and P. Woodward, J.C.S.Dalton, in press.
29. M. Green, G.N. Pain, and F.G.A. Stone, unpublished work.
30. W.D. Jones, M.A. White, and R.G. Bergman, J.Amer.Chem.Soc., 1978, 100, 6773.
31. W.A. Herrmann, C. Krüger, R. Goddard, and I. Bernal, J.Organometallic Chem., 1977, 140, 73; W.A. Herrmann, Angew.Chemie Internat.Edn., 1979, 17, 800.

RECEIVED December 4, 1980.

INDEX

K

L

M

W

Z